DNA Conjugates and Sensors

RSC Biomolecular Sciences

Titles in the Series:
 1: Biophysical and Structural Aspects of Bioenergetics
 2: Exploiting Chemical Diversity for Drug Discovery
 3: Structure-based Drug Discovery: An Overview
 4: Structural Biology of Membrane Proteins
 5: Protein–Carbohydrate Interactions in Infectious Disease
 6: Sequence-specific DNA Binding Agents
 7: Quadruplex Nucleic Acids
 8: Computational and Structural Approaches to Drug Discovery: Ligand–Protein Interactions
 9: Metabolomics, Metabonomics and Metabolite Profiling
10: Ribozymes and RNA Catalysis
11: Protein–Nucleic Acid Interactions: Structural Biology
12: Therapeutic Oligonucleotides
13: Protein Folding, Misfolding and Aggregation: Classical Themes and Novel Approaches
14: Nucleic Acid–Metal Ion Interactions
15: Oxidative Folding of Peptides and Proteins
16: RNA Polymerases as Molecular Motors
17: Quantum Tunnelling in Enzyme-Catalysed Reactions
18: Natural Product Chemistry for Drug Discovery
19: RNA Helicases
20: Molecular Simulations and Biomembranes: From Biophysics to Function
21: Structural Virology
22: Biophysical Approaches Determining Ligand Binding to Biomolecular Targets: Detection, Measurement and Modelling
23: Innovations in Biomolecular Modeling and Simulations: Volume 1
24: Innovations in Biomolecular Modeling and Simulations: Volume 2
25: Recent Developments in Biomolecular NMR
26: DNA Conjugates and Sensors

How to obtain future titles on publication:
A standing order plan is available for this series. A standing order will bring delivery of each new volume immediately on publication.

For further information please contact:
Book Sales Department, Royal Society of Chemistry, Thomas Graham House, Science Park, Milton Road, Cambridge, CB4 0WF, UK
Telephone: +44 (0)1223 420066, Fax: +44 (0)1223 420247,
Email: booksales@rsc.org, Visit our website at http://www.rsc.org/Shop/Books/

DNA Conjugates and Sensors

Edited by

Keith R Fox
Centre for Biological Sciences, University of Southampton, Southampton, UK

Tom Brown
School of Chemistry, University of Southampton, Southampton, UK

RSC Publishing

RSC Biomolecular Sciences No. 26

ISBN: 978-1-84973-427-1
ISSN: 1757-7152

A catalogue record for this book is available from the British Library

Published by The Royal Society of Chemistry,
Thomas Graham House, Science Park, Milton Road,
Cambridge CB4 0WF, UK

Registered Charity Number 207890

For further information see our web site at www.rsc.org

Printed in the United Kingdom by Henry Ling Limited, Dorchester, DT1 1HD, UK

Preface

Nucleic acids, DNA and RNA, are well known for their fundamental roles in the storage and transmission of genetic material. The ability to prepare synthetic oligonucleotides has therefore revolutionised our ability to manipulate gene sequences and has improved our understanding of the fundamental properties of DNA and RNA. In addition to their biological functions, nucleic acids can fold into defined structures that are endowed with some precise recognition properties, which have been exploited in diagnostic tools, drug molecules, biosensors and nanostructures. Their properties have been further enhanced by attaching other groups, which facilitate their detection, stabilise them against degradation and promote their enzymic and recognition properties.

In the post-genomic era there is a need for the detection of DNA and RNA targets with high sequence specificity and sensitivity. Since oligonucleotides can spontaneously form duplexes with their complementary strands, and do so with extremely high sequence specificity, they are excellent diagnostic tools for the identification and treatment of diseases, as well as for forensic applications.

The addition of fluorescent groups has facilitated the detection of very small quantities of genetic material and together with advances in biophysical instrumentation has enabled the measurement of the dynamics of single molecules. Attaching other DNA-binding groups has provided protection against nuclease digestion as well as enhancing the stability of DNA duplexes. Advances in chemistry have enabled the joining of single strands of DNA, cross-linking complementary strands, labelling oligonucleotides with reporter groups and immobilising DNA on surfaces. Such conjugates can be prepared with the standard nucleic acid backbone, and also with derivatives with modified backbones (such as LNA or PNA) or unusual bases. Other novel technologies have been developed that utilise oligonucleotides as artificial enzymes,

RSC Biomolecular Sciences No. 26
DNA Conjugates and Sensors
Edited by Keith R Fox and Tom Brown
© The Royal Society of Chemistry 2012
Published by the Royal Society of Chemistry, www.rsc.org

with precise recognition properties, or as folded structures that depend on the external environment.

The recognition of nucleic acids by short synthetic oligonucleotides offers a means to modulate the expression of specific genes, with applications in the treatment of viral infections, cancer and other diseases. Several oligonucleotide analogues are currently undergoing human clinical trials and there is optimism that some of these will make it to the market. Continued success in these areas will depend on synthetic efforts to improve their properties. Their conjugation with other functional groups is especially useful in this respect. These conjugates may improve the strength and specificity of hybridization, or may endow the oligonucleotide with entirely new properties, such as the ability to react with a target analyte.

This volume is written by leaders in the field, who describe the preparation, properties and applications of these DNA conjugates. Several have been used as sensors (aptamers, riboswitches and nanostructures), based on the ability of nucleic acids to adopt specific structures in the presence of ligands, whilst others contain reporter groups such as proteins or fluorophores that enable RNA or DNA to be used in detection or single molecule studies. This book will interest researchers in areas related to analytical chemistry, chemical biology, medicinal chemistry, molecular pharmacology, and structural and molecular biology.

Keith R. Fox and Tom Brown
Southampton, UK

Contents

RSC Biomolecular Sciences No. 26
DNA Conjugates and Sensors
Edited by Keith R Fox and Tom Brown
© The Royal Society of Chemistry 2012
Published by the Royal Society of Chemistry, www.rsc.org

**Chapter 8   Making Sense of Catalysis: The Potential of DNAzymes
as Biosensors   190**
Simon A. McManus, Kha Tram and Yingfu Li

**Chapter 9   Electrochemical Techniques as Powerful Readout Methods
for Aptamer-based Biosensors   211**
Bingling Li and Andrew D. Ellington

CHAPTER 1

Fluorophore-functionalised Locked Nucleic Acids (LNAs)

PATRICK J. HRDLICKA*[a] AND
MICHAEL E. ØSTERGAARD[b]

[a] Department of Chemistry, University of Idaho, P.O. Box 442343, Moscow, ID 83844-2343, USA; [b] Department of Medicinal Chemistry, Isis Pharmaceuticals, 2855 Gazelle Court, Carlsbad, CA 92010, USA
*Email: hrdlicka@uidaho.edu

1.1 Introduction

Fluorophore-modified oligonucleotides (ONs) are extensively used in mechanistic biological studies, molecular diagnostics, drug research, biotechnology and materials science.[1] Specific applications include their use to monitor the progress of real-time polymerase chain reaction (PCR),[2] detect cellular RNA,[3] detect single nucleotide polymorphisms (SNPs),[4] study RNA-folding,[5] monitor enzyme activities,[6] and generate self-assembled chromophore arrays.[7] These applications have forced chemists to develop fluorophore-functionalised building blocks with emission characteristics that are influenced by factors in their microenvironment.[8] For example, *hybridisation probes* enable detection of nucleic acid targets under conditions where excess probe cannot be washed away, by displaying low fluorescence emission in the absence of target, but prominent emission in the presence of target.[9]

In this chapter, we will focus on the synthesis, properties and applications of ONs modified with fluorophore-functionalised LNA (locked nucleic acid) monomers, since these materials display photophysical properties that are

RSC Biomolecular Sciences No. 26
DNA Conjugates and Sensors
Edited by Keith R Fox and Tom Brown
© The Royal Society of Chemistry 2012
Published by the Royal Society of Chemistry, www.rsc.org

difficult to mimic with more flexible monomers. As will be discussed in the following sections, these properties are linked to the unique structural characteristics of LNA-type monomers, which offer increased positional control of the fluorophore.

1.2 LNA – a Primer

As part of efforts directed toward developing high-affinity antisense oligonucleotides,[10,11] the Wengel[12] and Imanishi[13] groups independently developed LNA in the late 1990s. LNAs can formally be regarded as conformationally restricted analogues of 2'-O-methyl ribonucleotides in which the methyl group is connected to the 4'-position of the sugar ring (Figure 1.1). The resulting dioxabicyclo-[2.2.1]-heptane skeleton forces the five-membered furanose ring into a C3'-*endo* conformation, which resembles the conformation that is adopted by ribonucleotides in RNA duplexes (Figure 1.1).[14] Incorporation of LNA monomers into ONs gradually tunes the conformation of neighbouring 2'-deoxyribonucleotides from DNA-like C2'-*endo* conformations toward RNA-like C3'-*endo* conformations.[14] The effect is linked to the limited internal flexibility of LNA nucleotides, which restricts conformational interconversion of neighbouring nucleotides.[15] This influences the geometry of LNA-modified duplexes, which display greater RNA character than unmodified reference duplexes.[14,16] The four chiral centres in LNA nucleotides give rise to eight possible stereoisomers (the chirality of the 2'- and 4'-positions is interrelated owing to the oxymethylene ring). While six of the eight LNA stereoisomers result in improved RNA affinity relative to unmodified reference strands,[17] only

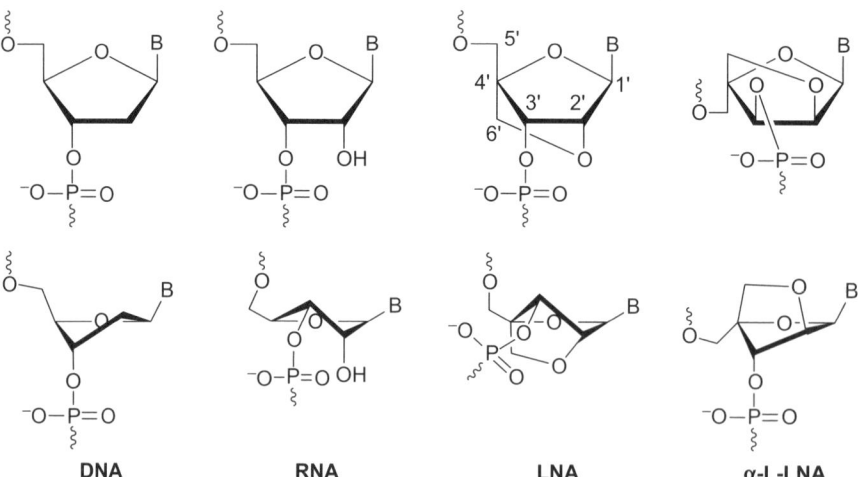

Figure 1.1   Structures of DNA, RNA LNA and α-L-LNA (*upper*) and their preferred sugar conformations (*lower*). Nucleoside numbering of the carbons in the bicyclic ring is shown for LNA.

α-L-LNA displays hybridisation properties that are comparable to those of LNA (Figure 1.1). Unlike LNA, which is an RNA mimic, α-L-LNA is considered a DNA mimic because duplexes between α-L-LNA-modified DNA strands and complementary DNA/RNA adopt geometries that globally resemble unmodified reference duplexes.[18,19]

Incorporation of LNA monomers into ONs results in significantly improved thermal affinity toward complementary DNA and RNA targets (ΔT_m/mod up to $+10\,°C$).[20] The stabilising effects of LNA monomers are sequence dependent and either entropy or enthalpy driven, suggesting that preorganisation of the LNA-modified strand or stronger base stacking interactions, respectively, contribute to stabilisation of the duplexes.[21] LNA-modified ONs, moreover, display excellent mismatch specificity.[22] ONs modified with α-L-LNA monomers have been less systematically studied owing to more limited commercial access, but generally display similar DNA/RNA affinity and mismatch discrimination as conventional LNAs.[23] The intriguing biophysical properties of LNA and α-L-LNA have led to the development of numerous LNA analogues.[24]

LNAs have found widespread use in fundamental research, biotechnology, diagnostics and drug development.[25] For example, their ability to stabilise interactions with RNA and provide protection from cellular nucleases has been widely explored in antisense technology.[26] Modulation of gene expression, through LNA-mediated targeting of messenger (m)RNA, pre-mRNA or micro (mi)RNA,[27–29] has accelerated gene function studies and led to the development of LNA-based antisense drug candidates against diseases of genetic origin.[20] Other applications of LNAs include their use as *in situ* hybridisation probes to monitor spatiotemporal expression patterns of miRNAs[30] and as primers to improve allele-selective PCR.[31] The readers are directed to other sources for additional background on LNA.[25,32]

1.3 Fluorophore-functionalised LNA – an Overview

1.3.1 Introduction

Many of the fluorophore-modified oligonucleotides (ONs) used in biotechnology are labelled at the termini through attachment of the fluorophore during solid-phase synthesis, *via* post-synthetic labelling, or through enzymatic incorporation.[33] While these labelling strategies have become largely routine, they often generate fluorescent probes that are insufficiently responsive to changes in their microenvironment and/or hybridisation state for certain applications.[8] This is largely due to their inherent flexibility, which leads to poor positional control of the fluorophore (Figure 1.2 – left). To address these limitations, a plethora of fluorophore-modified nucleotide monomers with lower inherent flexibility have been developed, which allow internal labelling of ONs and improved positional control of the fluorophore (Figure 1.2 – centre).[1,33] As will be discussed in Section 1.4, ONs modified with such

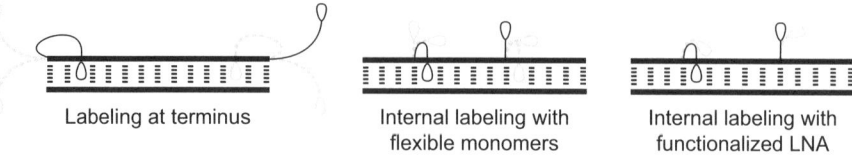

| Labeling at terminus | Internal labeling with flexible monomers | Internal labeling with functionalized LNA |

Figure 1.2   Interplay between monomer flexibility and positional control of fluorophore with different labelling approaches (shown for units with predominantly intercalative or groove binding modes). Solid and dashed lines
illustrate primary and alternative binding modes, respectively.

monomers display characteristics that have enabled the development of chromophore arrays and various diagnostic probes (representative monomers are
shown in Figure 1.3). Attachment of fluorophores to the conformationally
restricted LNA skeleton is poised to result in probes with even greater positional control of the fluorophore (Figure 1.2 – right).

An overview of the major classes of fluorophore-functionalised LNAs is
given in the following sections along with a brief discussion of their hybridisation properties and binding modes (Figure 1.4). The discussion of their
binding modes relies on indirect structural data, such as absorption and
fluorescence spectra and molecular modelling, owing to the absence of nuclear
magnetic resonance (NMR) solution or X-ray crystal structures. Fluorophores
have been conjugated to LNA-type monomers through: a) attachment to the
sugar moiety, b) attachment to the nucleobase moiety, or c) substitution of the
nucleobase (Figure 1.4). Pyrene-functionalised LNAs have been studied in
particular detail because:

- pyrene moieties engage in π-stacking with nucleobases or other pyrene
  moieties, enabling array formation, intercalation (stacking area: pyrene
  $\sim 184\,\text{Å}$ *vs.* A:T base pair $\sim 221\,\text{Å})^{[34]}$ and/or formation of pyrene–pyrene
  excimers[35]
- pyrene fluorescence is sensitive to the polarity of the microenvironment[36]
  and the nature of neighbouring nucleobases, which quench pyrene fluorescence *via* photoinduced electron transfer (guanine moieties are typically
  the strongest quenchers; $G > C \sim T > A$).[37]

1.3.2   N2′-Functionalised 2′-amino-LNA

Hybridisation properties. ONs that are modified with 2′-amino-LNA monomers
carrying small groups at the N2′-position (*e.g.* methyl; benzoyl; 2-aminoethyl;
amino acids) display similar affinity toward DNA/RNA targets as conventional
LNA.[38,39] Large N2′-fluorophores such as pyrene, perylene or coronene (Figure 1.4), on the other hand, have different impacts on target affinity, depending
on the orientation, steric bulk and linker chemistry of the fluorophore (Table
1.1).[38,40–44] For example, monomers in which the fluorophore is connected *via* a

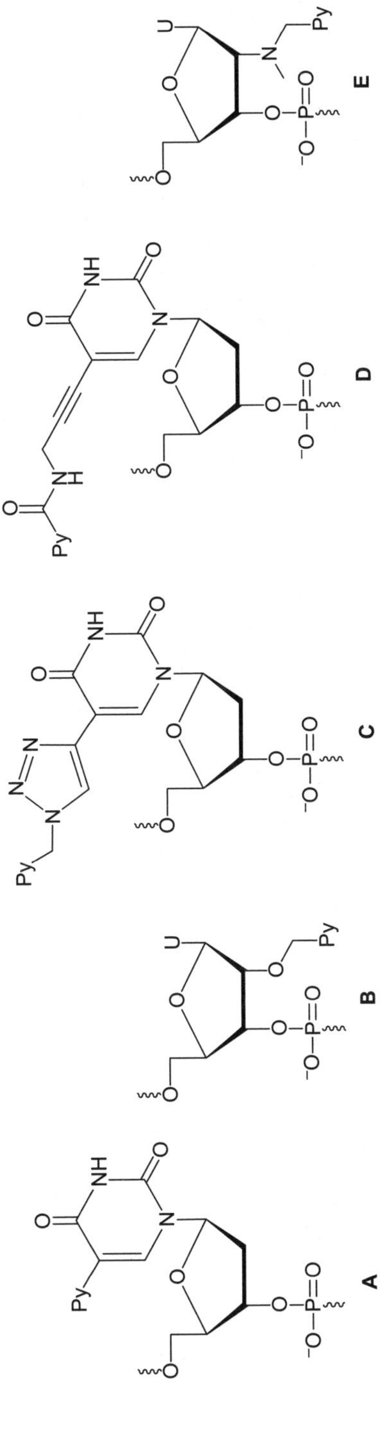

Figure 1.3 Pyrene-functionalised monomers with intermediate flexibility. Py = pyren-1-yl.

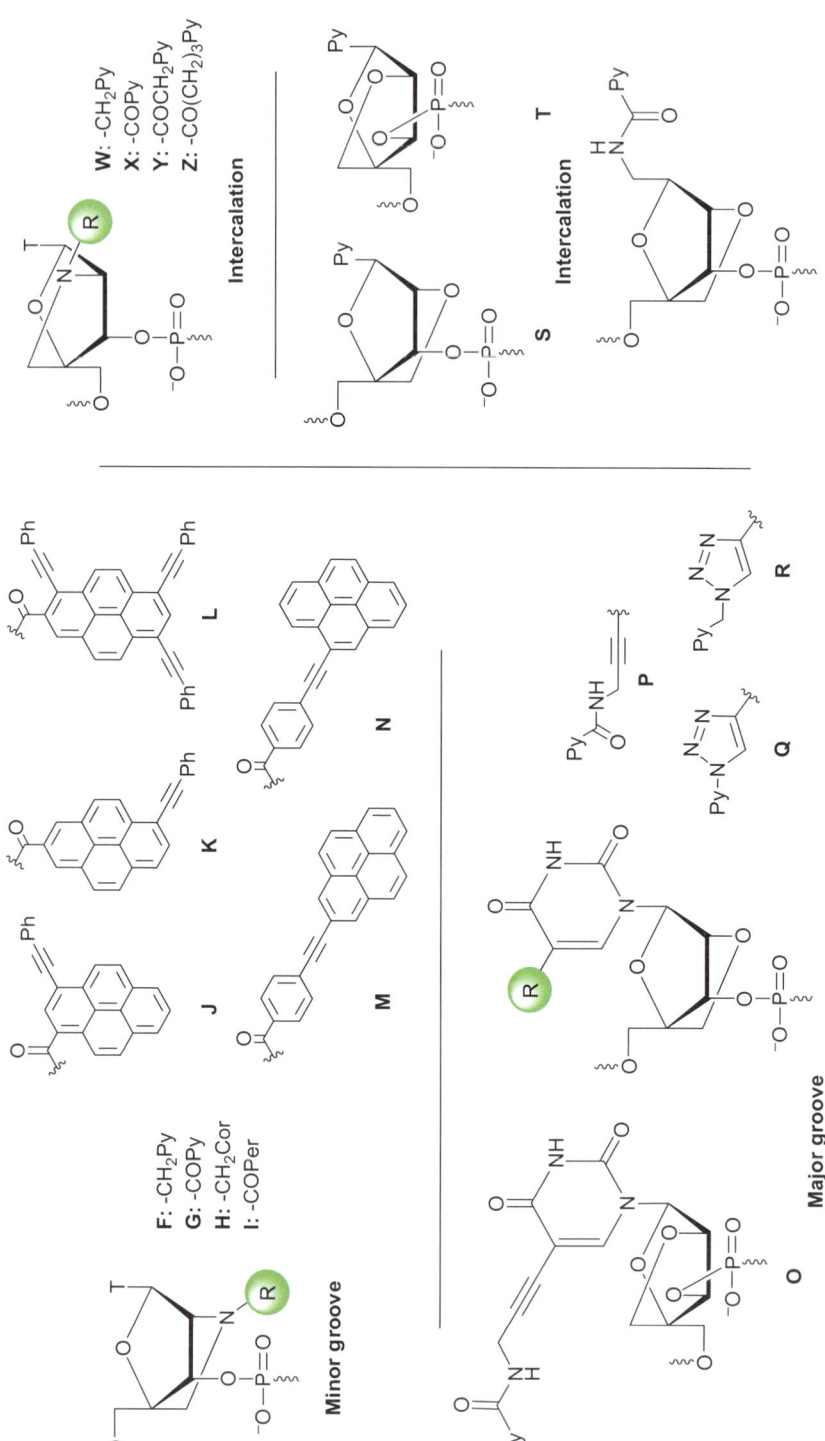

Figure 1.4 Fluorophore-functionalised LNA monomers. Py = pyren-1-yl, Per = perylen-3-yl, Cor = coronen-1-yl.

Table 1.1 Representative thermal denaturation temperatures of duplexes between N2′-fluorophore-functionalised 2′-amino-LNAs and complementary DNA or RNA.

| | ΔT_m (°C) | | | |
| | 5′-GTG A**B**A TGC | | 3′-CAC TA**B** ACG | |
Monomer	*vs. DNA*	*vs. RNA*	*vs. DNA*	*vs. RNA*
LNA-T[47,48]	+5.0	+9.5	+6.5	+9.5
F[45,49]	+3.0	+5.0	+1.0	nd
G[46]	+2.5	+7.0	+6.0	+9.5
H[44]	+3.0	+1.5	+2.0	+3.0
I[42]	+3.0	+6.5	+7.0	+7.5
J[41]	−8.0	−6.5	−6.5	0.0
K[41]	+5.5	+8.0	+6.0	+9.0
L[41]	+2.5	nt	nt	nt
M[43]	nd	nd	+6.5	+3.5
N[43]	nd	nd	+10.0	+9.5

Measured at 1 μM concentration of each strand in medium salt buffer ([Na⁺] = 110 mM, [Cl⁻] = 100 mM, pH 7.0 (NaH₂PO₄/Na₂HPO₄)). nd = not determined, nt = no transition. For structures of monomers, see Figure 1.4.

N2′-acyl linker generally result in greater duplex stabilisation than N2′-alkyl linked monomers. Fluorescence emission profiles (see Section 1.4.2) and molecular modelling studies suggest that the N2′-substituents are directed toward the minor groove (Figure 1.5).[38,45,46]

Synthesis. The most convenient synthetic route to N2′-functionalised 2′-amino LNA-T phosphoramidites initiates from commercially available diacetone-α-D-allose, which is converted into glycosyl donor **1** *via* an optimised multistep reaction sequence involving: O3-benzylation, regioselective 5,6-*O*-isopropylidene cleavage, oxidative cleavage of the resulting vicinal diol, crossed aldol condensation and Cannizzaro reduction, mesylation of the resulting diol, and acetolysis of the remaining isopropylidene group (Scheme 1.1).[50] Glycosylation of **1** under Vorbrüggen conditions,[51] followed by O2′-deacylation, O2′-mesylation and intramolecular nucleophilic displacement, results in the formation of anhydronucleoside **2**.[52] Opening under acidic conditions, followed by O2′-triflation of the resulting *threo*-configured nucleoside and installation of a 2′-azido group with inversion of configuration, affords nucleoside **3**. Azide reduction *via* a Staudinger reaction, and concurrent intramolecular substitution of the 6′-mesylate, affords 2′-amino-LNA derivative **4**.[52] A series of non-trivial protecting group manipulations converts **4** into partially protected amino alcohol **5**,[49,52] which is used as a substrate for N2′-functionalisation *via* reductive amination or chemoselective *N*-acylation. O3′-phosphitylation then affords phosphoramidites **6**, which are used in machine-assisted solid-phase DNA synthesis.[38] A variety of fluorophores (*e.g.* pyrene, perylene, coronene derivatives) have been attached in this manner despite the resource-intensive synthetic route (10–20% yield from diacetone-α-D-allose over ~20 steps).[38,41–44]

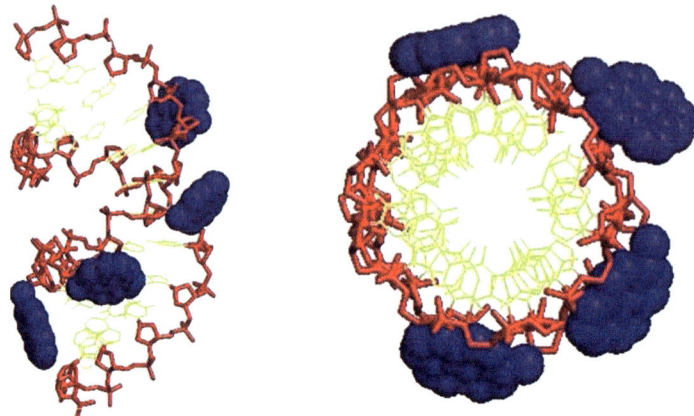

Figure 1.5 Position of fluorophores in DNA duplexes modified with N2′-functiona-
lised 2′-amino-LNA. Two representations of the lowest energy structure of
the duplex between 5′-d(TT**F** A**F**A **F**A**F** CAc G) and complementary
DNA, where **F** is 2′-*N*-(pyren-1-yl)methyl-2′-amino-LNA-T and *c* is 5-
methylcytosin-1-yl LNA (from ref. 45; copyright 2004 Royal Society of
Chemistry).

1.3.3   N2′-Functionalised 2′-amino-α-L-LNA

Hybridisation properties. ONs that are modified with 2′-amino-α-L-LNA
monomers carrying small non-aromatic units at the N2′-position, such as ethyl
or acetyl groups, have detrimental impact on duplex stability (ΔT_m down to
$-17\,°\mathrm{C}$ per modification).[53] In stark contrast, ONs modified with pyrene-
functionalised 2′-amino-α-L-LNA monomers (Figure 1.4) display exceptional
thermal affinity toward DNA targets (ΔT_m up to $+19\,°\mathrm{C}$ per modification,
Table 1.2), which far exceeds that of conventional α-L-LNA.[53] The linker
between the fluorophore and sugar skeleton has considerable influence on
duplex thermostability; short N2′-acyl linkers are favoured over N2′-alkyl and
longer N2′-acyl linkers (Table 1.2).

The pronounced DNA selectivity, along with hybridisation-induced bathochromic shifts of pyrene absorption maxima and increased intensity of circular dichroism (CD) signals in the pyrene region, strongly support an
intercalative binding mode for the pyrene moieties of monomers **W**–**Y**.[53] Closer
analysis of the molecular arrangement in these monomers reveals that the
attachment points of the nucleobase and pyrene moieties are restricted relative
to each other as a consequence of the 2-oxo-5-azabicyclo[2.2.1]heptane skeleton
(Figure 1.6). This, together with the short rigid linker between the bicyclic
skeleton and pyrene moiety, results in forced intercalation of the fluorophore
into the duplex core. Results from molecular modelling studies provide additional support for this hypothesis (Figure 1.6).

Synthesis. The synthesis of N2′-functionalised 2′-amino-α-L-LNA-T monomers initiates from inexpensive diacetone-α-D-glucose, which is converted into

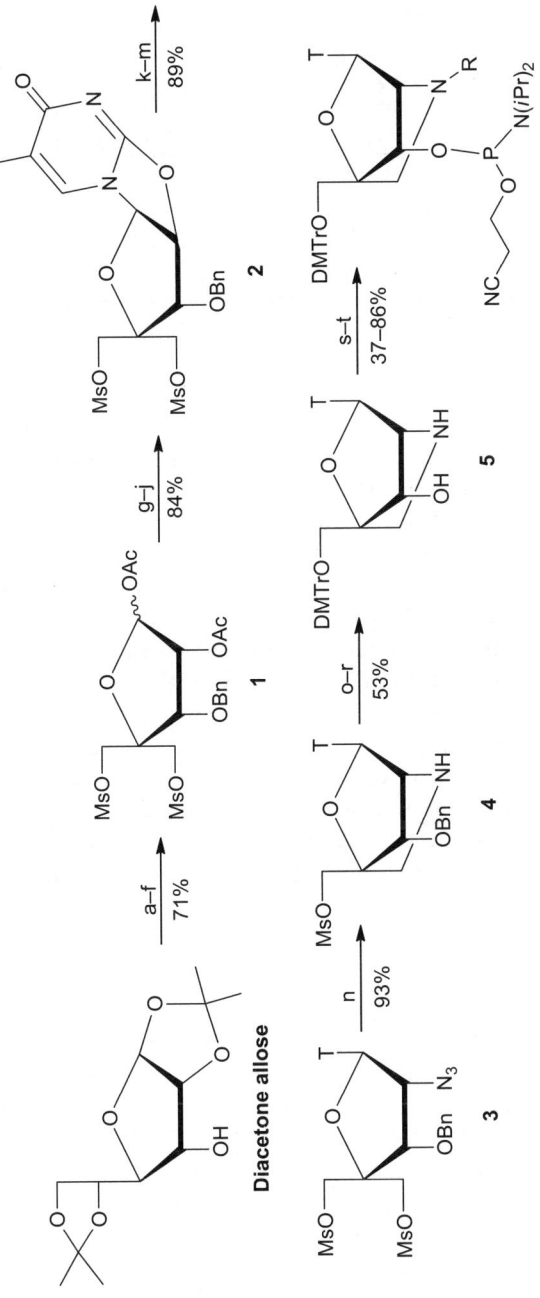

Scheme 1.1 Outline of synthetic route to N2′-functionalised 2′-amino-LNA-T phosphoramidites: (a) BnBr, NaH, THF; (b) 80% aq. AcOH; (c) NaIO₄, THF/H₂O; (d) HCHO, aq. NaOH, 1,4-dioxane; (e) MsCl, pyridine/CH₂Cl₂; (f) c. H₂SO₄, Ac₂O, AcOH; (g) thymine, BSA, TMSOTf, CH₃CN; (h) half sat. NH₃/MeOH; (i) MsCl, pyridine; (j) DBU, CH₃CN; (k) acetone, dil. aq. H₂SO₄; (l) Tf₂O, DMAP, pyridine/CH₂Cl₂; (m) NaN₃, DMF; (n) PMe₃, aq. NaOH, THF; (o) NaOBz, DMF; (p) sat. NH₃/MeOH; (q) DMTrCl, pyridine/CH₂Cl₂; (r) HCOONH₄, 20% Pd(OH)₂/C, EtOAc; (s) N2′-functionalisation (*e.g.* ArCOOH, HBTU, EtN(*i*-Pr)₂, DMF or ArCHO, NaBH(OAc)₃, ClCH₂CH₂Cl); (t) NC(CH₂)₂OP(Cl)N(*i*-Pr)₂, EtN(*i*-Pr)₂, CH₂Cl₂.

Table 1.2   Representative thermal denaturation temperatures of duplexes between N2′-fluorophore-functionalised 2′-amino-α-L-LNAs and DNA or RNA complements.

Monomer	ΔT_m (°C)			
	5′-GTG *A**B**A* TGC		3′-CAC *TA**B*** ACG	
	vs. DNA	*vs. RNA*	*vs. DNA*	*vs. RNA*
α-L-LNA-T	+ 6.0	+ 8.5	+ 8.0	+ 10.0
W	+ 14.0	+ 5.0	+ 15.5	+ 7.5
X	+ 19.0	+ 10.0	+ 19.5	+ 11.5
Y	+ 15.5	+ 9.5	+ 16.5	+ 12.0
Z	+ 6.0	+ 7.0	+ 6.5	+ 6.5

See Table 1.1 for experimental conditions. For structures of monomers, see Figure 1.4. T_m values are from ref. 53.

methyl furanoside **7** in a similar manner to that discussed for 2′-amino-LNA, except that the 1,2-*O*-isopropylidene group is cleaved using hydrogen chloride in methanol (Scheme 1.2).[54] Furanoside **7** is converted into an inseparable anomeric mixture of nucleoside **8** *via* a reaction sequence entailing O2-trifla-tion, installation of a C2-azido group with inversion of configuration (*i.e.* azido group pointing 'up'), acetolysis and Vorbrüggen glycosylation. Attempts to develop a route in which a C2′-azido group is installed at the nucleoside level were unsuccessful.[54] Treatment of **8** under Staudinger conditions results in an anomeric mixture of bicyclic nucleosides from which 2′-amino-α-L-LNA nucleoside **9** is isolated in moderate yield. Subsequent protecting group manipulations furnish key intermediate **10**, which is used as a substrate for N2′-functionalisation.[53] While a handful of N2′-pyrene- and coronene-functiona-lised 2′-amino-α-L-LNA monomers have been prepared in this manner, the time- and resource-intensive route has prevented full exploration of this compound class ($\sim$ 20 steps; <4% overall yield from diacetone-α-D-glucose).[53–55]

1.3.4   C5-Functionalised LNA

Hybridisation properties. ONs modified with C5-functionalised LNA-U monomers carrying small substituents (*e.g.* ethynyl, 3-aminopropyn-1-yl, amino acids), generally display higher affinity toward DNA/RNA targets than conventional LNAs.[47,48] In contrast, larger and more hydrophobic sub-stituents, such as fatty acids, cholesterol or pyrene derivatives (see Figure 1.4), are detrimental to duplex thermostability; similar observations have been made with the corresponding C5-functionalised α-L-LNA-U monomers (Table 1.3).[47,48,56] Fluorescence emission spectra suggest that the C5-substituent is directed toward the major groove (see Section 1.4.3).[56] We stipulate that interactions between H6 and H3′ (or H2′ in α-L-LNA monomers) hinder rotation around the glycosidic bond, resulting in greater positional control of

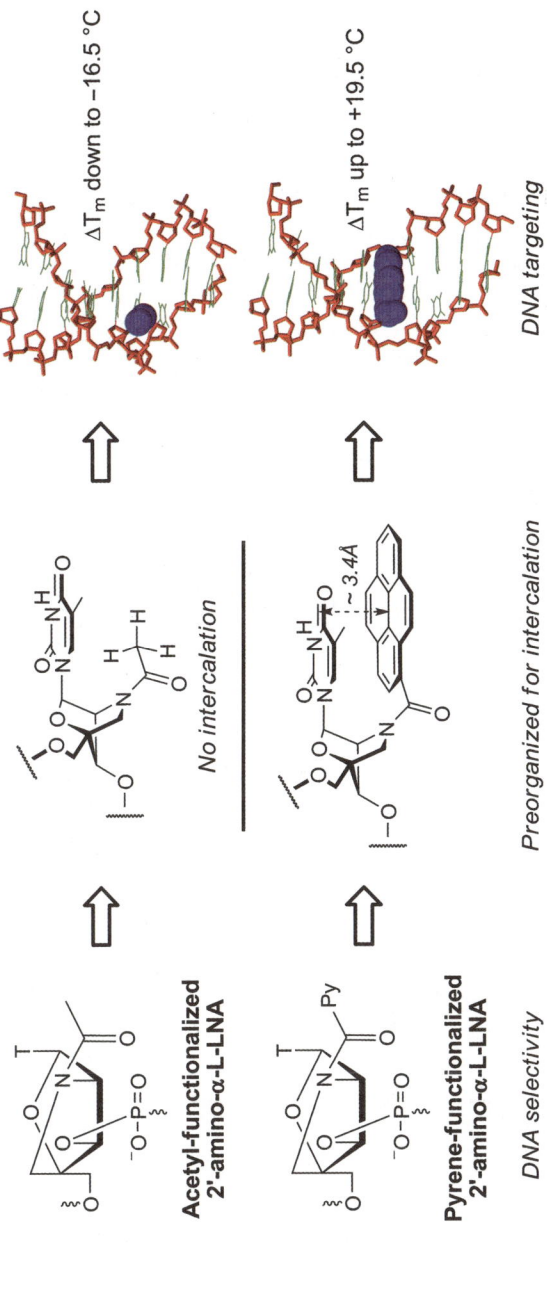

Figure 1.6 Binding modes of N2′-functionalised 2′-amino-α-L-LNA (reproduced with permission from ref. 53; copyright 2009 American Chemical Society).

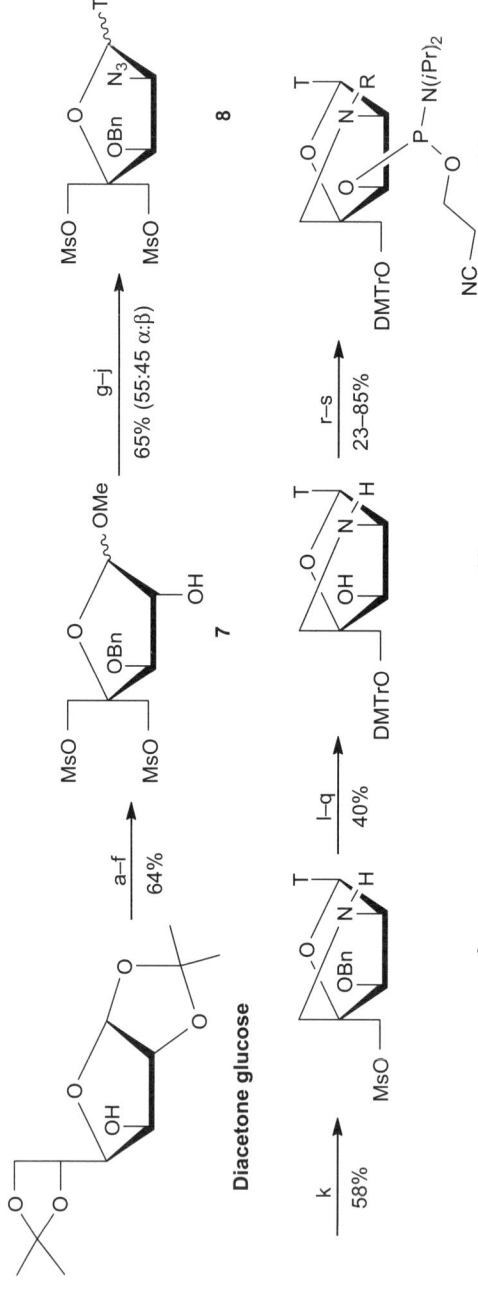

Scheme 1.2 Outline of synthetic route to N2′-functionalised 2′-amino-α-L-LNA-T phosphoramidites: (a) BnBr, *n*-Bu₄NI, NaH, THF; (b) 80% aq. AcOH; (c) NaIO₄, THF/H₂O; (d) HCHO, aq. NaOH, 1,4-dioxane; (e) MsCl, pyridine; (f) CH₃COCl, MeOH; (g) Tf₂O, pyridine/CH₂Cl₂; (h) NaN₃, 15-crown-5, DMF; (i) c. H₂SO₄, Ac₂O, AcOH; (j) thymine, BSA, TMSOTf, ClCH₂CH₂Cl; (k) PMe₃, aq. NaOH, THF; (l) (CF₃CO)₂O, pyridine/CH₂Cl₂; (m) KOAc, 18-crown-6, 1,4-dioxane; (n) sat. NH₃/MeOH; (o) BCl₃, hexane/CH₂Cl₂; (p) DMTrCl, DMAP, pyridine; (q) aq. NaOH, EtOH/pyridine; (r) N2′-functionalisation (*e.g.* PyCHO, NaB-H(OAc)₃, ClCH₂CH₂Cl or PyCOOH, HATU, EtN(*i*-Pr)₂, DMF); (s) NC(CH₂)₂OP(Cl)N(*i*-Pr)₂, EtN(*i*-Pr)₂, CH₂Cl₂.

Table 1.3 Representative thermal denaturation temperatures of duplexes between C5-fluorophore-functionalised LNA/α-L-LNA and DNA or RNA complements.

| | ΔT_m (°C) | | | |
| | 5'-GTG A**B**A TGC | | 3'-CAC TA**B** ACG | |
Monomer	vs. DNA	vs. RNA	vs. DNA	vs. RNA
LNA-T[47,48]	+ 5.0	+ 9.5	+ 6.5	+ 9.5
O[56]	− 1.0	+ 1.5	0	+ 2.5
P[56]	− 6.5	− 4.0	− 4.0	0.0
Q[47]	− 10.5	− 2.0	nd	nd
R[47]	− 5.5	− 1.5	nd	nd

See Table 1.1 for experimental conditions. For structures of monomers, see Figure 1.4.

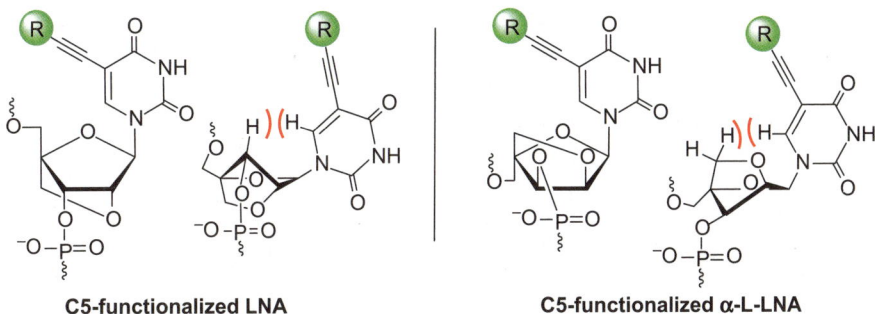

C5-functionalized LNA   **C5-functionalized α-L-LNA**

Figure 1.7 Illustration of interactions between H6 and H3′ (LNA) or H2′ (α-L-LNA) in C5-functionalised LNA/α-L-LNA, which hinder rotation about the glycosidic bond.

the fluorophore relative to corresponding C5-functionalised DNA monomers (Figure 1.7).

Synthesis. The synthesis of C5-functionalised LNA-U monomers initiates from glycosyl donor **1**,[50] which is converted into fully deprotected LNA uridine diol **12** *via*: a) Lewis acid-catalysed glycosylation with persilylated uracil, b) tandem O2′-deacylation and intramolecular nucleophilic displacement furnishing the LNA skeleton, and c) protecting group manipulations (Scheme 1.3).[57,58] C5-Iodination of **12** followed by O5′-dimethoxytritylation provides access to key intermediate **13**, which is coupled to fluorophore-modified terminal alkynes *via* the Sonogashira approach.[56,58] Subsequent O3′-phosphitylation provides desired phosphoramidite **14**. Interestingly, the synthesis of C5-functionalised LNA-U monomers only requires two additional steps relative to conventional LNA monomers, *i.e.* C5-iodination and C5-functionalisation. As a result, these monomers are the most readily accessible functionalised LNA monomers (∼15% overall yield from diacetone allose; ∼15 steps). The

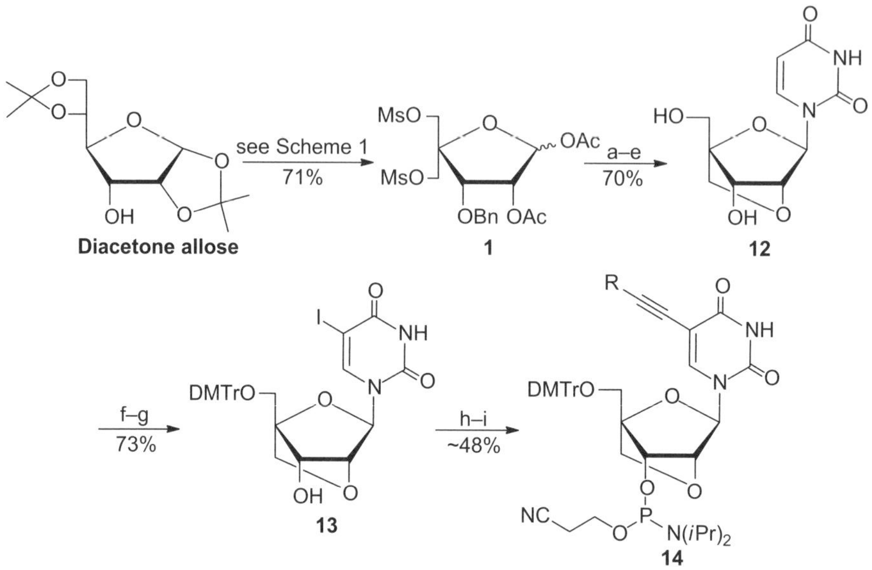

Scheme 1.3 Outline of synthetic route to C5-functionalised LNA-U phosphoramidites: (a) uracil, BSA, TMSOTf, CH$_3$CN; (b) aq. NaOH, 1,4-diox-
ane; (c) NaOBz, DMF; (d) aq. NaOH, THF; (e) 88% HCOOH, 20%
Pd(OH)$_2$/C, THF/MeOH; (f) I$_2$, CAN, AcOH; (g) DMTrCl, pyridine;
(h) C5′-functionalisation (*e.g.* PyCONH$_2$CH$_2$C≡CH, Pd(PPh$_3$)$_4$, CuI,
Et$_3$N, DMF); (i) NC(CH$_2$)$_2$OP(Cl)N(*i*-Pr)$_2$, EtN(*i*-Pr)$_2$, CH$_2$Cl$_2$.

synthesis of C5-fluorophore-functionalised α-L-LNA-U monomers proceeds
from diacetone-α-D-glucose *via* a similar route.[56]

1.3.5   LNA with Fluorescent Nucleobase Surrogates

Hybridisation properties. ONs modified with LNA or α-L-LNA based C-gly-
cosides **S**, **T** or **U** (see Figure 1.4) display strongly reduced thermal affinity
toward complementary DNA relative to unmodified reference strands (Table
1.4).[59–61] However, these probes display interesting universal hybridisation
characteristics, *i.e.* they exhibit virtually identical DNA/RNA target affinity
regardless of the nucleotide opposite of the modification site (Table 1.4). These
characteristics strongly suggest that the pyrene moieties of these *C*-glycosides
act as nucleobase surrogates, which force the opposing nucleotide out from the
duplex core in a similar manner to that reported for DNA-based monomers.[62,63] Development of universal hybridisation probes has been a long-
standing goal because of their potential application as degenerate PCR primers
and microarray probes when the identity of one or more nucleotides in a target
sequence is unknown.[64]

Synthesis. The synthetic route to the representative *C*-glycoside LNA
monomer **S** initiates from diacetone-α-D-allose (Scheme 1.4).[59] The starting

Table 1.4 Thermal denaturation temperatures (T_ms are shown) of duplexes between centrally modified ONs and DNA targets.

| | | | T_m (°C) | | |
| | | | 5'-GTG A**B**A TGC: 3'-CAC TYT ACG | | |
Monomer	Y:	A	C	G	T
DNA-T[59]		28	11	12	19
S[59]		18	17	18	19
T[60]		21	22	27	23
U[61]		21	20	19	21

See Table 1.1 for experimental conditions. For structures of monomers, see Figure 1.4.

material is converted to methyl furanoside **16** in an equivalent manner to that discussed for **7** (Scheme 1.2) with the exception that a *para*-methoyxybenzyl (PMB) group is used for protection of the O3-position rather than a regular benzyl group. PMB, which can be oxidatively cleaved using 2,3-dichloro-5,6-dicyano-1,4-benzoquinone (DDQ), is used to minimise cleavage of the benzylic O4–C1 bond during O3-deprotection. Base-induced cyclisation of methyl furanoside **16** followed by several protecting group manipulation steps provides dioxabicyclo-[2.2.1]-heptane **17**. Acidic hydrolysis of the unstable acetal generates γ-hydroxy-aldehyde **18**, which is stereospecifically converted into **19** upon treatment with Grignard reagents. Subsequent reformation of the bicyclic ring under Mitsunobu conditions followed by protecting group manipulations, including the aforementioned cleavage of the PMB group, provides phosphoramidite **20** (<5% overall yield, ~16 steps). Different aryl groups have been introduced in this manner.[59] The corresponding phosphoramidite of α-L-LNA monomer **T** is obtained in a related manner although additional inversion and protection/deprotection steps are needed.[60]

1.4 Applications of Fluorophore-functionalised LNA

1.4.1 Formation of Pyrene Arrays

The use of nucleic acids as scaffolds for programmable arrangement of chromophores has received considerable attention owing to the prospect of developing DNA-based light harvesting antenna systems.[7,65,66] Pyrene arrays have often been studied as simple model systems toward this end.

Formation of pyrene arrays in the major groove of DNA duplexes has been realised using ONs with five sequential incorporations of 5-(pyren-1-yl)-2'-deoxyuridine monomer **A** (see Figure 1.3), as evidenced by hybridisation-induced excitonic coupling of pyrene signals in CD spectra.[67] The presence of mismatched base pairs in the proximity of the array results in electronic decoupling of the pyrene moieties, which implies interesting diagnostic applications for detection of single nucleotide polymorphisms.

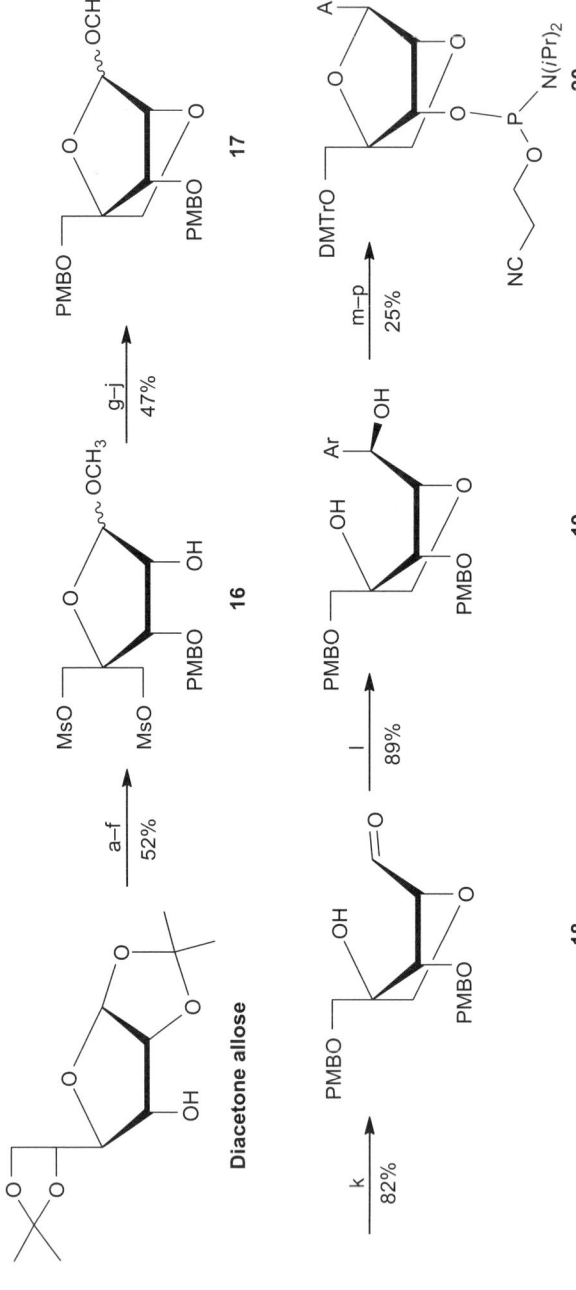

Scheme 1.4   Outline of synthetic route to LNA monomers with nucleobase surrogates: (a) *p*-MeOC$_6$H$_4$CH$_2$Cl, NaH, DMF; (b) 70% AcOH; (c) NaIO$_4$, H$_2$O; (d) HCHO, aq. NaOH, THF/H$_2$O; (e) MsCl, pyridine; (f) HCl/CH$_3$OH/H$_2$O; (g) NaH, DMF (isolation of major anomer); (h) KOAc, 18-crown-6, 1,4-dioxane; (i) sat. NH$_3$ in MeOH; (j) *p*-MeOC$_6$H$_4$CH$_2$Cl, NaH, THF; (k) 80% AcOH; (l) ArMgBr, THF; (m) TMAD, Bu$_3$P, benzene; (n) DDQ, CH$_2$Cl$_2$/H$_2$O; (o) DMTrCl, pyridine; (p) NC(CH$_2$)$_2$OP(Cl)N(*i*-Pr)$_2$, EtN(*i*-Pr)$_2$, CH$_2$Cl$_2$.

Formation of pyrene arrays in the minor groove has been realised using duplexes between RNA targets and 2′-*O*-methyl RNA strands that are sequentially modified with 2′-*O*-(pyren-1-yl)methyluridine monomer **B** (see Figure 1.3).[68] The use of RNA duplexes, rather than enzymatically more stable DNA duplexes, is necessary to minimise undesired intercalation of the pyrene moieties.[69] In an alternative approach, an interstrand array of pyrene moieties was generated in the minor groove of RNA duplexes, which are modified with 2′-*O*-(pyren-1-yl)methyl uridine/adenosine pairs.[70]

Several pyrene arrays, which are based on functionalised LNA monomers, have been developed. The first example involves DNA duplexes with interstrand arrangements of 2′-*N*-(pyren-1-yl)methyl-2′-amino-LNA monomer **F** (Figures 1.4 and 1.8).[45] The thermostability values of singly modified and reference DNA duplexes are virtually identical. Placement of two **F** monomers in a '−1 interstrand zipper arrangement' leads to considerable duplex stabilisation and excimer formation, whereas '+1 interstrand monomer arrangements' do not ($T_m = 35\,°C$ *vs.* $30\,°C$, respectively, Figure 1.8). These trends are augmented by incorporation of additional −1 zipper cassettes. The lowest energy structures from force-field calculations reveal structural features that are consistent with these characteristics, as pyrene moieties in −1 zipper

DNA Duplex	T_m (°C)	Excimer
5'-GTG ATA TGC 3'-CAC TAT ACG	28	-
5'-GTG ATA TGC 3'-CAC TA**F** ACG	29	-
5'-GTG ATA TGC 3'-CAC **F**AT ACG	27	-
5'-GTG A**F**A TGC 3'-CAC **F**AT ACG	35	strong
5'-GTG A**F**A TGC 3'-CAC TA**F** ACG	30	-
5'-TTT ATA TAT CAC G 3'-AAA TAT ATA GTG C	38	-
5'-TTT ATA TAT CAC G 3'-AAA **F**A**F** A**F**A GTG C	35	weak
5'-TT**F** A**F**A **F**A**F** CAC G 3'-AAA TAT ATA GTG C	45	weak
5'-TT**F** A**F**A **F**A**F** CAC G 3'-AAA **F**A**F** A**F**A GTG C	77	strong

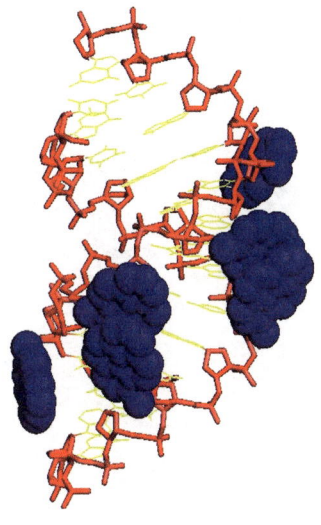

Figure 1.8 Formation of pyrene interstrand arrays using 2′-*N*-(pyren-1-yl)methyl-2′-amino-LNA monomer **F** (for structure, see Figure 1.4). *Left*: T_m-values and fluorescence characteristics of DNA duplexes modified with monomer **F**. *Right*: lowest energy structure from force field calculations on the duplex with three −1 interstrand arrangements of monomer **F** ($T_m = 77\,°C$) (from ref. 45; copyright 2004 RSC Publishing).

arrangements stack pair wise across the minor groove (Figure 1.8).[45] More brightly fluorescent versions of this 'communication system', based on N2'-(phenylethynyl)pyrene-functionalised 2'-amino-LNA monomers **M** or **N** (Figure 1.4), have been developed.[43] The predictability of the original communication system has been used to signal stepwise self-assembly of branched oligonucleotides into higher order structures.[71,72] The extraordinary duplex thermostability, directional preference and robustness of the communication system underscore the advantages of functionalised LNA.

In another example, 2'-*N*-(pyren-1-yl)acetyl-2'-amino-α-L-LNA monomer **Y** has been used to generate regular arrangements of pyrene moieties inside the core of DNA duplexes (Figure 1.9).[73] ONs modified with 5'-(**YΦ**)-steps, *i.e.* an arrangement where monomer **Y** is 3'-flanked by abasic site monomer **Φ**, display remarkable affinity toward DNA targets with abasic sites in the +1 interstrand position relative to **Y** (Figure 1.9). Thus, a 13-mer DNA duplex containing four of these 5'-(**YΦ**):3'-(**AΦ**)- cassettes, each separated by one base pair, displays a T_m of ~60 °C, whereas the corresponding reference duplex, where the 5'-(**YΦ**):3'-(**AΦ**)-units are replaced by 5'-(TA):3'-(AT) base pairs, has a T_m of ~35 °C (Figure 1.9). Hybridisation-induced bathochromic shifts of pyrene absorption maxima, along with data from thermodynamic studies and force field calculations, suggest that monomer **Y** forces the pyrene moiety into the void formed by the **Φ**:**Φ** pair, where it engages in efficient π–π stacking with neighbouring base pairs (Figure 1.9).[73] Similar, albeit far less thermostable, pyrene arrangements have been generated using the *C*-glycoside LNA monomer **U**.[61]

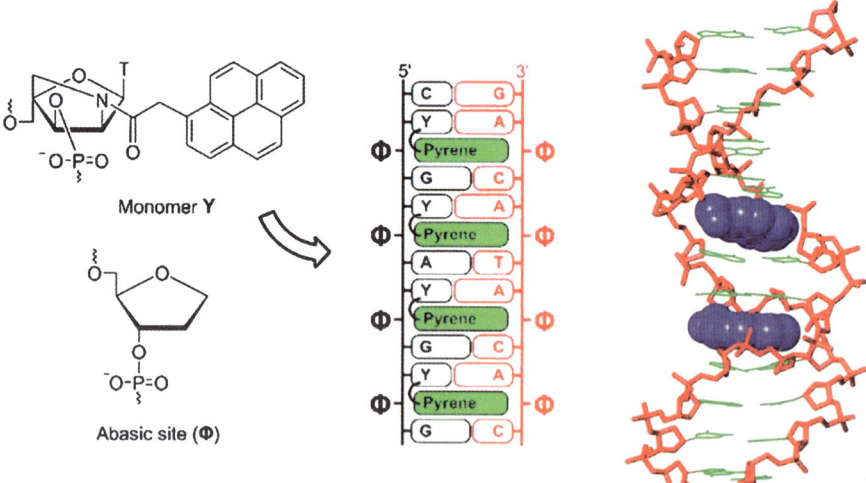

Figure 1.9   Regular arrangements of pyrene moieties in duplex cores using 5'-(**YΦ**): 3'-(**AΦ**)-units. Molecular modelling structure depicts 13-mer DNA duplex containing two separated 5'-(**YΦ**):3'-(**AΦ**)-units (reproduced with permission from ref. 73; copyright 2008 American Chemical Society).

1.4.2 Hybridisation Probes

Hybridisation probes are designed to display: a) low fluorescence in the absence of target, typically through quenching interactions between the fluorophore and nucleobase moieties,[37] and b) prominent fluorescence in the presence of target, by placing the fluorophore in a less quenching environment. Early examples of hybridisation probes include ONs modified with cyanine,[74,75] fluorescein[76,77] or pyrene moieties.[78] For example, the fluorescence intensity of RNA strands modified with 2′-*O*-(pyren-1-yl)methyluridine monomer **B** (see Figure 1.3) increases up to 30-fold upon hybridisation with RNA, while much smaller increases are observed with DNA targets. The fluorescence properties are moderately influenced by the nature of the 3′-flanking nucleobase [emission quantum yield $\Phi_F = 0.10/0.10/0.16/0.24$ for duplexes between complementary RNA and 5′-r(ACA **B**XC AGU GUU GAU) where X = G/A/U/C, respectively].[78] In contrast, incorporation of monomer **B** into DNA strands produces probes with far more variable emission profiles.[79,80] These differences are due to different pyrene binding modes; NMR solution structures of RNA:RNA and DNA:DNA duplexes modified with monomer **B** show that the pyrene moieties predominantly are located in the minor groove and duplex core, respectively.[69] This underlines just how interrelated are the photophysical properties and positional control of the fluorophore.

In contrast, ONs modified with C5-pyrene-functionalised triazole-linked 2′-deoxyuridine monomer **C** (see Figure 1.3) result in 9- to 23-fold increases in fluorescence upon hybridisation with DNA, as well as RNA, targets. Accordingly, the data suggest that placement of the pyrene moiety in the major groove[81] dominates over intercalative binding modes, which would result in quenching of fluorescence. These hybridisation probes only display mild sequence limitations; prominent hybridisation-induced increases in fluorescence intensity are observed when monomer **C** is flanked by A/C/G, while flanking Ts result in inadequate quenching of single stranded probes and low increases. The moderate quantum yields ($\Phi_F \leq 0.16$) are the main drawback of these probes. Interestingly, ONs modified with the corresponding C5-functionalised LNA monomer **R** (see Figure 1.4) display markedly larger hybridisation-induced intensity increases (up to 51- to 42-fold increases *vs.* DNA and RNA, respectively).[47] The formation of more brightly fluorescent duplexes suggests that the LNA skeleton plays an active role in directing the pyrene moiety into the non-quenching major groove. Similar observations have been made in comparative studies of other C5-functionalised DNA/LNA monomers (see Section 1.4.3).[56]

DNA strands modified with 2′-*N*-(pyren-1-yl)-carbonyl-2′-amino-LNA monomer **G** (see Figure 1.4) result in prominent increases in fluorescence intensity upon hybridisation with DNA and RNA targets (typically 2- and 10-fold), leading to the formation of brightly fluorescent duplexes ($\Phi_F = 0.28$–0.99, Figure 1.10).[46,82,83] Successful design of these 'Glowing LNA'[84] requires incorporation of at least two **G** monomers separated by at least one nucleotide, because this decreases the fluorescence intensity of single stranded probes

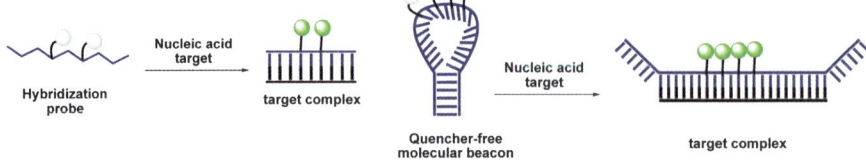

	Relative fluorescence emission quantum yield					
	5'-GCA T(U)A**G**CAC			5'-GCA **GAG**CAC		
Backbone	**SSP**	**DNA**	**RNA**	**SSP**	**DNA**	**RNA**
DNA	0.50	0.79	0.81	0.27	0.81	0.70
RNA	0.84	0.89	0.93	0.16	0.85	0.89
O2'-Me RNA	1.00	0.97	0.98	0.14	0.68	0.86

Figure 1.10   Principle of hybridisation probes and quencher-free molecular beacons modified with 2'-*N*-(pyren-1-yl)carbonyl-2'-amino-LNA-T monomer **G** (upper panel), and quantum yields of Glowing LNA probes with different backbone chemistries in the absence (SSP) or presence of complementary DNA/RNA (adapted with permission from ref. 83; copyright 2010 American Chemical Society).

through pyrene–nucleobase interactions (Figure 1.10). Molecular modelling studies suggest that the bicyclic skeleton and short rigid amide linker of monomer **G** force the pyrene moiety into the non-quenching minor groove of duplexes.[46] The importance of positional control is emphasised by the fact that ONs modified with the slightly more flexible 2'-*N*-(pyren-1-yl)-methyl-2'-amino-LNA monomer **F** do not display reliable hybridisation-induced increases in fluorescence intensity and result in the formation of far less brightly fluorescent duplexes. Incorporation of monomer **G** into RNA or 2'-*O*-methyl-RNA strands yields probes that display even larger increases in fluorescence intensity upon target binding, because the single stranded probes (SSPs) are more efficiently quenched, while duplexes display even higher quantum yields (Figure 1.10).[83]

Monomer **G** has also been evaluated as a building block in 'quencher-free' molecular beacons (MBs).[83] Unlike conventional MBs, which are end-functionalised with fluorophore–quencher pairs, leading to binding-induced dequenching of fluorescence emission,[85] quencher-free MBs rely on the presence of microenvironment-sensitive monomers in the target loop to signal binding-induced changes in the secondary structure.[86] Hybridisation of a 29-mer quencher-free MB with four separated **G** monomers in the 15-mer target loop (Figure 1.10) to complementary DNA/RNA targets results in 3- to 9-fold increases in signal intensity and formation of brightly fluorescent duplexes ($\Phi_F = 0.45$–0.62).[83] The quencher-free MBs display high biostability (>48 h) and greater thermal discrimination of singly mismatched DNA/RNA targets than their linear counterparts, which facilitates their use for imaging of cellular RNA.[83]

The interesting characteristics of multilabelled Glowing LNA stimulated the development of other N2'-fluorophore-functionalised 2'-amino-LNA monomers.[40-44] For example, ONs modified with 2′-*N*-(perylen-3-yl)carbonyl-2′-amino-LNA monomer **I** (see Figure 1.4) display a similar pattern of hybridisation-induced increases in fluorescence intensity to that of the original Glowing LNA probes.[42] While red-shifted emission relative to the original Glowing LNA probes is observed (~ 490 nm *vs.* ~ 400 nm), the emission increases and duplex quantum yields are lower ($\Phi_F = 0.09$–0.50). Optimised probes based on monomer **I** were, nonetheless, also used for cellular RNA imaging.

ONs modified with the larger 2′-*N*-4-(coronen-1-yl)methyl-2′-amino-LNA monomer **H** (see Figure 1.4) also display red-shifted emission (~ 430 nm) but only display minor changes in signal intensity upon hybridisation to DNA/RNA targets.[44] Loss of positional control, due to the more flexible linker connecting the coronene and sugar moieties, is one of the likely reasons for these photophysical characteristics.

ONs modified with various N2′-(phenylethynyl)pyrenecarbonyl-functionalised 2′-amino-LNA monomers **J–L** (see Figure 1.4) have been shown to display strongly red-shifted fluorescence emission (~ 420–520 nm, depending on the number of phenylethynyl substituents).[41] High duplex quantum yields and modest increases in signal intensity are observed for some of these probes, but the trends are less predictable than with the original Glowing LNA probes.

The pyren-1-ylcarbonyl moiety of Glowing LNA monomer **G** has been conjugated to ring-expanded LNA monomers, but multilabelled probes were not studied, which prevents direct comparison.[87]

In summary, the above examples demonstrate that attachment of the pyren-1-yl-carbonyl moiety to the 2′-amino-LNA skeleton (*i.e.* monomer **G**) produces hybridisation probes with desirable characteristics, presumably by striking an optimal balance between steric and electronic requirements.

1.4.3 Base Discriminating Fluorescent Probes

Fluorophore-functionalised LNA monomers have also found use as building blocks for base discriminating fluorescent (BDF) probes. Unlike hybridisation probes, BDF probes display duplex emission characteristics that are strongly dependent on the nature of the nucleotide *opposite* the BDF monomer (Figure 1.11).[88,89] Accordingly, these probes are useful for detection of SNPs, which are the most frequently occurring genetic variations in the human genome and important biomedical markers.[4]

ONs modified with C5-[3-(1-pyrenecarboxamido)propynyl]-functionalised 2′-deoxyuridine monomer **D** (see Figure 1.3) are interesting BDF probes, which have been used for detection of SNP sites in human breast cancers.[90,91] Prominent signal increases are observed upon hybridisation to complementary DNA (2- to 10-fold), while hybridisation to DNA strands with SNP-sites opposite monomer **D** results in much smaller increases (Figures 1.11 and 1.12). These trends presumably correspond to different fluorophore binding modes: a) in unhybridised probes, interactions between the fluorophore and nucleobase moieties ensure low emission levels; b) upon duplex formation with complementary DNA, the fluorophore is placed in the non-quenching major groove

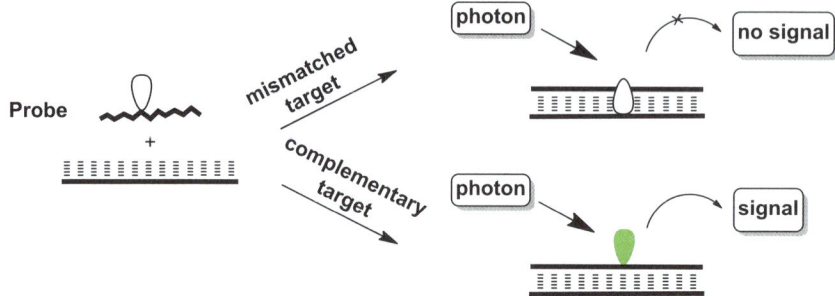

Figure 1.11   Principle of base discriminating fluorescent (BDF) probes.

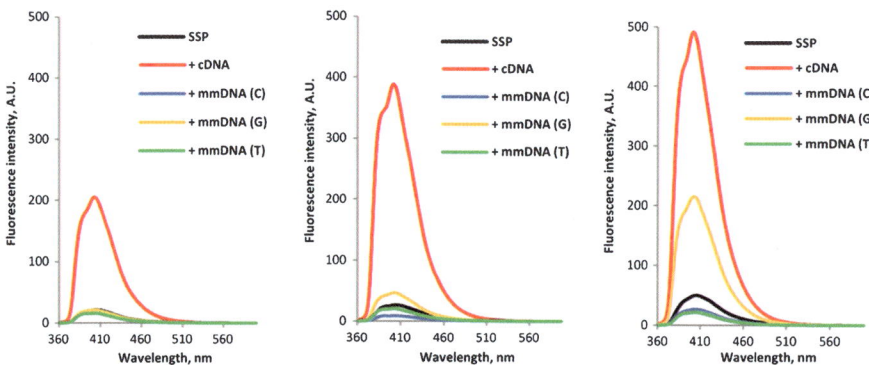

Figure 1.12   Fluorescence emission spectra of probes modified with C5-[3-(1-pyr-enecarboxamido)propynyl]-functionalised DNA (left), LNA (centre) or α-L-LNA (right) monomers. Probe sequence: 5′-d(CG CAA G*B*G ACC GC), where *B* = monomer **D**, **P** or **O**. Spectra are for probes in absence (SSP) or presence of complementary DNA (cDNA) or mismatched DNA (mmDNA, mismatched nucleotide across from modification listed in parenthesis). $\lambda_{ex} = 344$ nm, $T = 5\,^{\circ}\text{C}$ (adapted with permission from ref. 56; copyright 2011 Wiley).

resulting in high levels of emission; c) duplex formation with SNP-containing DNA targets results in a change in nucleobase orientation from the *anti* to *syn* conformation, resulting in intercalation of the fluorophore and nucleobase-mediated quenching.[90] Considering that guanine moieties are the most efficient quenchers of pyrene fluorescence, it is not surprising that the emission characteristics of these probes are influenced by neighbouring nucleotides. Probes in which monomer **D** is flanked by cytosine or guanine units discriminate SNPs more efficiently than probes with **ADA/TDT** contexts (emission decreases by 86–92% *vs.* 54–86% relative to matched duplexes).[56]

The corresponding LNA and α-L-LNA monomers **P** and **O** (see Figure 1.4) were explored on the basis of the hypothesis that the extreme sugar pucker of the bicyclic skeletons would influence the nucleobase orientation (see Figure 1.7) and, accordingly, the position of the fluorophore. Indeed, ONs modified with LNA monomer **P** display similar hybridisation-induced increases in emission (2- to 15-fold), 5–60% larger duplex quantum yields ($\Phi_F = 0.44$–0.67) and improved SNP-discrimination when the monomers are flanked by C/G (86–97% decreased emission relative to matched duplex, Figure 1.12).[56] ONs modified with α-L-LNA monomer **O** display slightly larger hybridisation-induced signal increases (4- to 10-fold), 20–50% larger duplex quantum yields ($\Phi_F = 0.50$–0.80) and, with the exception of G-mismatches, improved SNP-discrimination with probes having A<u>O</u>A or T<u>O</u>T sequence contexts (emission decreased by 65–93% relative to matched duplex).[56] The larger duplex quantum yields observed with LNA and α-L-LNA monomers **O** and **P** most likely reflect greater fluorophore occupancy in the major groove. Higher duplex quantum yields have also been observed for ONs modified with C5-pyrene-functionalised triazole-linked LNA monomers **Q** and **R** relative to their DNA counterparts (see Figure 1.4).[47]

1.4.4 Excimer-based Probes

Excimers, *i.e.* electronically excited π-stacking dimers of identical fluorophores, emit highly Stokes-shifted fluorescence (*e.g.* ∼120 nm for pyrene excimers[35]). The presence of excimer emission provides valuable structural insight as the fluorophores adopt co-planar arrangements with a separation of ∼3.4 Å. As discussed in previous sections, fluorophore-functionalised LNA monomers offer great positional control of fluorophores and have accordingly been used in the development of probes that rely on excimer signals for detection of nucleic acid targets. ONs with two next-nearest neighbour incorporations of 2′-*N*-(pyren-1-yl)acetyl-2′-amino-α-L-LNA monomer **Y** are an example of this (Figure 1.13).[92] Duplexes with matched DNA/RNA targets predominantly display pyrene monomer fluorescence, whereas duplexes with mismatched base pairs near the modified region (positions 4–7, Figure 1.13) display intense excimers (2- to 11-fold increase). Similar to BDF probes, and unlike hybridisation probes, this assay does not require stringent temperature control. In other words, SNPs are discriminated even if mismatched duplexes are formed. As discussed previously, the pyrene moiety of monomer **Y** is forced into the core of matched duplexes, which prevents excimer formation due to spatial separation of the pyrene moieties (Figure 1.13). Mismatched duplexes have a more dynamic duplex geometry, which allows the pyrene moieties to adopt extrahelical orientations suitable for formation of pyrene–pyrene excimers (Figure 1.13).[92]

Another class of probes that rely on excimers to signal the presence of SNPs are the so-called dual probes, where two end-labelled probes are assembled into a ternary complex upon target binding.[93] This brings the fluorophores into close proximity, leading to excimer formation (Figure 1.14). The use of two shorter probes for target binding, rather than a long probe, results in improved thermal discrimination of mismatches and reduces false positive

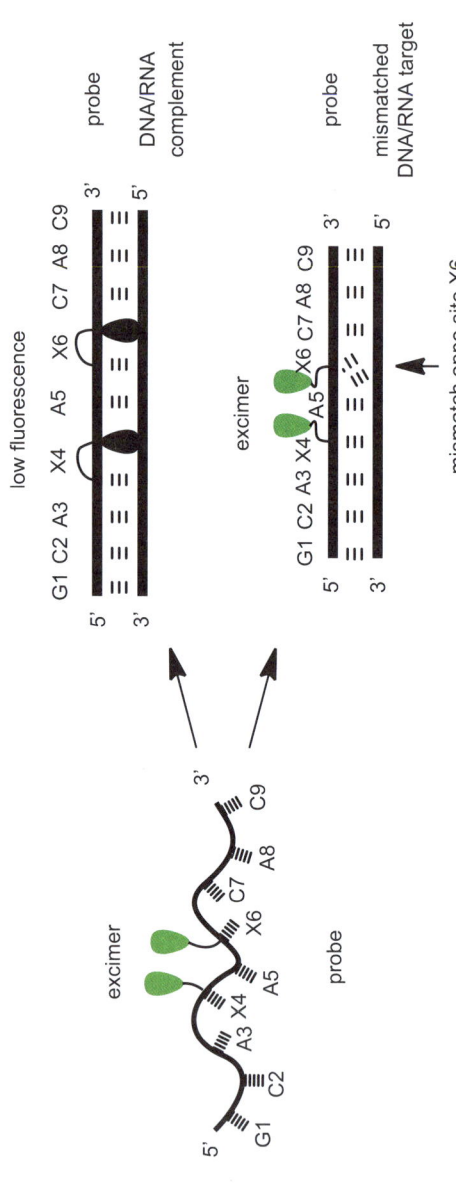

Figure 1.13 Principle of SNP detection using ONs with two next-nearest neighbour incorporation of 2′-*N*-(pyren-1-yl)acetyl-2′-amino-α-L-LNA monomer **Y** (see Figure 1.4 for structure of monomer) (reproduced with permission from ref. 92; copyright 2007 Wiley).

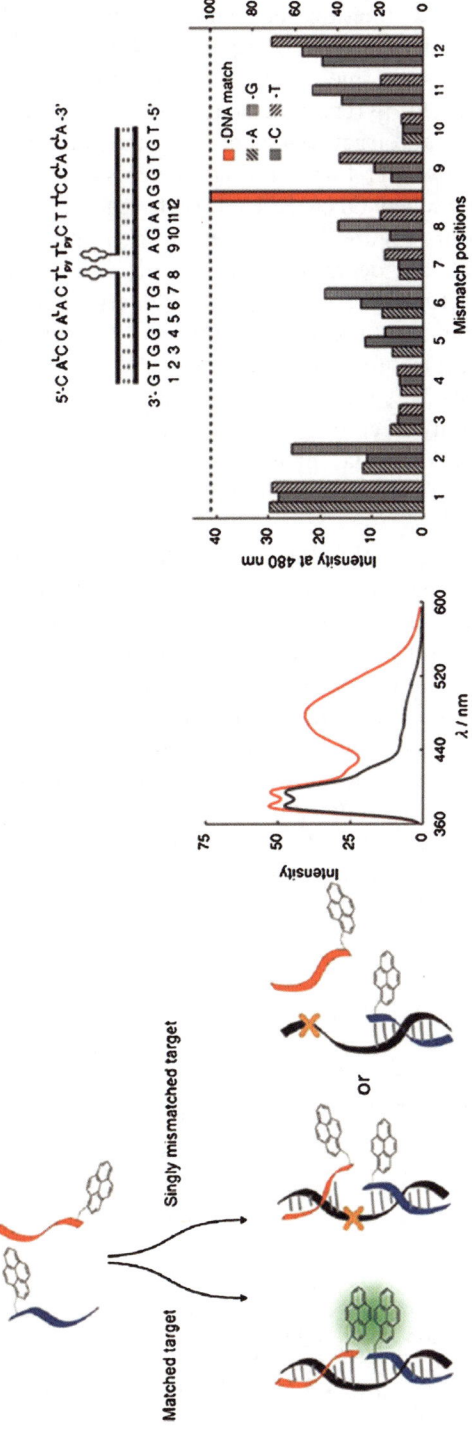

Figure 1.14 *Left:* principle of SNP-discrimination using excimer-forming dual probes. *Centre:* fluorescence emission spectra of duplexes between LNA-based dual probes end-functionalised with monomer **F** and complementary (*red*) or singly mismatched DNA target (*black*). *Right:* probe/target sequences; discrimination of SNPs in positions 1–12 ($\lambda_{em} = 480$ nm) (adapted with permission from ref. 94; copyright 2007 Wiley).

signals. Another advantage of this approach is that the large Stokes shifts allow for clear fluorescent distinction between unhybridised probes and target complexes. LNA-modified ONs, which are end-labelled with $2'$-N-(pyren-1-ylmethyl)-$2'$-amino-LNA monomer **F** (Figure 1.4), have been used in dual probe assays.[94] As expected, pyrene monomer fluorescence is predominantly observed in the absence of target, while binding to complementary DNA/ RNA targets results in excimer formation (Figure 1.14). In contrast, hybridisation to singly mismatched DNA/RNA targets leads to significantly less intense excimer emission (Figure 1.14). Thermal denaturation experiments have demonstrated that the decreased excimer intensity is a result of: a) mismatch-induced structural perturbation of the ternary complex leading to inefficient pyrene–pyrene stacking (positions 8–11, Figure 1.14), and/or b) lack of ternary complex formation due to efficient thermal mismatch discrimination (positions 2–7, Figure 1.14), which is augmented by the short probes and high content of conventional LNA.[94] In contrast, the corresponding 'covalently linked dual probe' only allows for fluorescent discrimination of SNPs in a narrow region (positions 7–10) through local disruption of excimers.[94]

Invader LNA probes, *i.e.* DNA duplexes modified with one or more $+1$ interstrand zipper arrangements of $2'$-N-(pyren-1-yl)methyl-$2'$-amino-α-L-LNA-T monomer **W** (see Figure 1.4), enable detection of double stranded DNA (dsDNA) targets through changes in excimer emission (Figure 1.15).[95,96] This specific monomer arrangement (termed an energetic hotspot) forces the pyrene moieties into the same region within a duplex core, leading to excimer formation, partial duplex unwinding and decreased thermostability.[95,96] Other interstrand monomer arrangements induce much higher duplex thermostabilisation, highlighting the importance of positional control to obtain functional control.[95,96] In contrast, the two strands that comprise an Invader LNA probe display very high affinity toward complementary DNA (ΔT_{m}/mod

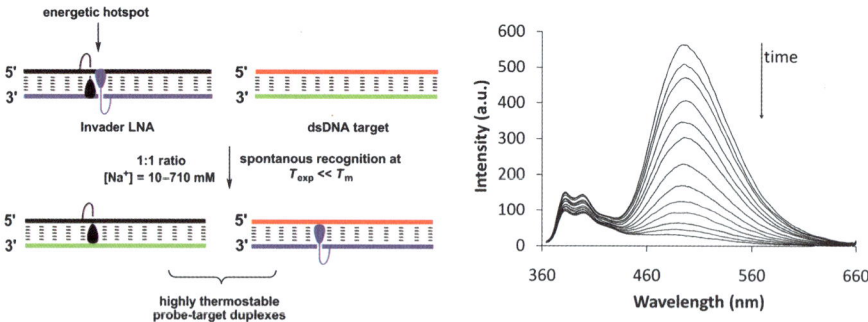

Figure 1.15   *Left*: illustration of Invader LNA concept. *Right*: time course of fluorescence emission upon addition of Invader LNA to a complementary dsDNA target ($\lambda_{\mathrm{ex}} = 335\,\mathrm{nm}$, $20\,^{\circ}\mathrm{C}$). Invader LNA: $5'$-d(GGT A**W**A TAT AGG C):$3'$-d(CCA TA**W** ATA TCC G) (from ref. 96; copyright 2010 RSC Publishing).

up to $+15\,^{\circ}\text{C}$).[53] The large differences in thermostability between probe–target and Invader probe duplexes generate an energetically favourable gradient for recognition of mixed-sequence dsDNA targets at physiologically relevant ionic strengths (Figure 1.15).[95,96] The recognition process is monitored by a decrease in excimer emission as the pyrene moieties are forced apart (Figure 1.15). Addition of a 13-mer Invader LNA to an equimolar quantity of complementary dsDNA target at 110 mM NaCl, pH 7 and low experimental temperatures results in $\sim$50% recognition within $\sim$40 min. Recognition of dsDNA even occurs in buffers of high ionic strength (710 mM NaCl, pH 7, $t_{50\%} \sim 125$ min). The use of Invader LNAs with more than one energetic hotspot accelerates dsDNA recognition. Addition of Invader LNAs to non-isosequential dsDNA targets results in much slower and less complete decay of the excimer signal, which demonstrates the specificity of the recognition process.[96] Subsequent studies have shown that the free energy for dsDNA recognition can be increased through changes in the linker chemistry of the Invader LNA monomers.[97]

1.5 Summary and Perspective

The following classes of fluorophore-conjugated LNAs have been highlighted in the previous sections:

- *N2'-functionalised 2'-amino-LNAs*, which direct fluorophores into the minor groove, without major impacts on duplex thermostability. These characteristics have led to the development of pyrene arrays; communication systems signalling self-assembly of higher order nucleic acid structures; brightly fluorescent hybridisation probes for RNA imaging; and dual probes for SNP-discrimination.
- *N2'-functionalised 2'-amino-α-L-LNAs*, which direct aromatic fluorophores into duplex cores, resulting in extraordinary stabilisation through π-stacking with neighbouring nucleobases. These monomers have been used in the development of regular pyrene arrangements inside duplex cores; excimer-based probes for discrimination of SNPs; and novel strategies for detection of mixed-sequence dsDNA under physiologically relevant conditions.
- *C5-functionalised LNAs and α-L-LNAs*, which direct fluorophores into the major groove, despite marked duplex destabilisation. These monomers have been used to generate base discriminating fluorescent probes for detection of SNPs.
- *C-glycoside LNAs and α-L-LNAs*, which force fluorescent nucleobase surrogates into the duplex core. Although strong duplex destabilisation ensues, these probes display interesting universal hybridisation characteristics and have been used to form regular pyrene arrangements.

A unifying element of these different classes of fluorophore-functionalised LNA is that functional control is achieved through positional control *via* the combined use of conformationally restricted bicyclic skeletons and short linkers for attachment of fluorophores (see Figure 1.2). Several classes of fluorophore-functionalised LNA remain unexplored. For example, attachment of fluorophores to the C8-position of LNA purines[98] is poised to afford building blocks that display interesting BDF-characteristics;[99] the extreme sugar pucker of the LNA skeleton will restrict the rotational freedom of the nucleobase, leading to stricter positional control of the fluorophore, when ompared with corresponding DNA monomers. LNA monomers that are functionalised *via* the extra ring are another class of interesting building blocks.[100–104] The attached fluorophores would be expected to point toward the minor groove, and potentially afford probes with similar characteristics to N2'-functionalised 2'-amino-LNA monomers. Moreover, recent reports on C5'-branched LNA and α-L-LNA monomers have demonstrated that small substituents are well tolerated in the major groove of nucleic acid duplexes, which suggests the 5'-position as another interesting attachment point for fluorophores.[105,106]

The long synthetic routes to fluorophore-functionalised LNA monomers, however, remain a major limitation, and development of synthetically more feasible analogues is therefore needed. As discussed herein, flexible fluorophore-functionalised monomers rarely display comparable target affinity and/or emission profiles relative to their LNA-based counterparts. In a rare exception to this, however, we recently demonstrated that O2'-intercalator-functionalised uridine and N2'-intercalator-functionalised 2'-*N*-methyl-2'-aminouridine monomers (see Figure 1.2) largely mimic the remarkable DNA hybridisation properties of N2'-pyrene-functionalised 2'-amino-α-L-LNAs (average ΔT_m/mod $\sim 8\,^\circ$C).[107] Conformational restriction of the sugar moiety presumably plays a less important role in this case because intercalation of the fluorophore is the dominating binding mode. Gratifyingly, these monomers have been demonstrated to be equipotent mimics of the original Invader LNA monomers, which will allow more extensive exploration of this novel strategy of dsDNA targeting.[97]

An interesting approach to more readily accessible 2'-amino-LNA monomers has been proposed by Wengel's group. Inspired by the ability of conventional LNA monomers increasingly to tune the conformation of neighbouring 2'-deoxyribonucleotides toward RNA-like C3'-*endo* conformations during duplex formation,[14] ONs were modified with 2'-*N*-pyren-1-ylmethyl-2'-*N*-methylaminouridine monomer **E** (see Figure 1.2) and surrounded by conventional LNA nucleotides.[108] Unfortunately, the probes display markedly lower target affinity than the corresponding reference DNA–LNA mixmers, presumably because LNA-induced conformational tuning of monomer **E** into a C3'-*endo* conformation is not energetically favourable. However, other combinations of fluorophore-functionalised DNA-monomers and LNA-backbones are likely to afford readily accessible mimics of fluorophore-functionalised LNAs in the future.[109]

The unique photophysical properties of fluorophore-functionalised LNAs will undoubtedly continue to stimulate development of new building blocks, leading to applications in life sciences, sensor technology and materials science.

References

1. U. Asseline, *Curr. Org. Chem.*, 2006, **10**, 491–518.
2. J. Wilhelm and A. Pingoud, *ChemBioChem*. 2003, **4**, 1120–1128.
3. D. P. Bratu, B.-J. Cha, M. M. Mhlanga, F. R. Kramer and S. Tyagi, *Proc. Natl. Acad. Sci. USA*, 2003, **100**, 13308–13313.
4. S. Kim and A. Misra, *Annu. Rev. Biomed. Eng.*, 2007, **9**, 289–320.
5. M. K. Smalley and S. K. Silverman, *Nucleic Acids Res.*, 2006, **34**, 152–166.
6. N. Dai and E. T. Kool, *Chem. Soc. Rev.*, 2011, **40**, 5756–5770.
7. V. L. Malinovskii, D. Wenger and R. Häner, *Chem. Soc. Rev.*, 2010, **39**, 410–422.
8. B. Juskowiak, *Anal. Bioanal. Chem.*, 2011, **399**, 3157–3176.
9. L. E. Morrison, *J. Fluoresc.*, 1999, **9**, 187–196.
10. J. Wengel, *Acc. Chem. Res.*, 1999, **32**, 301–310.
11. S. Obika, S. M. A. Rahman, A. Fujisaka, Y. Kawada, T. Baba and T. Imanishi, *Heterocycles*, 2010, **81**, 1347–1392.
12. A. A. Koshkin, S. K. Singh, P. Nielsen, V. K. Rajwanshi, R. Kumar, M. Meldgaard, C. E. Olsen and J. Wengel, *Tetrahedron*, 1998, **54**, 3607–3630.
13. S. Obika, D. Nanbu, Y. Hari, J. I. Andoh, K. I. Morio, T. Doi and T. Imanishi, *Tetrahedron Lett.*, 1998, **39**, 5401–5404.
14. M. Petersen, C. B. Nielsen, K. E. Nielsen, G. A. Jensen, K. Bondensgaard, S. K. Singh, V. K. Rajwanshi, A. A. Koshkin, B. M. Dahl, J. Wengel and J. P. Jacobsen, *J. Mol. Recognit.*, 2000, **13**, 44–53.
15. K. E. Nielsen and H. P. Spielmann, *J. Am. Chem. Soc.*, 2005, **127**, 15273–15282.
16. M. Petersen, K. Bondensgaard, J. Wengel and J. P. Jacobsen, *J. Am. Chem. Soc.*, 2002, **124**, 5974–5982.
17. V. K. Rajwanshi, A. E. Hakansson, M. D. Sorensen, S Pitsch, S. K. Singh, R. Kumar, P. Nielsen and J. Wengel, *Angew. Chem. Int. Ed.*, 2000, **39**, 1656–1659.
18. K. M. E. Nielsen, M. Petersen, A. E. Håkansson, J. Wengel and J. P. Jacobsen, *Chem. Eur. J.*, 2002, **8**, 3001–3009.
19. J. T. Nielsen, P. Stein and M. Petersen, *Nucleic Acids Res.*, 2003, **31**, 5858–5867.
20. T. Koch and H. Ørum, in *Antisense Drug Technology – Principles, Strategies and Applications*, ed. S. T. Crooke, CRC Press, Boca Raton, 2nd edn, 2008, pp. 519–564.
21. P. M. McTigue, R. J. Peterson and J. D. Kahn, *Biochemistry*. 2004, **43**, 5388–5405.
22. Y. You, B. G. Moreira, M. A. Behlke and R. Owczarzy, *Nucleic Acids Res.*, 2006, **34**, e60.
23. M. D. Sørensen, L. Kværnø, T. Bryld, A. E. Håkansson, B. Verbeure, G. Gaubert, P. Herdewijn and J. Wengel, *J. Am. Chem. Soc.*, 2002, **124**, 2164–2176.
24. T. P. Prakash, *Chem. Biodiv.*, 2011, **8**, 1616–1641.
25. H. Kaur, B. R. Babu and S. Maiti, *Chem. Rev.*, 2007, **107**, 4672–4697.

26. C. F. Bennett and E. E. Swayze, *Annu. Rev. Pharmacol. Toxicol.*, 2010, **50**, 259–293.
27. C. Wahlestedt, P. Salmi, L. Good, J. Kela, T. Johnsson, T. Hokfelt, C. Broberger, F. Porreca, J. Lai, K. K. Ren, M. Ossipov, A. Koshkin, N. Jacobsen, J. Skouv, H. Oerum, M. H. Jacobsen and J. Wengel, *Proc. Natl. Acad. Sci. USA*, 2000, **97**, 5633–5638.
28. J. Elmen, M. Lindow, S. Schutz, M. Lawrence, A. Petri, S. Obad, M. Lindholm, M. Hedtjarn, H. F. Hansen, U. Berger, S. Gullans, P. Kearney, P. Sarnow, E. M. Straarup and S. Kauppinen, *Nature*. 2008, **452**, 896–899.
29. M. A. Graziewicz, T. K. Tarrant, B. Buckley, J. Roberts, L. Fulton, H. Hansen, H. Ørum, R. Kole and P. Sazani, *Mol. Ther.*, 2008, **16**, 1316–1322.
30. E. Wienholds, W. P. Kloostermann, W. P. Misk, E. Alvarez-Saavedra, E. Berezikov, E. Bruijn, H. R. Horvitz, S. Kauppinen and R. H. A. Plasterk, *Nature*. 2005, **309**, 310–311.
31. D. Latorra, K. Campbell, A. Wolter and J. M. Hurley, *Hum. Mutat.*, 2003, **22**, 79–85.
32. M. A. Campbell and J. Wengel, *Chem. Soc. Rev.*, 2011, **40**, 5680–5689.
33. S. H. Weisbrod and A. Marx, *Chem. Commun.*, 2008, 5675–5685.
34. K. M. Guckian, B. A. Schweitzer, R. X. F. Ren, C. J. Sheils, D. C. Tahmassebi and E. T. Kool, *J. Am. Chem. Soc.*, 2000, **122**, 2213–2222.
35. F. M. Winnik, *Chem. Rev.*, 1993, **93**, 587–614.
36. K. Kalyanasundaram and J. K. Thomas, *J. Am. Chem. Soc.*, 1977, **99**, 2039–2044.
37. M. Manoharan, K. L. Tivel, M. Zhao, K. Nafisi and T. L. Netzel, *J. Phys. Chem.*, 1995, **99**, 17461–17472.
38. M. D. Sørensen, M. Petersen and J. Wengel, *Chem. Commun.*, 2003, 2130–2131.
39. M. W. Johannsen, L. Crispino, M. C. Wamberg, N. Kalra and J. Wengel, *Org. Biomol. Chem.*, 2011, **9**, 243–252.
40. D. Lindegaard, A. S. Madsen, I. V. Astakhova, A. D. Malakhov, B. R. Babu, V. A. Korshun and J. Wengel, *Bioorg. Med. Chem.*, 2008, **16**, 94–99.
41. I. V. Astakhova, V. A. Korshun and J. Wengel, *Chem. Eur. J.*, 2008, **14**, 11010–11026.
42. I. V. Astakhova, V. A. Korshun, K. Jahn, J. Kjems and J. Wengel, *Bioconj. Chem.*, 2008, **19**, 1995–2007.
43. I. V. Astakhova, D. Lindegaard, V. A. Korshun and J. Wengel, *Chem. Commun.*, 2010, 8362–8364.
44. P. Gupta, N. Langkjær and J. Wengel, *Bioconj. Chem.*, 2010, **21**, 513–520.
45. P. J. Hrdlicka, B. R. Babu, M. D. Sørensen and J. Wengel, *Chem. Commun.*, 2004, 1478–1479.
46. P. J. Hrdlicka, B. R. Babu, M. D. Sørensen, N. Harrit and J. Wengel, *J. Am. Chem. Soc.*, 2005, **127**, 13293–13299.
47. M. E. Østergaard and P. J. Hrdlicka, unpublished data.

48. M. E. Østergaard, P. Kumar, B. Baral, D. J. Raible, T. S. Kumar, B. A. Anderson, D. C. Guenther, L. Deobald, A. J. Paszczynski, P. K. Sharma and P. J. Hrdlicka, *ChemBioChem.* 2009, **10**, 2740–2743.

49. B. R. Babu and J. Wengel, personal communication.

50. H. M. Pfundheller and C. Lomholt, *Curr. Protoc. Nucleic Acid Chem.*, 2002, **35**, 4.12.1–4.12.16.

51. A. A. Koshkin, J. Fensholdt, H. M. Pfundheller and C. Lomholt, *J. Org. Chem.*, 2001, **66**, 8504–8512.

52. C. Rosenbohm, S. M. Christensen, M. D. Sørensen, D. S. Pedersen, L. E. Larsen, J. Wengel and T. Koch, *Org. Biomol. Chem.*, 2003, **1**, 655–663.

53. T. S. Kumar, A. S. Madsen, M. E. Østergaard, S. P. Sau, J. Wengel and P. J. Hrdlicka, *J. Org. Chem.*, 2009, **74**, 1070–1081.

54. T. S. Kumar, A. S. Madsen, J. Wengel and P. J. Hrdlicka, *J. Org. Chem.*, 2006, **71**, 4188–4201.

55. I. V. Astakhova, T. S. Kumar and J. Wengel, *Collect. Czech. Chem. Commun.*, 2011, **76**, 1347–1360.

56. M. E. Østergaard, P. Kumar, B. Baral, D. C. Guenther, B. A. Anderson, F. M. Ytreberg, L. Deobald, A. J. Paszczynski, P. K. Sharma and P. J. Hrdlicka, *Chem. Eur. J.*, 2011, **17**, 3157–3165.

57. T. S. Kumar, P. Kumar, P. K. Sharma and P. J. Hrdlicka, *Tetrahedron Lett.*, 2008, **49**, 7168–7170.

58. P. Kumar, M. E. Østergaard and P. J. Hrdlicka, *Curr. Protoc. Nucleic Acid Chem.*, 2011, **44**, 4.43.1–4.43.22.

59. B. R. Babu, A. K. Prasad, S. Trikha, N. Thorup, V. S. Parmar and J. Wengel, *J. Chem. Soc. Perkin Trans.*, 2002, **1**, 2509–2519.

60. Raunak, B. R. Babu, M. D. Sørensen, V. S. Parmar, N. H. Harrit and J. Wengel, *Org. Biomol. Chem.*, 2004, **2**, 80–89.

61. C. Verhagen, T. Bryld, M. Raunkjaer, S. Vogel, K. Buchalova and J. Wengel, *Eur. J, Org. Chem.*, 2006, 2538–2548.

62. T. J. Matray and E. T. Kool, *J. Am. Chem. Soc.*, 1998, **120**, 6191–6192.

63. S. P. Sau and P. J. Hrdlicka, *J. Org. Chem.*, 2012, **77**, 5–16.

64. D. Loakes, *Nucleic Acids Res.*, 2001, **29**, 2437–2447.

65. R. Varghese and H. A. Wagenknecht, *Chem. Commun.*, 2009, 2615–2624.

66. T. N. Nguyen, A. Brewer and E. Stulz, *Angew. Chem. Int. Ed.*, 2009, **48**, 1974–1977.

67. E. Mayer-Enthart and H. A. Wagenknecht, *Angew. Chem. Int. Ed.*, 2006, **45**, 3372–3375.

68. M. Nakamura, Y. Shimomura, Y. Ohtoshi, K. Sasa, H. Hayashi, H. Nakano and K. Yamana, *Org. Biomol. Chem.*, 2007, **5**, 1945–1951.

69. M. Nakamura, Y. Fukunaga, K. Sasa, Y. Ohtoshi, K. Kanaori, H. Hayashi, H. Nakano and K. Yamana, *Nucleic Acids Res.*, 2005, **33**, 5887–5895.

70. M. Nakamura, Y. Murakami, K. Sasa, H. Hayashi and K. Yamana, *J. Am. Chem. Soc.*, 2008, **130**, 6904–6905.

71. D. Lindegaard, B. R. Babu and J. Wengel, *Nucleos. Nucleot. Nucleic Acids*, 2005, **24**, 679–681.

72. K. Pasternak, A. Pasternak, P. Gupta, R. N. Veedu and J. Wengel, *Bioorg. Med. Chem.*, 2011, **19**, 7407–7415.

73. T. S. Kumar, A. S. Madsen, M. E. Østergaard, J. Wengel and P. J. Hrdlicka, *J. Org. Chem.*, 2008, **73**, 7060–7066.

74. T. Ishiguro, J. Saitoh, H. Yawata, M. Otsuka, T. Inoue and Y. Sugiura, *Nucleic Acids Res.*, 1996, **24**, 4992–4997.

75. J. B. Randolph and A. S. Wagonner, *Nucleic Acids Res.*, 1997, **25**, 2923–2929.

76. A. O. Crockett and C. T. Wittwer, *Anal. Chem.*, 2001, **290**, 89–97.

77. D. J. French, C. L. Archard, T. Brown and D. G. McDowell, *Mol. Cell Probes.*, 2001, **15**, 363–374.

78. K. Yamana, H. Zako, K. Asazuma, R. Iwase, H. Nakano and A. Murakami, *Angew. Chem. Int. Ed.*, 2001, **40**, 1104–1106.

79. K. Yamana, R. Iwase, S. Furutani, H. Tsuchida, H. Zako, T. Yamaoka and A. Murakami, *Nucleic Acids Res.*, 1999, **27**, 2387–2392.

80. S. S. Iversen and P. J. Hrdlicka, unpublished data.

81. M. E. Østergaard, D. C. Guenther, P. Kumar, B. Baral, L. Deobald, A. J. Paszczynski, P. K. Sharma and P. J. Hrdlicka, *Chem. Commun.*, 2010, 4929–4931.

82. M. E. Østergaard, J. Maity, B. R. Babu, J. Wengel and P. J. Hrdlicka, *Bioorg. Med, Chem. Lett.*, 2010, **20**, 7265–7268.

83. M. E. Østergaard, P. Cheguru, M. R. Papasani, R. A. Hill and P. J. Hrdlicka, *J. Am. Chem. Soc.*, 2010, **132**, 14221–14228.

84. D. Auld and A Simeonov, *Assay Drug Dev. Technol.*, 2005, **3**, 581–593.

85. K. Wang, Z. Tang, C. J. Yang, Y. Kim, X. Fang, W. Li, Y. Wu, C. D. Medley, Z. Cao, J. Li, P. Colon, H. Lin and W. Tan, *Angew. Chem. Int. Ed.*, 2009, **48**, 856–870.

86. N. Venkatesan, Y. J. Seo and B. H. Kim, *Chem. Soc. Rev.*, 2008, **37**, 648–663.

87. D. Honcharenko, C. Zhou and J. Chattopadhyaya, *J. Org. Chem.*, 2008, **73**, 2829–2842.

88. D. W. Dodd and R. H. E. Hudson, *Mini-Rev. Org. Chem.*, 2009, **6**, 378–391.

89. A. Okamoto, Y. Saito and I. Saito, *J. Photochem. Photobiol., C.* 2005, **6**, 108–122.

90. A. Okamoto, K. Kanatani and I. Saito, *J. Am. Chem. Soc.*, 2004, **126**, 4820–4827.

91. A. Okamoto, K. Tainaka, Y. Ochi, K. Kanatani and I. Saito, *Mol. Biosyst.*, 2006, **2**, 122–126.

92. T. S. Kumar, J. Wengel and P. J. Hrdlicka, *ChemBioChem.* 2007, **8**, 1122–1125.

93. D. M. Kolpashchikov, *Chem. Rev.*, 2010, **110**, 4709–4723.

94. T. Umemoto, P. J. Hrdlicka, B. R. Babu and J. Wengel, *ChemBioChem.* 2007, **8**, 2240–2248.

95. P. J. Hrdlicka, T. S. Kumar and J. Wengel, *Chem. Commun.*, 2005, 4279–4281.

96. S. P. Sau, T. S. Kumar and P. J. Hrdlicka, *Org. Biomol. Chem.*, 2010, **8**, 2028–2036.
97. S. P. Sau and P. J. Hrdlicka, unpublished data.
98. P. J. Hrdlicka, P. Kumar and Michael E. Østergaard, PCT/US2010/048520.
99. Y. Shinohara, K. Matsumoto, K. Kugenuma, T. Morii, Y. Saito and I. Saito, *Bioorg. Med. Chem. Lett.*, 2010, **20**, 2817–2820.
100. P. P. Seth, G Vasquez, C. A. Allerson, A. Berdeja, H. Gaus, G. A. Kinberger, T. P. Prakash, M. T. Migawa, B. Bhat and E. E. Swayze, *J. Org. Chem.*, 2010, **75**, 1569–1581.
101. P. P. Seth, J. Yu, C. R. Allerson, A. Berdeja and E. E. Swayze, *Bioorg. Med. Chem. Lett.*, 2011, **21**, 1122–1125.
102. J. Xu, Y. Liu, C. Dupouy and J. Chattopadhyaya, *J. Org. Chem.*, 2009, **74**, 6534–6554.
103. Q. Li, F. Yuan, C. Zhou, O. Plashkevych and J. Chattopadhyaya, *J. Org. Chem.*, 2010, **75**, 6122–6140.
104. S. Kumar, M. H. Hansen, N. Albaek, S. I. Steffansen, M. Petersen and P. Nielsen, *J. Org. Chem.*, 2009, **74**, 6756–6769.
105. P. P. Seth, C. R. Allerson, A. Siwkowski, G. Vasquez, A. Berdeja, M. T. Migawa, H. Gaus, T. P. Prakash, B. Bhat and E. E. Swayze, *J. Med. Chem.*, 2010, **53**, 8309–8318.
106. P. P. Seth, C. R. Allerson, M. E. Østergaard and E. E. Swayze, *Bioorg. Med. Chem. Lett.*, 2012, **22**, 296–299.
107. S. Karmakar, B. A. Anderson, R. L. Rathje, S. Andersen, T. B. Jensen, P. Nielsen and P. J. Hrdlicka, *J. Org. Chem.*, 2011, **76**, 7119–7131.
108. N. Kalra, B. R. Babu, V. S. Parmar and J. Wengel, *Org. Biomol. Chem.*, 2004, **2**, 2885–2887.
109. I. V. Astakhova, A. V. Ustinov, V. A. Korshun and J. Wengel, *Bioconj. Chem.*, 2011, **22**, 533–539.

CHAPTER 2

Fluorophore Conjugates for Single Molecule Work

ROHAN T. RANASINGHE AND DAVID KLENERMAN*

Department of Chemistry, University of Cambridge, Lensfield Road,
Cambridge CB2 1EW, UK
*Email: dk10012@cam.ac.uk

2.1 Introduction

We begin this chapter by introducing some inherent properties of single molecule fluorescence spectroscopy (SMFS) that have been key to its success in studies of DNA. In genomics (Section 2.3.1.1), single molecule methods have delivered unrivalled speeds of data acquisition, reducing the time required to sequence a human genome from several years to one day. This remarkable acceleration has been possible because the spatial and temporal resolution of SMFS allows detection in a massively parallel way, where millions of small scale polymerisation reactions are monitored simultaneously. The efficiency with which information is generated from reagents and time is therefore dramatically better than is possible using bulk techniques. In biophysical studies (Section 2.3.2), the measurement of single molecules reveals the nature of dynamic processes that are inaccessible to bulk methods without synchronisation (Figure 2.1A) and permits the resolution of multiple populations of species that would be hidden or misrepresented by ensemble-averaged techniques (Figure 2.1B).

This section illustrates the optical methods that have most enabled the study of DNA at the single molecule level by fluorescence spectroscopy. The detailed

RSC Biomolecular Sciences No. 26
DNA Conjugates and Sensors
Edited by Keith R Fox and Tom Brown
© The Royal Society of Chemistry 2012
Published by the Royal Society of Chemistry, www.rsc.org

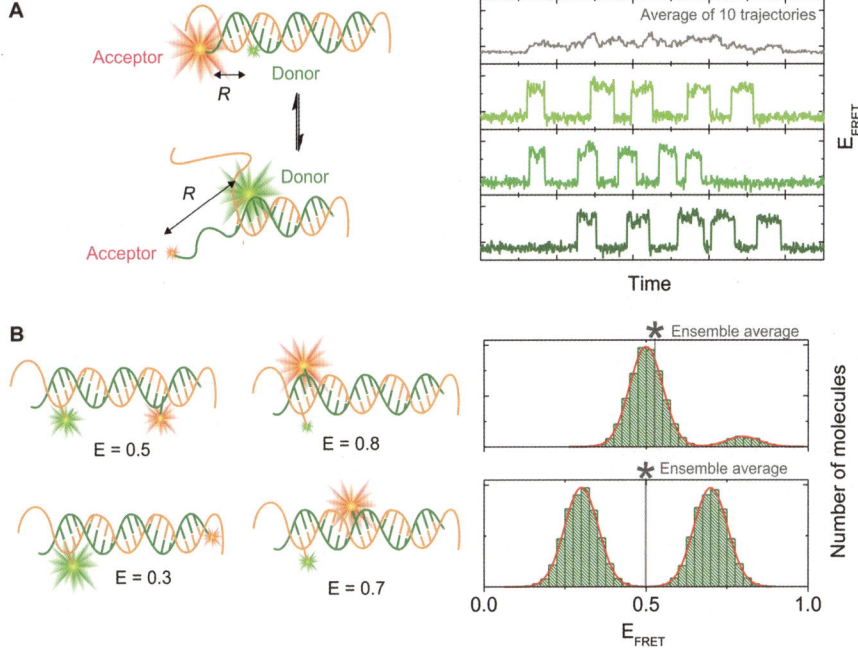

Figure 2.1 Advantages of single molecule detection. (A) Studying dynamics: observations of transitions of a DNA duplex between two structures with different Förster resonance energy transfer efficiencies (E_{FRET}). Simulated data from three single molecules show individual transitions, while the average of as few as 10 time traces (or trajectories) obscures the resolution of the individual events because they are not synchronised. (B) Revealing heterogeneity: the presence of a small fraction (10%) of a species with high FRET efficiency (0.8) in an excess of DNA with $E_{FRET} = 0.5$ can be observed as a bimodal distribution in a FRET histogram, while in a bulk measurement the result is a small shift in the measured value (0.53), which may not be considered significant. Ensemble averaging completely misrepresents the nature of a sample consisting of an equimolar mixture of species with $E_{FRET} = 0.3$ and 0.7, indicating a single FRET efficiency of 0.5. Both histograms are generated from simulated Gaussian distributions ($n = 1000$).

development of SMFS and its application to studying biological questions have been reviewed extensively by leading practitioners in the field.[1–7] In particular, the article by Moerner and Fromm provides a comprehensive survey of experimental techniques used for SMFS.[3] As a result, we only describe the basic principles underlying the analytical applications described in this chapter.

2.1.1 Single Molecule Detection Techniques

The measurement of single fluorophores is now a routine practice in many laboratories as a result of technological advances that have occurred in the

20 years since the first observation of single Rhodamine 6G molecules dissolved in water at room temperature.[8] The availability of commercial instruments, from sequencing platforms such as the HeliScope Single Molecule Sequencer by Helicos to microscope systems for basic research, such as the MicroTime 200 by PicoQuant, underlines the maturity of the technology.

In order for single molecule detection (SMD) to be achieved, several conditions must be met. To maximise the signal, the detector must be very sensitive [typically avalanche photodiodes (APDs) or charge-coupled device (CCD) cameras are used] and emitted photons must be collected very efficiently, usually by microscope objectives with high numerical aperture (NA). The minimisation of background noise resulting from fluorescent impurities or Raman scattering from solvent molecules or bulk solutes is also of critical importance. This goal can be achieved by reducing the observation volume, because signals from bulk species increase with the volume sampled, while those from the single fluorophore of interest do not. There are three strategies for minimising the observation volume of particular importance for the applications described here: confocal illumination, total internal reflection fluorescence microscopy (TIRFM) and the use of zero-mode waveguides (ZMWs), each of which is described below.

In the confocal setup (Figure 2.2A), excitation light from a laser source is focused by a microscope objective to a diffraction-limited spot with a diameter of $\sim 1 \, \mu m$,[9] whose radial intensity profile obeys a Gaussian function. Emitted photons are collected by the same objective, separated from excitation light by a dichroic mirror, and a pinhole (typically $50 \, \mu m$ in width) is positioned in the conjugate image plane to reject out-of-focus light. This arrangement of optical components defines an elliptical observation volume of $\sim 0.5 \, fL$ (Figure 2.2B). Even when sampling this minute volume, it is necessary to work at low concentrations of analyte to ensure observation of one molecule at a time. Simple calculation tells us that a concentration of 3.3 nM would lead to an average of one molecule occupying this detection volume at a given time, and to ensure that discrete bursts from single molecules are observed, it is necessary to work at analyte concentrations of 10–100 pM.[10] In this single molecule regime, it is possible to obtain multiple parameters such as fluorescence intensities, lifetimes, anisotropy and Förster resonance energy transfer (FRET) efficiency,[11] though intensity and FRET are the most widely used in SMFS of DNA. Finally, if analytes are moving, either by Brownian motion or in flowing samples, it is necessary to collect data with high temporal resolution, so that signals from individual molecules are observed as discrete 'bursts' of fluorescence.

In contrast to discrete burst analysis, fluorescence correlation spectroscopy (FCS) relies on measuring temporal fluctuations in intensity, where the detection volume can be occupied by multiple fluorophores. It is most useful in analysis of more concentrated samples, in the 1–100 nM range.[4,12] Since these fluctuations are due to molecules diffusing in and out of the detection volume, the autocorrelation function, $G(\tau)$, that is generated enables measurement of diffusion coefficients and molecular dynamics, for example in single stranded

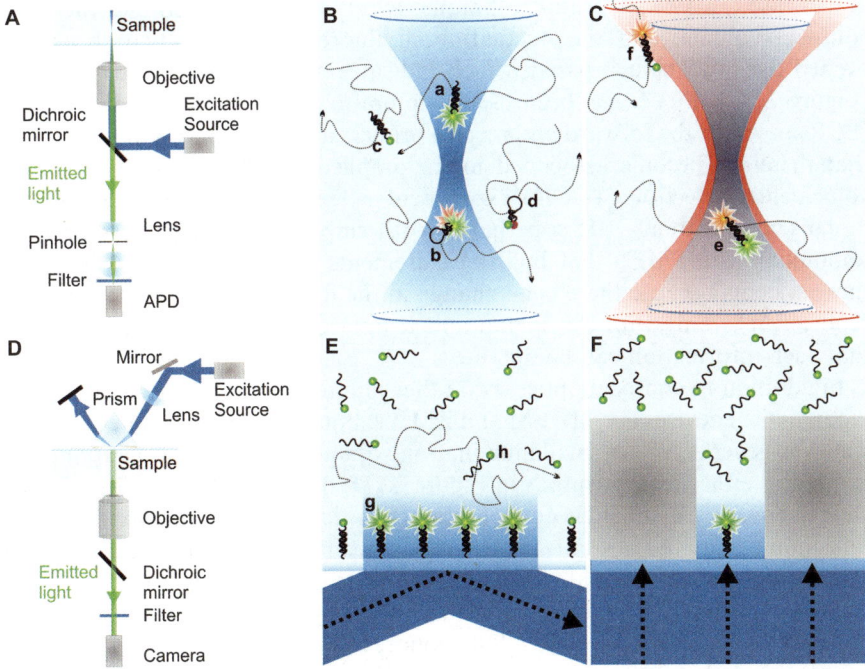

Figure 2.2 Configurations for SMFS. (A) Confocal microscope setup. (B) Detection of single molecules diffusing through a confocal illumination volume ($\sim 1\,\mu m$ diameter, $\sim 0.5\,fL$). Molecules that are labelled with a fluorophore that are excited by the laser light give rise to a fluorescence burst in one colour (molecule a), or two colours (molecule b) if undergoing Förster resonance energy transfer (FRET). When energy transfer is measured using SMFS, the experiment is often called smFRET (single molecule FRET) or spFRET (single pair FRET). If the path of the molecule does not take it through the illuminated volume, no emission occurs (molecules c and d). (C) Detection using two overlapped laser beams. Dual-labelled species that encounter the overlapped region during Brownian motion exhibit fluorescence from both reporters (molecule e), while some molecules are only excited by one laser because of imperfect overlap (molecule f). (D) Prism-based TIRFM setup. (E) Detection of fluorescence by TIRFM on a microscope slide. Due to the evanescently decaying illumination field (depth $\sim 100\,nm$), only surface-bound species (*e.g.* molecule g) are excited for long enough to generate a significant signal above background, whereas mobile species (*e.g.* molecule h) encounter the illuminated volume only briefly owing to Brownian motion and therefore do not build up a significant signal above noise in any given pixel when averaged over the duration of the measurement. (F) Detection of a single molecule inside a nanofabricated ZMW (diameter 20–100 nm), with an illuminated volume of 10–20 zL.

(ss-) and double stranded (ds)DNA.[13] FCS is most effective at distinguishing species with very large differences in molecular weight (and therefore hydrodynamic radius), which limits its usefulness. For this reason, fluorescence cross correlation spectroscopy (FCCS), first proposed in 1994,[14] was developed and first realised in 1997.[15] In this technique, two lasers are focused into the same confocal volume, and the fluctuation in fluorescence from two separately excited dyes can be compared, which indicates whether they are co-localised (Figure 2.2C). FCCS has been used to monitor a polymerase chain reaction (PCR) in which the forward and reverse primers are labelled with different dyes that therefore become associated in the amplicon.[16] The related two colour coincidence detection (TCCD) experiment extends the sensitivity to femto-molar concentrations.[17] It is perhaps worth emphasising for readers who are unfamiliar with SMFS that bulk measurements cannot reveal association in these systems because there is no change in the fluorescence intensity of either reporter; it is only with the spatial and temporal resolution offered by SMD that such information can be obtained.

In addition to confocal optics, two other illumination methods important in bioanalytical devices are TIRFM and ZMWs, both of which rely on evanescent waves to generate a very small illumination volume. In TIRFM, excitation light is directed either by a prism (as in Figure 2.2D) or a microscope objective to a slide or coverslip at an angle greater than the critical angle, so that total internal reflection occurs. As a result, the excitation beam does not propagate into the sample, but an exponentially decaying evanescent wave at the interface pene-trates the sample, generating a thin film of illumination (~ 100 nm) where its intensity is sufficient to excite fluorophores (Figure 2.2E).[18,19] ZMWs are nanofabricated cylindrical holes in a metal film, whose diameter (20–100 nm) is much smaller than the wavelength of the excitation light (Figure 2.2F). Under these conditions there are no propagating modes, and an exponentially decaying evanescent field produces an illuminated volume of ~ 20 zL. In con-trast to confocal optics, an average of one molecule in the detection volume is produced by an analyte concentration of 83 µM and single molecule detection of fluorophores at 10 µM can be achieved.[20]

2.2   Labelling DNA for Single Molecule Detection

2.2.1   Selection of Fluorophores

The intrinsically limited signal-to-noise ratio obtained in SMD places exacting demands on the properties of the reporter fluorophore. The number of photons emitted by a reporter within a given observation period is ultimately deter-mined by the number of excitation/emission cycles it can complete, which in turn depends on its fluorescent lifetime, because the molecule must return to its ground state before another excitation photon can be absorbed. In order to overcome the effects of background and shot noise it is necessary to collect ~ 25 photons, implying a minimum measurement time of 100 ns, given a fluorescence lifetime of 4 ns. However, while high NA objectives can collect a relatively large fraction of emitted photons (e.g. 26% of photons for NA = 1.3), once ineffi-ciencies in the detector and other optical components are taken into account, an overall efficiency as low as $\sim 1\%$ is achieved.[21] The quantum yield of the label further limits the photon emission rate, and the high illumination intensities used in SMD to ensure that the fluorophore is excited rapidly after relaxation tend to give rise to photon-driven processes that compete with fluorescence, such as bleaching or blinking (see Section 2.2.2). When these factors are taken into account, we arrive at a realistic minimum acquisition period of around 50–100 µs. This represents the temporal resolving power of the technique for studying the dynamics of, for example, protein–DNA interactions, and its inverse defines the throughput possible in analytical applications. In super-resolution microscopy (see Section 3.2.1), the precision of localisation is determined by the square root of the number of photons collected.[22] As a result, researchers who practise SMFS constantly strive to maximise the rate of photon emission, selecting only the reporters that display the largest extinction coefficients, highest quantum yields and greatest photostability at high laser power. Fluorescent nucleobase analogues, which have been very useful for monitoring structural changes of DNA in bulk experiments, have been employed with only limited success in SMFS owing to their photophysical properties.[23,24]

Since laser induced fluorescence (LIF) is the most suitable excitation mode for SMD, the dyes that have been most widely used are those efficiently excited by most common lasers; among these are the argon ion (488 nm), frequency doubled Nd:YAG (532 nm), krypton ion (568 nm) and HeNe (633 nm) laser sources, although modern solid state lasers are becoming available at multiple wavelengths. The 19th-century dyestuffs (xanthenes, rhodamines and cyanines) dominate the field, though the original chromophores have been finely tuned by synthetic chemistry to generate modern versions with superior properties. These are available from many suppliers (Life Technologies, ATTO TEC, GE Healthcare, Glen Research, Lumiprobe and Dyomics among others), each of which offers distinct patented fluorophore structures (Figure 2.3, Table 2.1). In the first demonstration of SMD, 80–100 fluorescein molecules were conjugated to a single γ-globulin molecule *via* a polyethyleneimine linker and detected following excitation by an Ar ion laser.[23] Despite its widespread use in applications from real time PCR to retinal angiography, fluorescein is not now widely used in SMFS because of its susceptibility to photobleaching (of most importance in imaging applications), and the pH-dependence of its quantum yield, which is due to ionisation of the phenolic group ($pK_a = 6.4$). In order to improve fluorescence at neutral pH, the 2,7-difluoro derivative of fluorescein, Oregon Green, was synthesised. The electron withdrawing effect of the fluorine atoms reduces the pK_a to 4.8, making this dye suitable for detection at or below pH 7.[24] Dyes from the rhodamine family are among those most commonly studied by SMFS, and many derivatives are available. The first detection of single fluorophores in aqueous conditions 20 years ago was of a 100 fM solution of Rhodamine 6G excited at 532 nm,[8] and sulfonated/elaborated derivatives of that dye, Alexa Fluor 532 and ATTO 532, remain popular in the present day.[25] For excitation at 488 nm, Rhodamine Green (Rhodamine 110) or its sulfonated derivatives Alexa Fluor 488 or ATTO 488 are widely used; these labels are preferred to fluorescein derivatives, because they are completely insensitive to

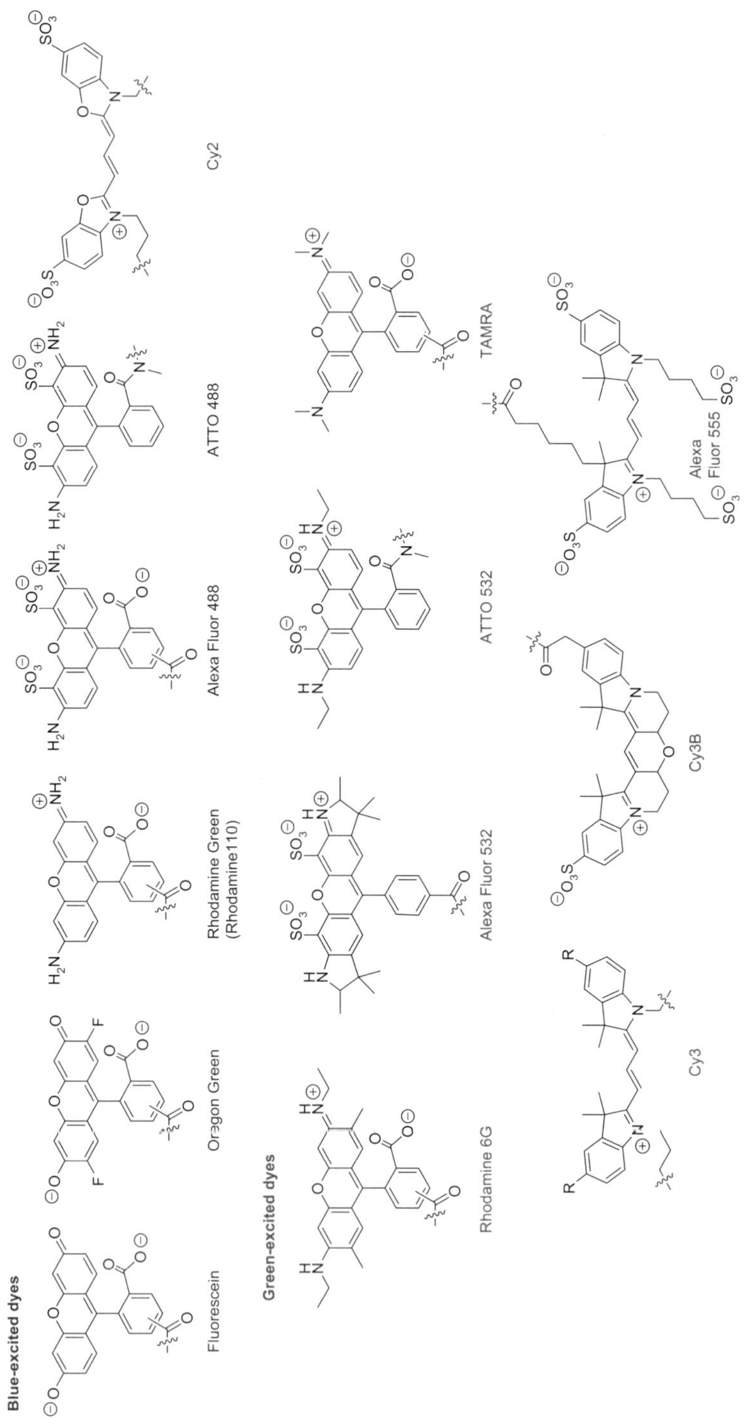

Blue-excited dyes

Fluorescein

Oregon Green

Rhodamine Green
(Rhodamine110)

Alexa Fluor 488

ATTO 488

Cy2

Green-excited dyes

Rhodamine 6G

Alexa Fluor 532

ATTO 532

TAMRA

Cy3

Cy3B

Alexa Fluor 555

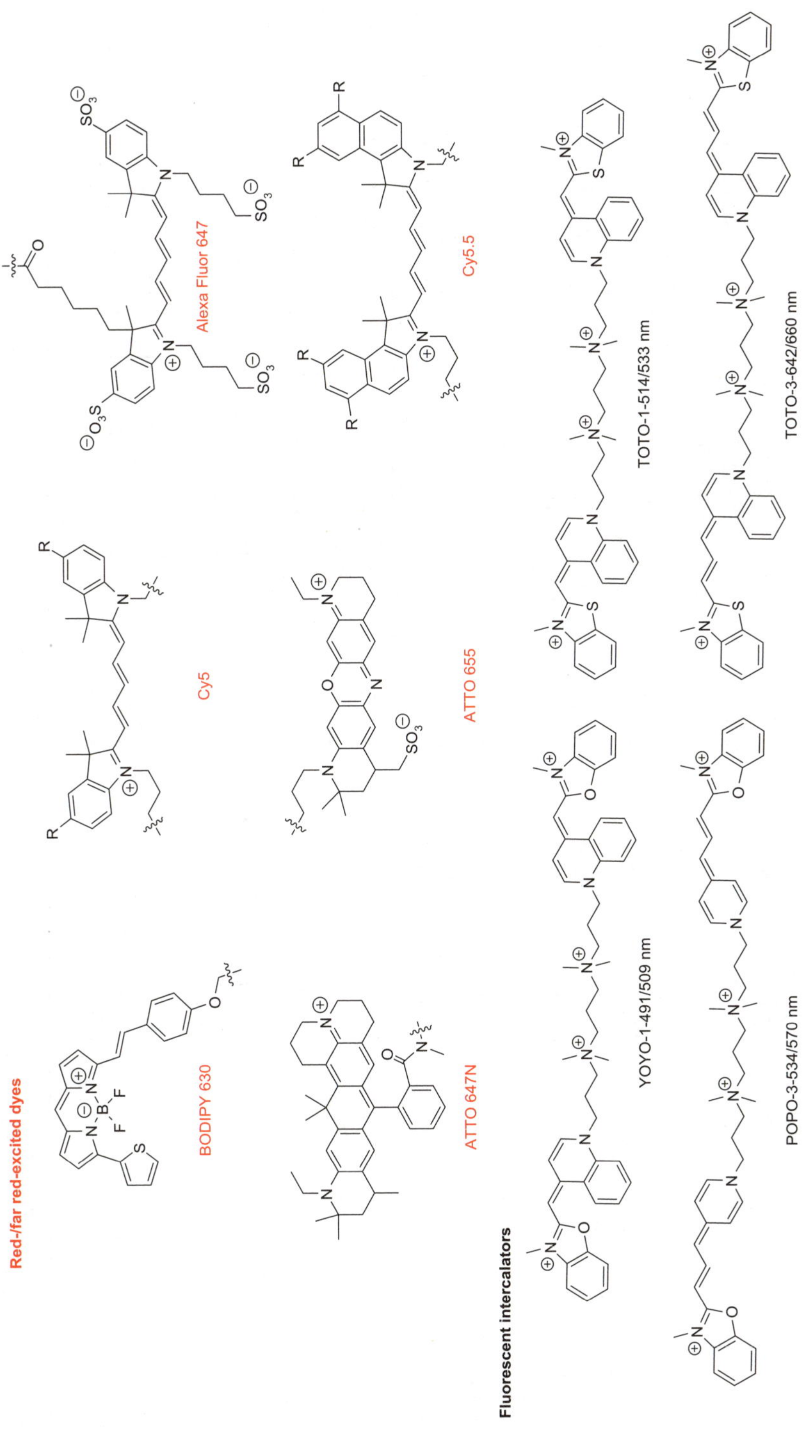

Figure 2.3 Chemical structures of fluorophores commonly used in SMFS. For phosphoramidite derivatives of Cy3, Cy5 and Cy5.5, R = H; for other reactive derivatives, R = SO$_3^-$.

Table 2.1 Fluorescence properties of organic fluorophores commonly used in SMFS. Parameters are reported for the free dyes and may be significantly altered upon conjugation to nucleic acids.

Dye	λ_{ex} (nm)	λ_{em} (nm)	ε ($M^{-1}\,cm^{-1}$)	Φ	τ (ns)	Available reactive derivatives
Fluorescein[R,39]	490	514	75 000	0.92	4.1	–NCS, –CO$_2$H, NHS ester, –NH$_2$, –N$_3$, maleimide, phosphoramidite
Oregon Green[S,I]	490	514	82 400	0.97	4.1	–CO$_2$H, NHS ester, –NH$_2$, maleimide, –N$_3$, –alkyne, –iodoacetamide
Rhodamine Green (Rhodamine 110)[S,I]	504[S,I]	532[S,I]	73 000[S,I]	0.92[R,40]	4.2[R,41]	–CO$_2$H, NHS ester
Alexa Fluor 488[S,I]	495	519	73 000	0.92	4.1	–CO$_2$H, NHS ester, –NH$_2$, maleimide, –N$_3$, –alkyne
ATTO 488[S,A]	501	523	90 000	0.80	3.2	–CO$_2$H, NHS ester, –NH$_2$, maleimide, –N$_3$, –iodoacetamide
Cy2	489[S,G]	506[S,G]	~150 000[S,G]	>0.12[S,G]	0.6[R,42]	Bis NHS ester
Rhodamine 6G	524[S,I]	552[S,I]	92 000[S,I]	0.95[R,40]	4.08[R,43]	–CO$_2$H, NHS ester
Alexa Fluor 532[S,I]	531	554	81 000	0.61	2.5	NHS ester, maleimide
ATTO 532[S,A]	532	553	115 000	0.90	3.8	–CO$_2$H, NHS ester, –NH$_2$, maleimide, –N$_3$, –iodoacetamide
TAMRA[R,39]	547	574	77 000	0.35	2.2	–NCS, –CO$_2$H, NHS ester, –NH$_2$, –N$_3$, alkyne, maleimide, iodoacetamide, phosphoramidite
Cy3[R,44]	548	562	150 000	0.04	<0.3	–CO$_2$H, NHS ester, –NH$_2$, maleimide, –N$_3$, –iodoacetamide phosphoramidite
Cy3B[R,44]	558	572	130 000	0.70	2.8	–CO$_2$H, NHS ester, maleimide,
Alexa Fluor 555[S,I]	555	565	150 000	0.10	0.3	–CO$_2$H, NHS ester, –NH$_2$, maleimide, –N$_3$, –alkyne, –iodoacetamide
Cy5[S,G]	646	664	250 000	0.27	1.0	–CO$_2$H, NHS ester, –NH$_2$, maleimide, –N$_3$, –iodoacetamide, phosphoramidite
Alexa Fluor 647[S,G]	650	668	270 000	0.33	1.0	–CO$_2$H, NHS ester, –NH$_2$, maleimide, –N$_3$, –alkyne, –iodoacetamide
ATTO 647N[S,A]	644	669	150 000	0.65	3.4	–CO$_2$H, NHS ester, –NH$_2$, maleimide, –N$_3$, –iodoacetamide
ATTO 655[S,A]	663	684	125 000	0.30	1.8	–CO$_2$H, NHS ester, –NH$_2$, maleimide, –N$_3$, –iodoacetamide
Cy5.5[S,G]	673	692	190 000	0.23	1.0	–CO$_2$H, NHS ester, –NH$_2$, maleimide, –N$_3$, –iodoacetamide phosphoramidite

^{S}Spectroscopic data provided by supplier (^{I} = Life Technologies, ^{A} = ATTO-TEC, ^{G} = GE Healthcare).
^{R}Spectroscopic data taken from literature (^{n} = Reference number).

pH in the useful analytical range, 4–8. The introduction of sulfonate groups is a common strategy for improving the performance of fluorophores,[26,27] because the reduced hydrophobicity results in increased solubility and a lower tendency to form aggregates in water. While aggregate formation is not usually a major concern at the low concentrations typically used for SMFS, dye molecules may interact with each other intramolecularly if bioconjugates are multiply labelled, or otherwise with hydrophobic amino acid side chains or nucleobases, adversely affecting fluorescence properties.[28–30]

Cyanine dyes, whose origins lie in the photographic industry, are also widely used in SMFS. DNA conjugates of Cy2, Cy3, Cy3B, Alexa Fluor 555, Cy5, Alexa Fluor 647, Cy5.5 and Cy7 are particularly useful in multicolour FRET experiments owing to their well-matched absorption and emission spectra, while cyanine-derived intercalators are alternative labels for SMD of DNA, obviating the need for chemical derivatisation of oligonucleotides. Dimeric dyes, such as YOYO-1 and TOTO-1, are usually preferred for SMFS because of their high affinity for dsDNA, large fluorescence enhancements upon binding and the availability of many derivatives whose absorption spectra enable efficient excitation by widely used laser sources.[31]

Quantum dots, luminescent core-shell semiconductor nanocrystals of diameter 2–10 nm, have been the subject of much interest since the first descriptions of their use in biological imaging in the late 1990s.[32,33] The most popular core materials in analytical applications are CdSe and CdTe, among the first to be synthesised. Commercial suppliers (*e.g.* Quantum Dot Corporation, now owned by Life Technologies, Evident Technologies and Crystalplex) can provide capped core-shell nanoparticles, and particles functionalised with groups such as streptavidin, amines and carboxylic acids, to enable conjugation to oligonucleotides. Their popularity in SMFS stems largely from their enhanced brightness, broad excitation spectra and resistance to photobleaching, but intermittent emission at the single molecule level caused by photoblinking can cause problems.[34] Careful optimisation of surface chemistry to ensure stability and solubility in aqueous media is also required. Reviews comparing quantum dots with organic dyes,[35] as well as describing their use in biomolecular assays and single molecule detection have recently been published,[36–38] so we do not replicate this material here.

2.2.2 Dye Photophysics

The suitability of a fluorophore for any given SMFS application largely depends on its photophysical properties. In order for fluorescence emission to occur, a chromophore in its ground state (S_0) must first be excited by absorption of a photon to its first excited singlet state (S_1), whereupon vibrational relaxation is followed by a transition back to the ground state, accompanied by emission of a photon (Figure 2.4A). The rate constant for photon absorption (k_{01}) is determined by the product of the excitation photon flux (photons cm^{-2} s) and the absorption cross-section (σ) of the chromophore,

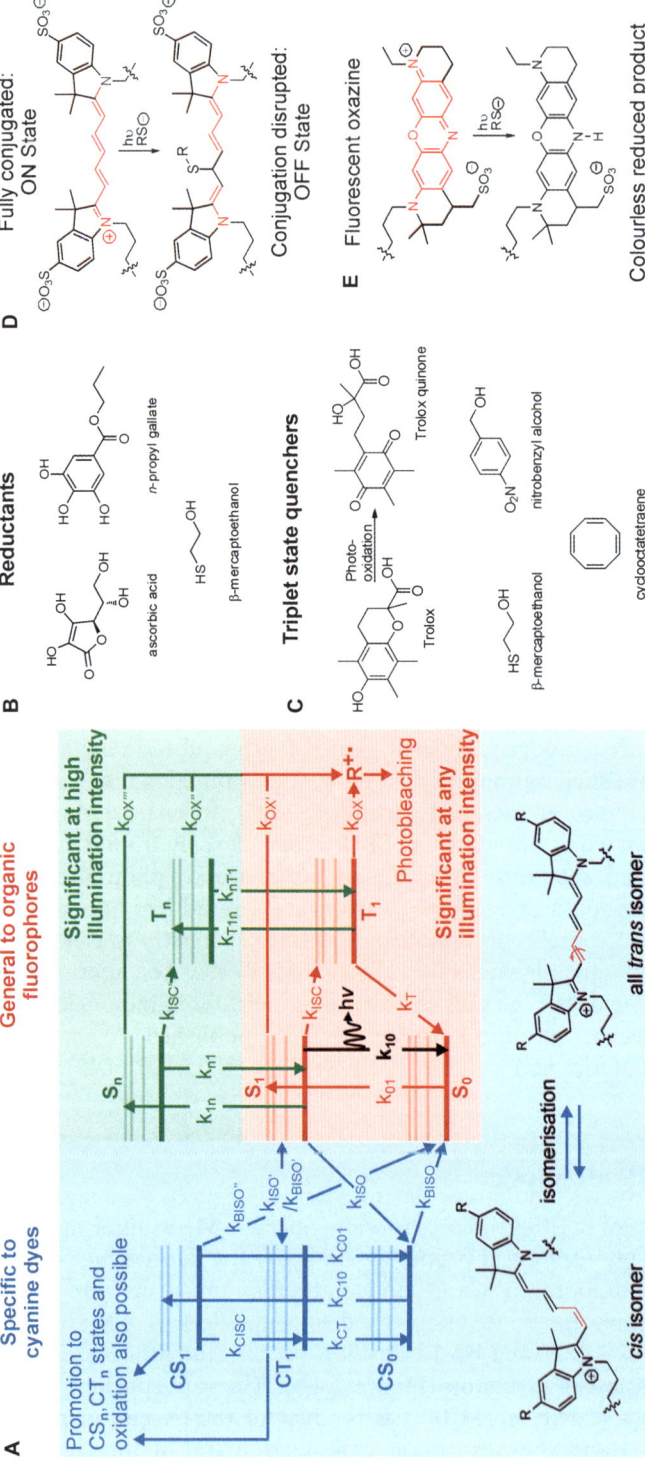

Figure 2.4　Photophysics of organic fluorophores. (A) Modified Jablonski diagram including excitation to higher excited electronic states and photoisomerisation. (B) Reductants used as additives in SMFS experiments. (C) Structures of widely used triplet state quenchers. (D) Photoswitching of Cy5 by formation of a thiol adduct. (E) Photochemical reduction of the oxazine ATTO655 chromophore.

from which its extinction coefficient derives. One can visualise this process as bullets (representing photons) being fired at a target (the dye) of a fixed size (the absorption cross-section). One might therefore consider firing bullets more rapidly at the target by increasing the illumination intensity to increase the rate of fluorescence emission which, as described previously, is critically important to SMFS. However, inspection of Figure 2.4A reveals that there are many possible fates of the excited S_1 state that do not give rise to fluorescence emission. First, even in the simplest three-state model (red part of Figure 2.4A), two processes compete with fluorescence: oxidation of the S_1 state to a radical cation (R^+) and intersystem crossing to the triplet T_1 state, which may relax back to the ground state S_0 or be oxidised to a radical cation. The oxidised species formed by either pathway may then undergo chemical reaction, leading to permanent destruction of the chromophore, known as photobleaching. Because the rate constants (k_{10}, k_{OX}, k_{ISC} and k_{OX}) for all of these relaxation pathways do not depend on the laser power, absorption of an excitation photon in this model is associated with a fixed probability of chromophore destruction (around 10^{-6}), sometimes called the quantum yield of photobleaching. On average then, increasing the laser power within this regime leads to emission of the same number of photons at a higher rate before bleaching prevails. The tradeoff is therefore between temporal/structural resolution and measurement longevity.

Under certain conditions, however, the simple three-state model breaks down because higher electronically excited states (S_n, T_n, *etc.*) become accessible (green part of Figure 2.4A). Since population of these states requires absorption of a second excitation photon by a transiently excited chromophore, they become significant only at high illumination intensities ($> 10^3$ W cm^{-2}, depending on the electronic structure of the dye).[45] Depending on the resonant frequency of the transition, the second absorbed photon may be of the same wavelength as that used for $S_0 \rightarrow S_1$ excitation, or if multicolour labelling is being used, it may be of the wavelength required to excite a different fluorophore. For example, when used as acceptors in smFRET experiments, several far-red dyes (such as Cy5, Alexa Fluor 647, ATTO 647N and ATTO 655) have been shown efficiently to absorb a second green or blue photon (used to excite the donor), while in the S_1 excited state.[46,47] As oxidation of higher excited states opens up new pathways for photobleaching that are not present in the three-state model, the quantum yield of photobleaching increases under these circumstances, reducing the number of photons that can be emitted before destruction of the chromophore. The probability of photobleaching TAMRA following excitation to S_1 within the three-state model (appropriate for excitation intensities $\leq 5 \times 10^4$ W cm^{-2}) is 2.5×10^{-6}, but increasing the power density 10-fold results in a proportionate increase in the probability of photobleaching.[45] As a result, the mean number of emitted photons decreases 10-fold, from 4×10^5 to 4×10^4. Given a collection efficiency of 1%, this means that all the information on the molecules under study must be extracted from analysis of a total of either 4000 or 400 photons, the latter being emitted at a faster rate. The optimal experimental conditions depend on the application: a high laser power may be selected for rapid quantitation of DNA duplexes in a diagnostic test using a confocal detection system, whereas a lower intensity would be better suited for monitoring a multistep enzymatic modification of DNA taking place over tens of seconds using TIRFM. A mean occupation time of around 100 µs is typical for short DNA duplexes diffusing through a confocal detection volume,[48] which may be even shorter if the analytes are being carried in a microfluidic flow. In this scenario, all the fluorescence from one molecule is typically collected in a single burst and ~ 25 photons are sufficient for detection, so illumination densities of $\geq 10^4$ W cm^{-2} are common, with up to 10^6 W cm^{-2} used for molecules undergoing fast flow.[49] In contrast, intensities of 10^2–10^3 W cm^{-2} are more typical for wide-field TIRF excitation, the most popular configuration for studies of molecular dynamics.[50]

The polymethine chain present in cyanine dyes provides another possible fate for the S_1 state: *trans-cis* isomerisation of the lowest energy all *trans* configuration (blue part of Figure 2.4A). The relatively low rates of the back isomerisation processes mean that a photostationary equilibrium can be reached relatively quickly under the conditions used for SMFS, in which $\sim 50\%$ of the chromophores are in a photoisomerised *cis* state.[51] When conjugated to oligonucleotides, sticking of dyes to DNA limits their rotational flexibility, so that unsticking becomes rate limiting for isomerisation, extending the lifetime of all isomers.[52–54] Isomerisation reactions can occur directly between ground states with thermal activation ($CS_0 \rightarrow S_0$),[55] *via* excited singlet states (*e.g.* $S_1 \rightarrow CS_0$, $CS_1 \rightarrow S_0$)[51] or *via* the *cis* triplet state ($S_1 \leftrightarrow CT_1$).[56] Since the *cis* isomers are weakly or non-fluorescent, they are essentially invisible, leading to the generation of a large 'zero peak' in smFRET histograms, or a 'blinking' event in a fluorescence trajectory. To eliminate effects due to photoisomerisation, Cy3B, a conformationally locked analogue of Cy3, has been developed.[44] This modification has a dramatic effect on the quantum yield, which increases from 0.04 for Cy3 to 0.7 for Cy3B. For detection in the far red part of the spectrum the carbopyronine dye ATTO 647N,[57] and the oxazine ATTO 655, offer conformationally rigid alternatives to the carbocyanines Cy5 and Alexa Fluor 647.

The presence of two unpaired electrons (occupying the $2p\pi$ molecular orbitals) in the $^3\Sigma_g^-$ ground state of molecular oxygen and its ubiquitous nature mean that it is a major player in the photobleaching of organic dyes *via* reaction with species bearing unpaired electrons, such as triplet states or radical cations, generating singlet oxygen in the process. To retard bleaching by oxidative processes, one option is to include a reductant such as ascorbic acid, *n*-propyl gallate or a thiol in the measurement buffer (Figure 2.4B), which can rescue the chromophore from a radical cation state before bleaching occurs, although the antioxidant concentration must be optimised to avoid reduction of the native dye.[58] A refinement of this approach uses both a reductant and an oxidant to recycle triplet states rapidly to singlet states by two successive electron transfers.[59]

Enzymatic removal of O_2 from SMFS media using the glucose oxidase–catalase system[60] is an alternative strategy frequently used to prolong single molecule measurements. The disadvantage that this approach suffers is that O_2

is also capable of quenching triplet states, regenerating S_0 without producing singlet oxygen; because the $T_1 \rightarrow S_0$ transition is spin-forbidden, the lifetime of T_1 can be relatively long in the absence of oxygen, increasing from $< 100\,\mu s$ in air to 4–$100\,ms$ when embedded in a polymer matrix in a vacuum in the case of the cyanine dye $DiIC_{12}$.[61] In the absence of oxygen then, a fluorophore may be observed to 'blink' for an extended period, which could be mistaken for a conformational change in certain experiments, *e.g.* in smFRET trajectories. Therefore the ideal medium for collecting extended single molecule trajectories is depleted of oxygen, but contains an additive that quenches triplet states, such as the vitamin E derivative Trolox, nitrobenzyl alcohol or a thiol (Figure 2.4C).[58] In particular, the use of Trolox in combination with glucose oxidase–catalase[62] has proved valuable in studies of protein–DNA interactions. Recently, the conjugation of Trolox and other triplet quenchers to oligonucleotides has been proposed as a method to increase the photostability of cyanine dyes located on a complementary strand.[63] The mode of action of Trolox remains unclear, with some workers favouring a collision-based triplet quenching mechanism while others attribute its function to the presence of a reducing and oxidising system (ROXS) consisting of the native α-tocopherol and its quinone congener (Figure 2.4C), formed by spontaneous oxidation in buffer.[64]

As a result of the advent of super-resolution microscopy (see Section 2.3.2.1), there has been a dramatic upturn in interest in photochemical reactions that produce non-fluorescent forms of widely used dyes, many of which were initially discovered by photophysical studies aimed at minimising the population of such dark states. For example, the inclusion of a thiol as a triplet quencher in deoxygenated buffer was found to induce long-lived dark states in several cyanine dyes (Cy5, Cy5.5 and Cy7, but not Cy3) lasting several seconds,[65] which has been attributed to the formation of a thioether adduct of the polymethine chain, causing disruption of the conjugated π-system (Figure 2.4D).[66] The oxazine dye ATTO 655 can be photoswitched under similar conditions, brought about by reduction of the chromophore (Figure 2.4E).[67]

2.2.3 Labelling Strategies

The method chosen to label DNA fluorescently for SMFS is heavily application dependent: according to the experiment it may be most appropriate to use a reactive derivative of the dye, a labelled dNTP or even a fluorophore that makes non-covalent interactions with duplex DNA (Figure 2.5). Generally, the key requirements for labelling reactions are high yields of the pure conjugate without any deleterious effect on the photophysical properties of the chromophore. In principle, SMD can tolerate low labelling efficiencies and the presence of free dye, because the fully labelled species are implicitly selected by the experiment. For example, detection of dual-labelled DNA duplexes by TCCD is possible in a 1000-fold excess of singly labelled oligonucleotides.[17] In practice, however, high labelling efficiencies are highly desirable because they increase the proportion of molecules from which useful information can be obtained. In experiments using four labelled strands,[68] an increase in labelling

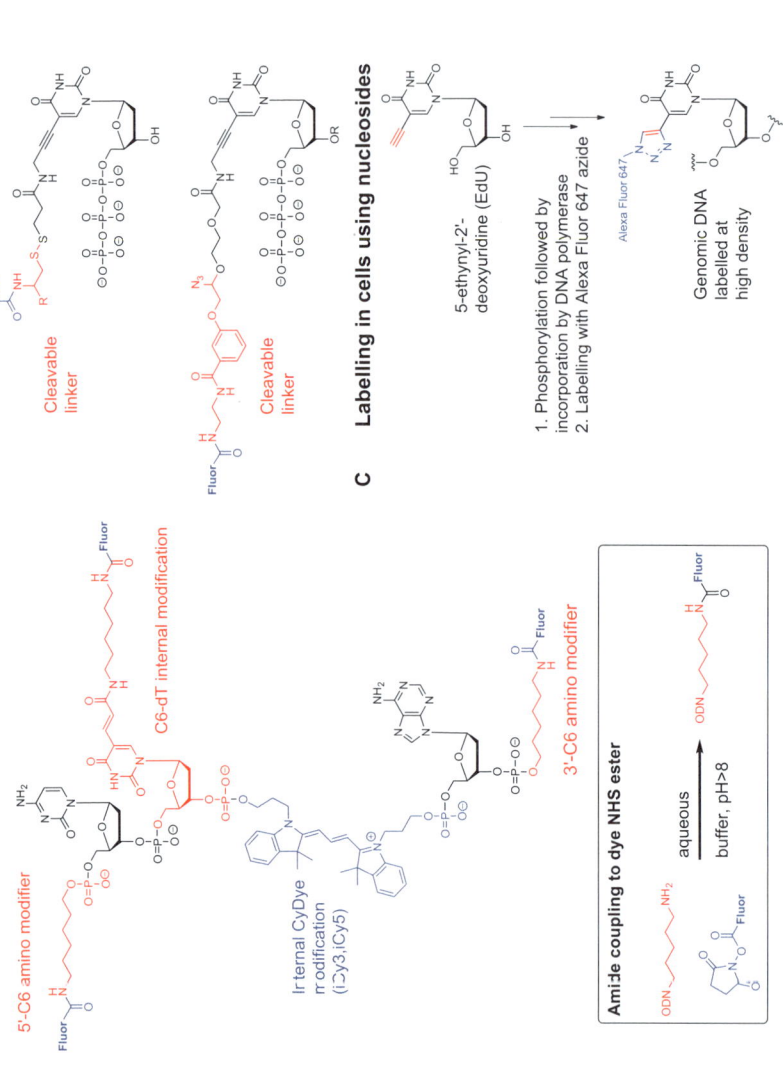

Figure 2.5 Locations of fluorescent labels used in SMFS of DNA: dye conjugates of oligodeoxynucleotides for use in genetic analysis and probing DNA structure and function (A), reversible dye terminator dNTPs (where R is a group that blocks chain extension, see Figure 2.6B and C) used in DNA sequencing (B) and labelling of genomic DNA (gDNA) in cells using an alkynyl-functionalised nucleoside and a fluorophore bearing an azide (C).

efficiency from 85% to 95% would lead to an increase in the maximum fraction of fully labelled species from 52% to 81%. In order to maximise the purity of labelled oligonucleotides, they are typically purified by one or two high-pressure liquid chromatography (HPLC) steps before use in SMFS and sometimes ethanol precipitation is used to remove traces of free dye.[54] When multi-strand DNA systems are being studied, it is sometimes useful to form the structures of interest by hybridisation and then carry out a PAGE (polyacrylamide gel electrophoresis) purification to recover the fully labelled complex.[69]

Fluorophore–oligonucleotide conjugates are required for DNA detection by hybridisation and for *in vitro* studies of DNA structure and protein–DNA interactions (Sections 2.3.1.2 and 2.3.2). Reactive dye derivatives available for the labelling of oligonucleotides include phosphoramidites for incorporation during automated synthesis, carboxylic acids and active esters for reaction with amino-modified oligonucleotides, maleimides and iodoacetamides for labelling thiolated nucleic acids, and amines for coupling to carboxylic acids and electrophiles (see Table 2.1). In addition to these reagents, azide- and alkyne-functionalised fluorophores are becoming increasingly available to take advantage of the efficient conjugation possible *via* Cu(I)-catalysed click chemistry, which is becoming a well-established methodology for synthetic manipulations of nucleic acids.[70] Labelling by phosphoramidite chemistry, if possible, is preferable because of the high coupling yields obtained. However, many chromophores suitable for SMFS (particularly rhodamines and cyanines) are unstable to oligonucleotide deprotection by heating in concentrated aqueous ammonia or other basic media, in which case post-synthetic modification is necessary. For example, deprotection of oligonucleotides containing the rhodamine chromophore introduced during automated synthesis has to be carried out using $0.05\,M\ K_2CO_3$ or *t*-butylamine–methanol–water $(1:1:2)$.[71] As a result, the predominant method for labelling is post-synthetic coupling of an active ester derivative of the fluorophore to an aliphatic amine introduced into the oligonucleotide using a modified phosphoramidite. The amino group is usually introduced at the 3′- or 5′-terminus of the oligonucleotide or internally within the sequence *via* the 5-position of pyrimidines (Figure 2.5A). Several oligonucleotide synthesis companies are able to incorporate any of these modifications and carry out post-synthetic labelling, so SMFS researchers can simply purchase DNA that is tailored to the needs of their experiment. High water solubility is particularly important for efficient post-synthetic labelling of deprotected oligonucleotides, though not for incorporation in oligonucleotide synthesis, where acetonitrile is the solvent of choice. For this reason the sulfonate groups of Cy3, Cy5 and Cy5.5 are absent from their phosphoramidite derivatives (see Figure 2.3). These commercially available cyanine monomers enable incorporation of the fluorophore internally within an oligonucleotide or attached to the 5′ end *via* a three carbon chain. The short linker promotes stacking of dyes located at the 5′ end upon the terminal base pair of duplexes formed by the modified oligonucleotides, which can affect the results of smFRET experiments by orientating two cyanine fluorophores at each end of a DNA duplex.[72]

Several next-generation sequencing technologies rely on synthetic dNTPs conjugated to SMFS-compatible fluorophores to label the growing strand transiently in order to read the sequence of the template strand (see Section 2.3.1.1, Figure 2.6).[73] In most cases the dye is attached to the nucleobase by a propargylamino group to enable efficient incorporation by DNA polymerases, while the 3'-OH group is blocked from extension by a cleavable group attached either directly to the oxygen atom or by a bulky group linked to another part of the dNTP (Figure 2.5B).[74]

Different strategies are required for labelling DNA inside cells, which is of renewed interest owing to the advent of super-resolution microscopy. This application requires dense dye functionalisation of the nucleic acid to maximise the amount of structural information retrieved, but must also use selective chemistry so that other cellular components are not labelled. Labelling of genomic DNA of HeLa cells at high density with Alexa Fluor 647 has recently been achieved *via* 'click' reaction between an alkynyl group enzymatically incorporated into the DNA and an azido-functionalised dye derivative (Figure 2.5C).[75]

2.3   Applications

The applications of SMFS to the study of DNA are now too numerous to cover exhaustively in this chapter. Instead, we select a few examples that illustrate the scope of the field and the principles underpinning the respective technologies, directing the reader to more detailed review articles where appropriate.

2.3.1   DNA Sequence Analysis

The impact of the polymerase chain reaction (PCR) on the fields of molecular biology, DNA sequencing, mRNA expression studies, genetic analysis, molecular diagnostics and forensic science has been revolutionary; the ability to amplify nucleic acid sequences of interest exponentially prior to their analysis has been central to countless applications in the last quarter of a century.[76] However, the introduction of SMFS and the availability of lab-on-a-chip devices have led to technologies that do not require amplification of nucleic acids. These adaptations can slash the economic cost and the time required to analyse nucleic acid sequences without compromising sensitivity.

2.3.1.1   Next-generation Sequencing Technologies

The most conspicuous application of SMFS techniques has been in the field of DNA sequencing. Since the completion of the Human Genome Project in 2003, several methods that generate sequence data in a massively parallel manner from a few or even individual DNA molecules have been developed. These are collectively called next-generation sequencing platforms and have been developed by Roche 454 Life Sciences, Illumina, Applied Biosystems, Helicos Biosciences and Pacific Biosciences among others (Figure 2.6). These

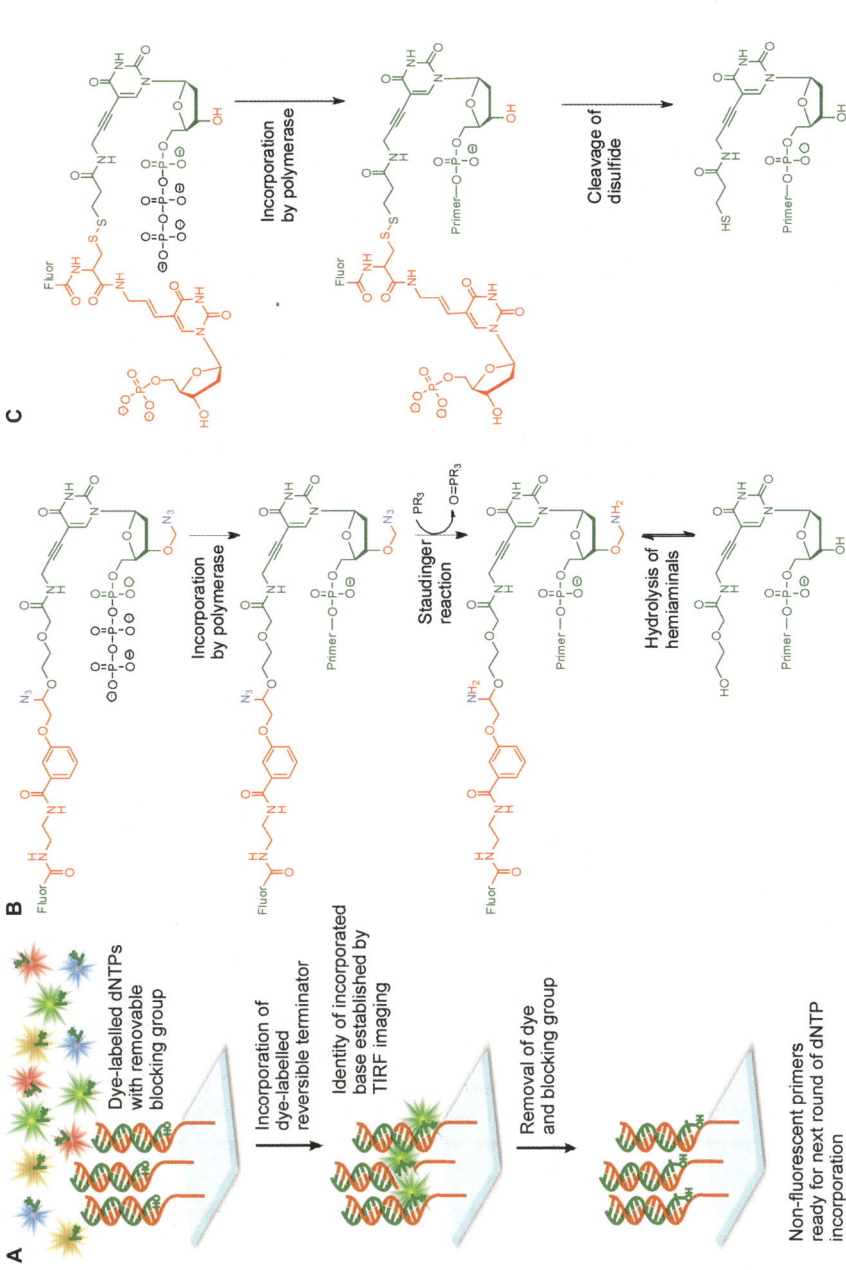

Figure 2.6 Next-generation sequencing-by-synthesis platforms. (A) Key steps in Illumina sequencing using dye-labelled reversible terminators. Chemical reactions leading to cleavage of dye molecules and 3'-blocking groups in the methods used by Illumina (B) and Helicos (C).

innovations can make dramatic time and cost savings when compared with Sanger sequencing; while the Human Genome Project took several years to complete and cost several hundred million US dollars, the latest instrument marketed by Illumina as of January 2012, the HiSeq 2500, is able to sequence a human genome in around a day at a reagent cost of < US$5000. A recent review provides a comprehensive overview of next generation platforms.[73]

In Illumina's sequencing-by-synthesis method, clonal arrays of ~ 1000 copies of the molecule to be sequenced are constructed on the surface of an optically transparent flow cell by polymerase amplification.[77,78] Sequence information is then generated by extension of a primer using reversible dye terminator chemistry with four fluorescent dNTPs each labelled with a different colour (Figure 2.6A). The 3′-oxygen of the fluorescent dNTP is blocked from further extension by the presence of an azidomethyl group. Unincorporated dNTPs are then washed out of the flow cell and the identity of the attached nucleotide established by imaging using prism-based TIRFM (see Figure 2.2D) before cleavage of the fluorophore and 3′-protecting group. This is achieved using a Staudinger reaction between a water-soluble phosphine and the azides, unmasking two hemiaminals which are rapidly hydrolysed (Figure 2.6B). The released 3′-OH is then available for reaction in another round of single base extension.

This sequencing-by-synthesis strategy is shared by the Helicos 'true Single Molecule Sequencing' (tSMS) method, where the fluorescence of a single primer strand elongated on an immobilised template is imaged in the presence of additives for scavenging of O_2 and radical species, as well as for quenching of triplet states.[79] The dye-labelled reversible terminators consist of a Cy5 label and a bulky nucleosidic group conjugated to the dNTP by a linker containing a disulfide (Figure 2.6C).[74] Since the same fluorophore is used for detection of all four nucleotides, each dNTP must be added in a separate cycle. Cleavage of both moieties is accomplished by reduction with a water soluble phosphine, enabling further elongation.

Pacific Biosciences' single molecule real time (SMRT) sequencing method takes a more direct approach.[80] Here, the processive incorporation of nucleotides by an individual polymerase is monitored (Figure 2.7A). The fluorescent label is attached to the ε-phosphorus atom of a modified 2′-deoxynucleotide pentaphosphate (dN5P) and is thus ejected with the leaving group in the extension reaction (Figure 2.7C). While this takes place, the dye is held in close proximity to the polymerase for a few hundred milliseconds, generating a 'pulse' of colour specific to each base (Figure 2.7B). The product is an unmodified DNA strand which can continue to participate in polymerisation. Since the K_M of the DNA polymerase is in the micromolar range, tremendous spatial resolution is required to ensure that only triphosphates undergoing reaction give rise to signal. This is achieved by the use of nanofabricated zero-mode waveguide (ZMW) structures which can illuminate an incredibly small volume of 10–20 zeptolitres $(10–20 \times 10^{-21} \, L)$.[81]

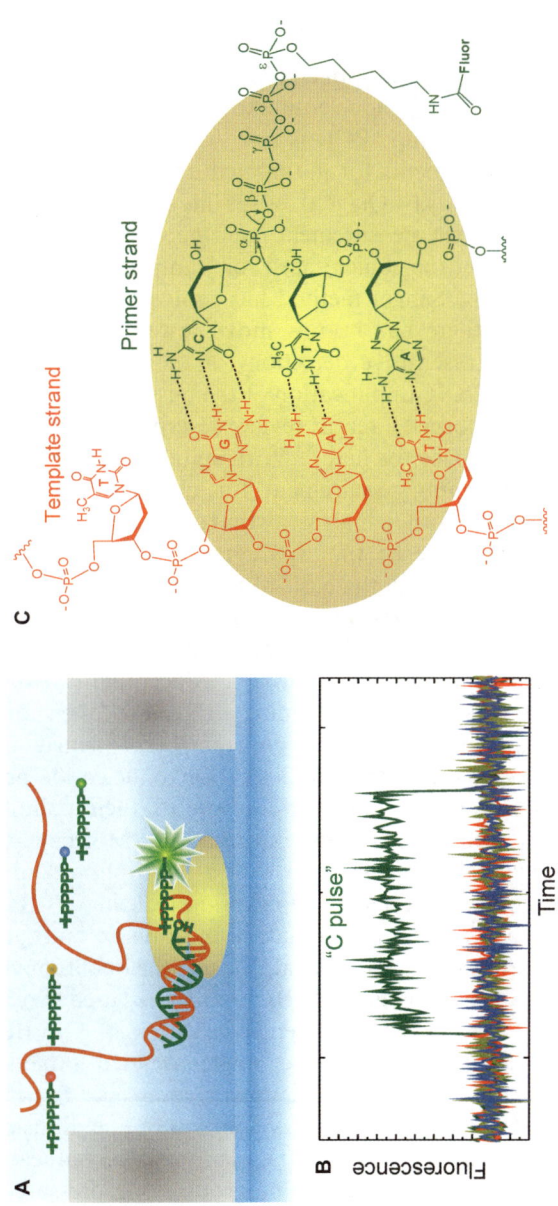

Figure 2.7 Single molecule real time sequencing. (A) Extension of a primer by a single polymerase immobilised within a zero-mode waveguide (ZMW). (B) Simulated data trace showing observation of a fluorescence pulse during extension, indicating incorporation of dCTP. (C) Chemical mechanism of polymerisation reaction using fluorescently labelled 2'-deoxynucleotide pentaphosphate (dN5P).

2.3.1.2   Ultrasensitive Detection of Nucleic Acids by Hybridisation

Despite the advances in DNA sequencing technologies, the detailed information they generate is unnecessary for small scale single nucleotide polymorphism (SNP) analysis or pathogen detection. Fluorescence-based real time PCR has been the method of choice for these applications in recent years.[82] These assays typically rely on a fluorogenic oligonucleotide for recognition of a sequence of interest in the amplicon. Real time PCR is now widely used for nucleic acid-based diagnostics, and instruments for automated preparation of assay-ready samples (*e.g.* the QIAsymphony by Qiagen), or for integrated sample processing, amplification and fluorescence measurement (*e.g.* the GeneXpert by Cepheid) have been marketed to meet the clinical need for 'sample-in–answer-out' capability. However, just as the Sanger method has evolved into next-generation sequencing technologies, there has been a move towards sequence analysis platforms that rely on ultrasensitive detection for the development of rapid assays that do not require PCR amplification of the target DNA. As with next-generation sequencing platforms, a recent review describes these approaches in more detail than we have space for here.[83] With suitable instrumentation and labels available, the remaining requirement for ultrasensitive detection of unlabelled DNA of biological origin is an assay that generates a detectable signal in the presence of a specific sequence. In the paragraphs below we illustrate some of these assays, separating them on the basis of the type of readout observed in the measurement: fluorescence intensity, co-localisation or FRET.

Detection based on an increase in fluorescence intensity in the presence of target DNA is perhaps the simplest approach conceptually. In the single molecule regime, this typically involves defining a threshold, then counting the number of events (either bursts of fluorescence or pixels, if imaging is used) whose intensity exceeds this level (Figure 2.8A). Oligonucleotide probes that exhibit increased fluorescence upon hybridisation to their complementary sequence can provide sequence-specific information. Molecular Beacons, hairpin-structured oligonucleotides labelled with a fluorophore and quencher first described by Tyagi and Kramer for monitoring real time PCR reactions, are one such probe type.[84] When hybridised, a fluorophore and quencher at either end of the stem sequence become distal, causing the quantum yield of the reporter to increase significantly (Figure 2.8B). Zhang *et. al* used two Molecular Beacons, labelled with either Oregon Green/Iowa Black or Cy5/BHQ-3, for detection and discrimination of synthetic single-stranded targets at concentrations down to 70 pM by counting bursts of fluorescence from molecules undergoing Brownian motion in a stationary solution.[85] The diffusion-limited encounter rate of Molecular Beacons with the micron-sized confocal detection volume has been improved upon by the same group using pressure-driven or electrokinetic flow, improving the data acquisition rate so that the time required to detect 70 pM unlabelled single stranded DNA hybridised to a complementary Molecular Beacon with 99% confidence dropped from ∼200 s to <4 s.[86] Another assay, reported by Castro *et. al.*, concentrated fluorescence

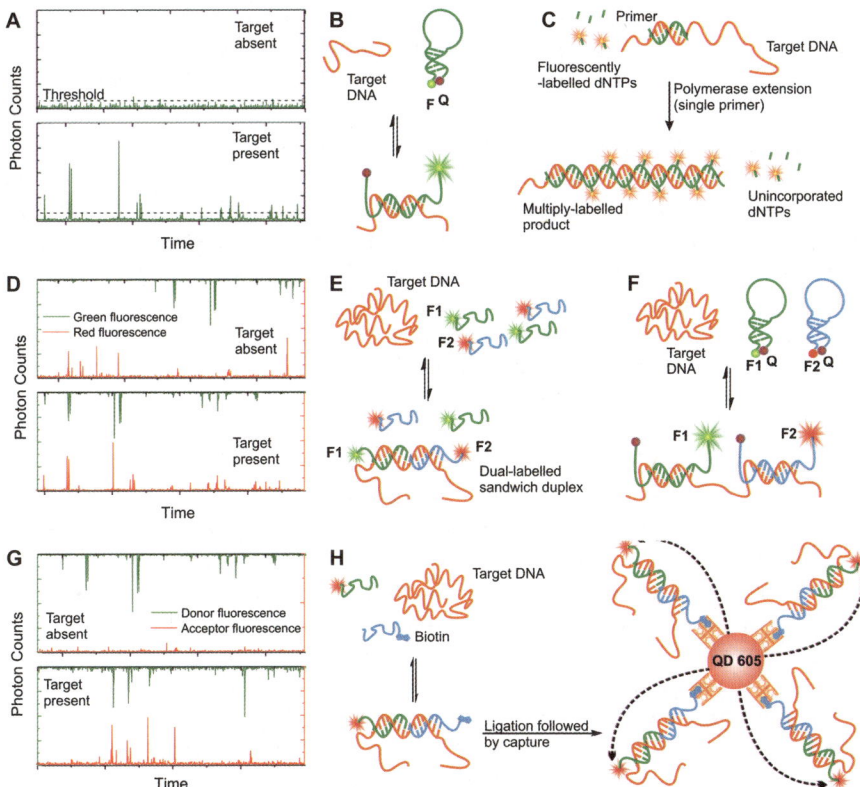

Figure 2.8 Sequence-specific detection of DNA sequences using SMFS. (A) Intensity-based detection of nucleic acid targets. A greater number of fluorescence bursts above threshold are observed upon hybridisation.* (B) Mode of action of Molecular Beacons. (C) Detection of target DNA using enzymatic labelling with multiple fluorophores. (D) Co-localisation based detection of nucleic acid sequences. In the absence of the target sequence, bursts in each channel are uncorrelated. After probe binding, simultaneous bursts of fluorescence are observed in both channels.* (E) Generation of coincidence events by binding of two labelled oligonucleotides to unlabelled target DNA. (F) Co-localisation of two fluorescently labelled Molecular Beacons on a target. (G) FRET-based detection of DNA. In the absence of target, there are no acceptor signals. Upon hybridisation, excitation of the donor results in bursts of fluorescence from donor and acceptor simultaneously.* (H) Generation of dual-labelled species by target-dependent ligation of dye- and biotin-labelled probes followed by capture to a quantum dot. *Panels (A), (D) and (G) show idealised burst data from a confocal detection experiment.

from the solution onto the probe–target duplex, so that binding events were observed as highly emissive species above a constant background.[87] This was achieved by incorporation of a dUTP derivative labelled with a TAMRA/BODIPY FRET pair during polymerase extension of a single primer hybridised to a polynucleotide fragment (Figure 2.8C). The resulting duplex

contained ~50 labels, enabling a 1.2 pM concentration of plasmid to be discriminated from the background due to unincorporated dUTP at a concentration of 10 nM using burst counting in flow following a 200-fold dilution.

As outlined in Figure 2.2C, one popular implementation of SMFS involves
using fluorophores attached to separate interacting biomolecules, leading to co-
localisation of the dye molecules upon binding. When the chromophores are
then simultaneously excited in the observation volume resulting from two
overlapped laser beams, co-localisation gives rise to a simultaneous burst of
emitted photons from both dyes, sometimes called a coincidence event (Figure
2.8D). This approach has advantages over the use of intensity alone for
detection of nucleic acids, because background coincidence events are not
caused by free dye, unless there is significant crosstalk between the two dyes
(caused by 'leaking' of emission between detection channels), or if two dye
molecules enter the excitation volume simultaneously, either by chance or non-
specific association. Similarly, autofluorescent impurities are less likely to be
detected in both channels. The most convenient way of harnessing this technique for detection of non-fluorescent endogenous DNA involves the use of
two fluorescent probes which co-localise upon hybridisation to the same
unlabelled target molecule, analogous to the 'sandwich' method used in
immunoassays (Figure 2.8E). This approach was first presented by Castro and
Williams for amplification-free detection of phage λ gDNA, which was detected
using two 15mer peptide nucleic acid (PNA) probes, labelled with Rhodamine
Green and BODIPY-TR,[88] and later for detection of gDNA from *B. anthracis*,
the causative agent of anthrax.[89] One disadvantage with homogeneous 'sandwich'-hybridisation methods in the detection of unamplified gDNA arises
because excess unhybridised fluorescent probes are not removed before analysis. As a result, the total concentration of fluorescent molecules must not
exceed the single molecule regime (≤250 pM), because the probability of
chance coincidence events rises.[90,91] However, it would be desirable for high
concentrations of fluorescent probes to be used in the rapid detection of
unamplified nucleic acids to accelerate rates of hybridisation. The problem of
increased background signal in this scenario can be solved by using a pair of
Molecular Beacons to bind the same target strand (Figure 2.8F).[92] This combination of intensity and co-localisation offers advantages over both separate
methods: owing to the efficient quenching in the 'closed' form, unbound probes
are unlikely to give rise to coincident signals even if two (or more) unbound
probes co-localise in the detection volume by chance.

The use of SMD also allows the study of individual FRET pairs (Figure
2.2B). When a single donor is excited in sufficient proximity to a suitable
acceptor, photons are emitted by the acceptor. If the FRET efficiency is less
than 100%, emission can be detected from both the donor and the acceptor
(Figure 2.8G), and coincidence analysis can be applied.[93] Indeed, smFRET
assays can be considered to be a subset of co-localisation methods in which the
donor and acceptor are extremely close in space. In analytical assays for nucleic
acids, there are advantages over both intensity-based and coincidence detection: the dependence of energy transfer on the inverse of the sixth power of the

interfluorophore separation [see eqn (2), below] means that it is not enough for two dyes to occupy the detection volume (diameter $\sim 1\,\mu m$) simultaneously to generate a chance FRET signal, rather they must maintain a proximity in the nanometre range for a significant fraction of the measurement time. The probability of this occurring by chance is vanishingly small even at relatively high concentrations. Furthermore, as only one excitation wavelength is used, there is no loss of detection efficiency in confocal measurements due to imperfect overlap of excitation sources (about 30% overlap is possible for 488 nm and 633 nm lasers),[17] although the overlap of detection volumes defined by separate pinholes must still be optimised. However, direct excitation of the acceptor at the donor's excitation wavelength and spectral crosstalk from donor emission can lead to weak signals in the acceptor emission channel. These possible sources of false positives have to be filtered out by thresholding. Since a high degree of spectral overlap between donor emission and acceptor excitation is required for efficient FRET, the suppression of direct acceptor excitation requires a donor fluorophore with a large Stokes shift. Another disadvantage of smFRET detection is that the total signal from the acceptor will be lower than that obtained by excitation at its absorption maximum because the energy transfer is less than 100% efficient, which could make binding events difficult to resolve from the background fluorescence. The use of a multivalent quantum dot as a donor for multiple organic fluorophores could address both of these issues, and take advantage of the highly efficient energy transfer reported between quantum dots and organic dyes in flow.[94] The assay developed by Zhang and co-workers (Figure 2.8H)[95] uses a streptavidin-functionalised quantum dot (QD605) as a FRET donor for Alexa Fluor 647. The excitation of the quantum dot at 488 nm elicits negligible emission from Alexa Fluor 647 ($\lambda_{max} = 650\,nm$), and the capture of ~ 50 acceptor duplexes per quantum dot ensures high acceptor fluorescence, enabling a high threshold to be used. Detection in pressure-driven flow allowed a limit of detection of 4.8 fM unlabelled target DNA, a 10-fold improvement over a Molecular Beacon assay in a head-to-head comparison. This concept has recently been extended to capture of two target sequences on single quantum dots followed by detection using a combination of coincidence and FRET.[96]

Unlike DNA sequencing platforms, these promising research methods are still some distance from finding real-world application in diagnostics or mutation analysis. A key remaining challenge is the integration of devices for sample preparation (*e.g.* cell lysis, nucleic acid extraction and purification), hybridisation and detection;[97] most of the assays described here rely on the use of pre-processed, purified DNA samples. Micro total analysis systems ('µTAS'), in which several steps are carried out in one automated device, are one attractive solution, particularly for 'point-of-care' applications.[98] As discussed above, interfacing a microfluidic system with SMD optics is likely to be essential to achieving rates of data acquisition that are competitive with PCR-based methods, and this is an active area of research.[99] Finally, the production of a cheap, compact and robust optical system capable of SMD in a primary care setting is an engineering problem that is yet to be solved. Whether confocal systems with their higher signal-to-noise ratios or wide-field imaging setups with the potential for greater volumetric throughput are more suitable for these applications remains an open question.

2.3.2 Elucidating DNA Structure and Function

Cornish and Ha have observed that the number of publications in the biological sciences containing the phrase "single molecule" in the title has grown exponentially since 2000, and they projected that in 25 years, every such paper would concern single molecule techniques.[6] This tongue-in-cheek forecast contains within it the more realistic implication that every biological scientist will have access to instruments capable of studying single molecules within this timeframe. Characterisation of single molecules might thereby become as commonplace a practice as, say, gel electrophoresis. This is already the case in large genome sequencing facilities, while commercial microscope systems aimed at biophysical research already offer the possibility of taking these studies outside of the realm of physical science laboratories equipped with home-built optical setups. In this section, we highlight a few examples of SMFS applied to the study of DNA structure and function, with the aim of providing a flavour of the field. The scope of this chapter excludes studies of RNA[100] and mechanical manipulation of nucleic acids,[101] which can be combined with SMFS.[102]

2.3.2.1 Determining DNA Structures

The major advantages of using SMFS over experimental methods such as X-ray crystallography, electron microscopy or bulk techniques such as nuclear magnetic resonance (NMR), circular dichroism (CD), ultraviolet (UV) or ensemble-averaged fluorescence spectroscopy to elucidate structures and functions of DNA are that rare species and dynamic processes can be observed under physiologically relevant solvation conditions (see Figure 2.1). The FRET efficiency (E_{FRET}), the probability of an excited state donor transferring energy to an acceptor fluorophore, has been the most useful parameter for obtaining structural information about DNA at length scales of ~ 1–10 nm. It is calculated from experimental data using eqn (1):

$$E_{FRET} = \frac{I_A}{I_A + \gamma I_D} \tag{1}$$

where I_A and I_D are the intensities of acceptor and donor fluorescence and γ is a factor that corrects for differences in quantum yields of the donor and acceptor, and instrumental detection efficiency of photons emitted by the two fluorophores. The usefulness of FRET efficiency primarily stems from its well-characterised dependence on the separation of donor and acceptor fluorophore according to eqn (2):[103,104]

$$E_{FRET} = \frac{1}{1 + \left(\dfrac{R}{R_0}\right)^6} \tag{2}$$

where R is the interfluorophore separation and R_0 is the donor–acceptor distance at which the efficiency is 50% (the Förster radius). If this radius is known, it is therefore possible to measure absolute donor–acceptor distances. R_0 is determined by eqn (3):

$$R_0 = \sqrt[6]{\frac{9\Phi_D(\ln 10)\kappa^2 J}{128\pi^5 n^4 N_A}} \tag{3}$$

where Φ_D is the quantum yield of the donor in the absence of the acceptor, κ^2 is the dipole orientation factor, J is the spectral overlap integral of donor emission and acceptor absorption, n is the refractive index of the medium and N_A is Avogadro's number.

The availability of synthetic oligonucleotides and the choice of labelling locations they offer (see Figure 2.5A) make FRET a versatile tool for probing DNA structures. Indeed, the first demonstration of smFRET was between two rhodamine dyes (TAMRA and Texas Red) arranged on a dsDNA scaffold.[105] In some cases the labelling strategy must be taken into account when interpreting experimental data, *e.g.* κ^2 may deviate from 2/3 (the value for freely rotating dyes) if the fluorophores interact with duplex DNA, as occurs for Cy3 and Cy5 attached to the 5'-end of DNA duplexes *via* a three-carbon spacer.[72] The Cy3–Cy5 pair is probably the one most commonly used to label DNA for FRET measurements, despite the low quantum yields of both dyes (see Table 2.1) and their complicated photophysical behaviour (see Section 2.2.2). The popularity of Cy5 (and the related carbocyanine dye Alexa Fluor 647) mainly derives from its large extinction coefficient, which leads to a high overlap integral with donors whose emission spectra are well separated from its own, such as Cy3, TAMRA or Rhodamine Green derivatives (*e.g.* Alexa Fluor 488 and ATTO 488). This property enables a high Förster radius to be maintained without significant crosstalk, *i.e.* donor emission leaking into the acceptor channel. Another advantage of the sulfonated Cy3–Cy5 pair is that transient short-range (<3 nm) dye–dye interactions are minimised. Di Fiori and Meller have shown that these interactions can lead to significant intermittent quenching of both fluorophores if rhodamine–rhodamine (TAMRA/ATTO 647N) or rhodamine–cyanine (Cy3/ATTO 647N and TAMRA/Cy5) pairs are used instead of Cy3–Cy5.[106]

Both confocal and TIRFM configurations have been used in smFRET studies of DNA, each offering distinct advantages. Because TIRFM is an imaging approach (illuminating an area of $\sim 0.05\,\text{mm}^2$ to a depth of ~ 100–$200\,\text{nm}$), it offers higher throughput than confocal measurement, which uses point detection: hundreds of molecules may be monitored simultaneously but discretely using TIRFM. On the other hand, freely diffusing analytes can be studied by confocal methods whereas TIRFM requires that the DNA be tethered to a microscope slide (Figure 2.9A and B), which may perturb the process being probed; as a result, chemical coating of the slide surface (*e.g.* with polyethylene glycol, PEG) to resist non-specific adsorption is an important requirement. In addition, the higher signal-to-noise ratios achieved in confocal detection deliver

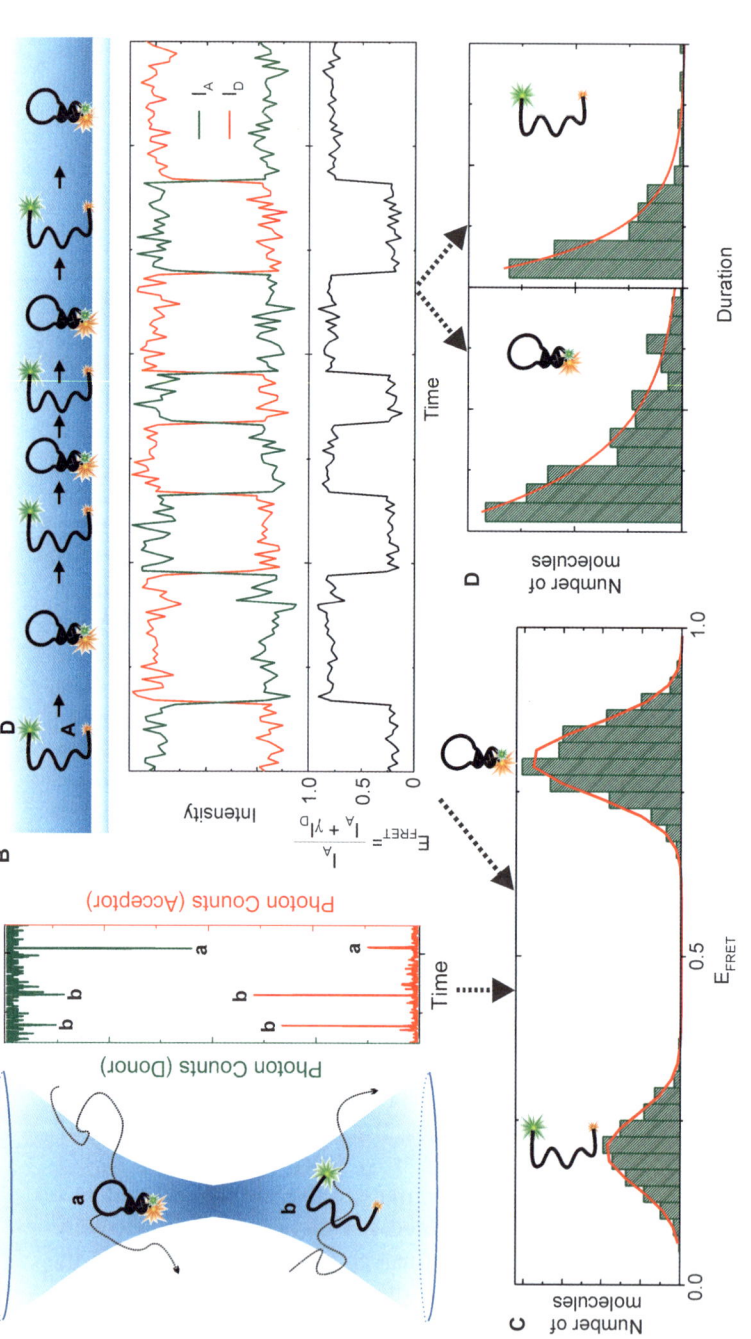

Figure 2.9   Structural information obtained using different optical configurations for SMFS. (A) Idealised two-colour fluorescence burst data obtained from DNA hairpins freely diffusing through a confocal detection volume, where closed (molecule a) and open (molecule b) hairpins exhibit high and low FRET states respectively. (B) Simulated donor and acceptor fluorescence trajectories of an immobilised hairpin undergoing repetitive opening/closing transitions observed using TIRFM. The FRET efficiency (E_{FRET}) can be calculated from I_D, I_A and γ using eqn (1). (C) Simulated histogram ($n = 1500$) of FRET efficiencies obtained from either confocal or TIRFM data, yielding equilibrium concentrations of open and closed forms of the hairpin. (D) Simulated duration distributions of the two species, which can be used to calculate rate constants for the opening and closing processes. If the transit time of the molecules through the detection volume is short compared to the lifetime of the open and closed states, such data cannot be recovered from confocal measurements.

increased temporal resolution because sufficient photons to overcome shot noise can be collected more quickly. Either method allows the concentrations of relatively long-lived species to be measured (Figure 2.9C), and analysing the persistence time of different molecular conformations allows microscopic rate constants for their interconversion to be parsed directly under equilibrium conditions (Figure 2.9D). These kinetic data can be obtained by applying a hidden Markov model to FRET trajectories of immobilised molecules,[107] or extracted from FRET fluctuations of diffusing analytes.[108] If the species being analysed travel through the confocal detection volume (either by Brownian motion or in flow) on a timescale that is much faster than their lifetime, it is necessary to perform non-equilbrium time-course experiments to characterise kinetics. Scanning confocal systems have been developed to monitor single immobilised molecules with high signal-to-noise ratio, but the throughput is limited to one reaction at a time.[109] TIRFM is the more popular choice for probing DNA structural dynamics as a result of these considerations, but confocal measurements have been extremely useful for some applications. Readers interested in setting up and running TIRFM smFRET experiments in their own laboratories are directed to the very useful practical guide published by Taekjip Ha's group, which includes advice on data analysis, selection of commercial optical components and reagents, and a protocol for coating microscope slides with PEG.[50]

The simplest dynamic process involving DNA is perhaps the conformational flexibility of single stranded nucleotides. This process has been characterised using the E_{FRET} distributions of a Cy3–Cy5 pair tethered at each end of a dT_{10-70} chain (Figure 2.10A), measuring the salt-dependence of the persistence length ssDNA.[110–113] The structural features of the B-DNA duplex can also been observed by FRET, where the Clegg model for interfluorophore separation takes account of the geometry of the double helix.[114] Deniz *et al.*'s demonstration in 1999 that this model could be tested at the single molecule level using spFRET (Figure 2.10B) was a landmark in SMFS.[111] More recently, precise measurement of kinks and bends in DNA duplexes has been achieved using confocal measurement of spFRET in conjunction with anisotropy and dye quenching.[115] In addition to characterising the structure of dsDNA, SMFS has also been useful in elucidating the kinetics of DNA hairpin opening and closing, where FRET fluctuations of freely diffusing species were monitored.[108] Higher order DNA structures whose dynamics have been probed at the single molecule level by FRET include Holliday junctions[69] and G-quadruplexes.[116] Experiments using smFRET are not restricted to two fluorophores; the development of three-colour (using Cy3, Cy5 and Cy5.5)[117] and four-colour (using Cy2, Cy3, Cy5 and Cy7)[118] versions of the method were demonstrated using Holliday junctions as model samples.

While smFRET is ideally suited to examining DNA–DNA and protein–DNA interactions at ~ 1–10 nm resolution, the structure of DNA at the ~ 10–200 nm length scale is key to understanding the function of genomes and DNA nanomaterials.[119] Because E_{FRET} depends on $\frac{R_0^6}{R^6}$ [eqn (2)], it is very insensitive to changes at these distances, where $R \gg R_0$ even for fluorophores with the

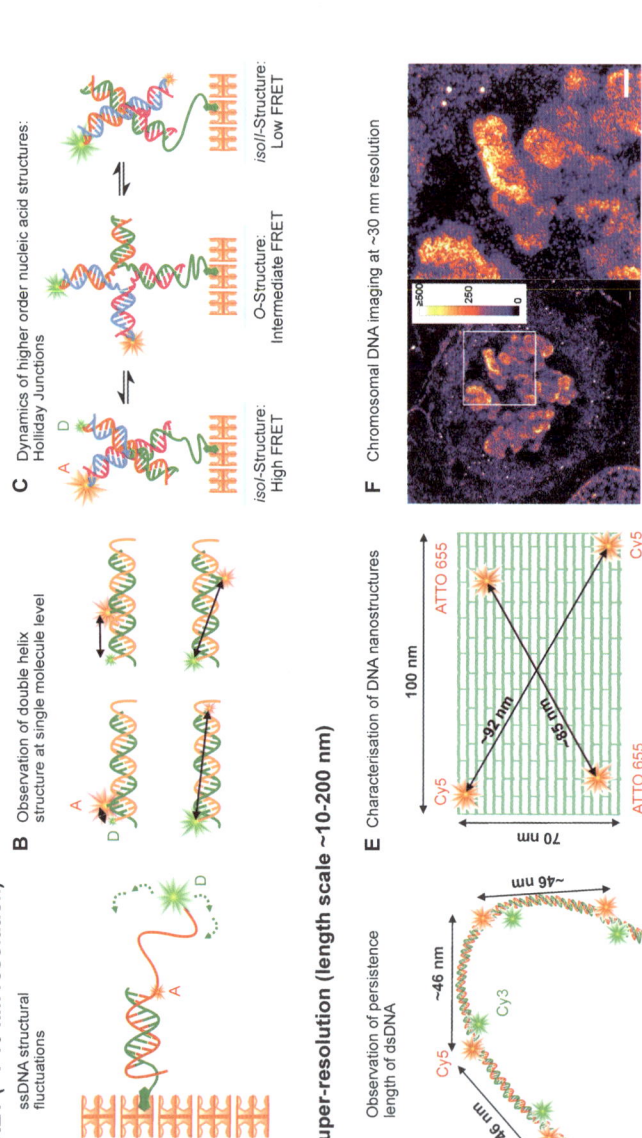

Figure 2.10 Studies of DNA structures using SMFS. (A) Characterising the flexibility of ssDNA using smFRET monitored by TIRFM.[110] (B) Observing the helical geometry of the B-DNA duplex by measuring the dependence of E_{FRET} on base pair separation using confocal optics and smFRET between a 3′-TAMRA donor and Cy5 acceptor attached to the base of a T-analogue.[111] (C) Elucidating the structural dynamics of Holliday junctions using spFRET between Cy3 and Cy5.[69] (D) Synthetic dsDNA used for super-resolution microscopy based on photoswitching of Cy3–Cy5 pairs separated by 135bp. (E) DNA origami used as a nanoscopic ruler labelled with either two Cy5 dyes or two ATTO 655 fluorophores (in separate experiments).[112] The theoretical dimensions of the rectangular DNA structure (100×70 nm) and the dye–dye distances measured by super-resolution microscopy are shown. (F) Super-resolution image of gDNA labelled with POPO-3 inside the nucleus of a fixed HeLa cell. The square in the left panel indicates the zoomed-in area shown in the right panel. Scale bar (in both images) = 1 μm. Adapted with permission from P. D. Simonson, E. Rothenberg, and P. R. Selvin, *Nano Lett.*, 2011, **11**, 5090–5096.[113] Copyright 2011 American Chemical Society.

largest spectral overlap. According to the Abbe diffraction limit, far-field light microscopy should not be effective in this regime either, because it is not possible to resolve objects in the focal plane that are separated by less than the distance, d, determined by eqn (4):

$$d = \frac{\lambda}{2n\sin\theta} \tag{4}$$

where λ is the wavelength of detected light, n is the refractive index of the medium and θ is the half-angle of the cone of light collected by the objective. The parameter $n\sin\theta$ is equivalent to the numerical aperture (NA). Therefore, an optical system with NA of 1.3 detecting green fluorescence at 520 nm has a maximum theoretical resolution of 200 nm. Despite this constraint, it has been known for ~25 years that sub-diffraction objects can be localised with precision beyond the diffraction limit using fluorescence microscopy, by calculating the centre of the point spread function (PSF), which defines the response of an imaging system to a point source.[120] Provided sufficient photons can be collected from lone fluorophores, their position can be determined at nanometre resolution with standard error $\langle(\Delta x)^2\rangle$ determined by eqn (5):[22]

$$\langle(\Delta x)^2\rangle = \frac{s^2}{N^2} \tag{5}$$

where s is the standard deviation of the PSF and N is the number of photons collected. Therefore, two or more fluorophores separated by distances smaller than the diffraction limit can still be localised precisely by SMFS as long as they can be observed in temporally separated image frames, *e.g.* by undergoing photoswitching between emissive and dark states (see Section 2.2.2, Figure 2.6D and E). Each frame can then be combined to reconstruct an image where the positions of all the fluorophores are known with high precision. This realisation has triggered an explosion of subtly different SMFS methods in the last 5–10 years that exploit this phenomenon. These techniques, collectively known as super-resolution microscopy, provide an alternative to electron microscopy that can be used in live cells; they have been reviewed recently and we deliberately gloss over the experimental details here.[121–123]

DNA structures have mostly been used as model samples in super-resolution microscopy for proof-of-concept experiments, *e.g.* using the persistence length of dsDNA to demonstrate super-resolution by photoswitching of a Cy3–Cy5 pair in the presence of a thiol (Figure 2.10D),[124] staining polynucleotides with YOYO-1 for imaging on a glass surface[125,126] or using DNA origami labelled with Cy5 or ATTO 655 as a nanoscopic calibrant (Figure 2.10E).[112] In the future though, it is likely that super-resolution techniques will begin to reveal information about DNA, characterising the structure or motion of new DNA-based nanomachines or elucidating the function of gDNA in live cells. The latter application has so far been hampered by the unavailability of high-density DNA-labelling chemistry for use *in vivo*. Intercalators (*e.g.* SYTO

16, POPO-3) can be used to stain DNA inside fixed cells (Figure 2.10F),[113] with the advantage that binding to nucleic acids enhances their quantum yield, reducing background fluorescence. However, their diminished brightness compared with other synthetic fluorophores limits the localisation precision to ~ 35 nm, because the number of photons collected affects the uncertainty in the measured position [eqn (5)]. High-density labelling of gDNA in fixed cells with Alexa Fluor 647 by click chemistry (see Figure 2.5C) has recently enhanced the resolution obtained to 10 nm.[75]

2.3.2.2   Protein–DNA Interactions

The processes by which proteins manipulate DNA with atomic precision are fundamental to life, regulating DNA repair, recombination, replication and transcription. SMFS, in particular smFRET, provides a means to observe these events without synchronising large numbers of reactions. Although confocal detection can provide detailed structural information about protein–DNA interactions, such as the 'scrunching' of promoter DNA by RNA polymerase,[127] the dynamic nature of these processes means that prolonged observation of immobilised molecules by TIRFM can yield more useful data. The fact that synthetic oligonucleotides can be efficiently labelled with bright, photostable dyes (see Sections 2.2.1 and 2.2.3) and immobilised to the passivated surface of a microscope slide has been crucial to this field of study. Because the timescales of the processes of interest are in the order of tens of seconds, management of dye photophysics using additives for removal of oxygen and quenching of triplet states has also been of critical importance (see Section 2.2.2).

Site-specific labelling of proteins with small molecule dyes is more difficult to achieve than conjugation to synthetic oligonucleotides. Automated phosphoramidite chemistry enables incorporation of a variety of groups whose reactivity is orthogonal to native DNA at predetermined locations (see Figure 2.5A). In contrast, labelling proteins with electrophilic groups (*e.g.* activated esters or maleimides) is unlikely to be selective if there are multiple nucleophilic amino acid residues (*e.g.* lysine or cysteine) present in the sequence. The resultant heterogeneity in the sample makes interpretation of SMFS data extremely difficult. As a result, the preferred labelling strategy for studying protein–DNA interactions with spFRET is to conjugate both fluorophores to the nucleic acid component so that their relative positions are altered by proteins acting on the DNA. Helicase enzymes, motor proteins fuelled by ATP-hydrolysis that unwind double stranded nucleic acids, have been well studied by this method. Ha *et al.* reported the first study of protein–DNA interaction dynamics using smFRET in 2002 using the Rep helicase from *E. coli*;[128] by 2007, unwinding of Cy3–Cy5-labelled dsDNA by the NS3 helicase from hepatitis C virus in discrete 3 bp bursts could be observed as successive decreases in E_{FRET} (Figure 2.11A).[129] Binding of two prokaryotic proteins important in DNA metabolism, RecA and SSB (single-stranded DNA-binding

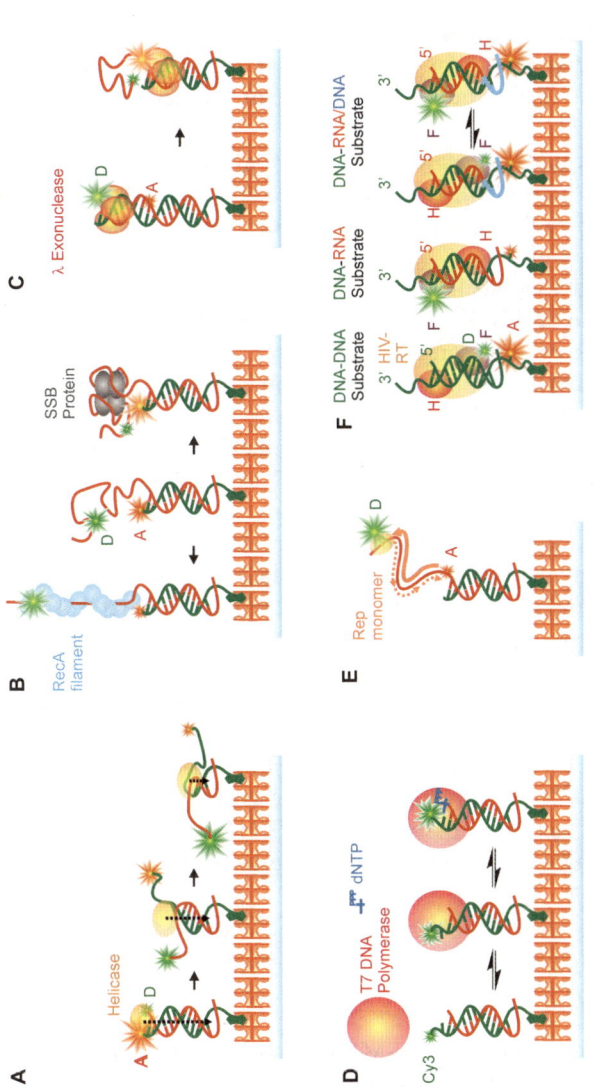

Figure 2.11 Studies of protein–DNA interactions using SMFS. (A) Unwinding of dsDNA by NS3 helicase leading to successive increases in the separation of a donor (Cy3) and acceptor (Cy5) attached at the ends of each strand of the duplex. (B) Changes in E_{FRET} between an acceptor attached to a single stranded overhang and at the end of the adjacent duplex, caused by protein–DNA interactions. RecA filament formation along ssDNA leads to a decrease in the transfer efficiency, whereas wrapping DNA around the four subunits of SSB increases E_{FRET}. (C) Increase in E_{FRET} between Cy3 and iCy5 (Figure 2.5A) attached to the same DNA strand upon degradation of its unlabelled complement by λ exonuclease. (D) Conformational changes of Cy3 attached to a primer strand upon formation of a ternary complex with a DNA polymerase and subsequent binding of a dNTP, monitored by increase in fluorescence intensity. (E) Observing repetitive shuttling of Cy3-labelled Rep helicase along a Cy5-labelled strand. (F) Binding orientations of HIV reverse transcriptase (RT) with different nucleic acid substrates. Binding of RT to a DNA–DNA duplex places the DNA polymerase domain F at the 3'-end of the primer strand, while a DNA–RNA substrate binds the RNaseH domain H at the 3'-end of the primer. If an RNA primer strand containing a polypurine tract (the native sequence for priming DNA synthesis) is used, substitution of two deoxynucleotides at the 3'-end leads to dynamic flipping between the two orientations.

protein), has been monitored in real time using oligonucleotides labelled with Cy5 at the end of a duplex and with Cy3 on an adjacent single-stranded overhang. This arrangement of fluorophores leads to an increase upon SSB-binding and a decrease in E_{FRET} upon RecA filament formation (Figure 2.11B);[130] eventually it was shown with three-colour FRET that SSB diffusion on ssDNA promotes formation of RecA filaments.[131] Introducing both fluorophores into the same strand of a duplex enabled the hydrolysis of the complementary unlabelled strand by λ exonuclease to be monitored (Figure 2.11C), revealing an intial distributive component of the degradation process.[132]

Dual-labelling of oligonucleotide substrates is not the only way in which protein–DNA interactions can be probed. In the simplest case, intensity fluctuations of a single fluorophore could report back on conformational changes. This approach was used to monitor the formation of a ternary complex between T7 DNA polymerase, a Cy3-labelled primer and a template, as well as subsequent dNTP binding (Figure 2.11D).[133] Each step was accompanied by an increase in the quantum yield of the cyanine fluorophore, consistent with increased restriction of the flexibility of its polymethine chain, thereby supressing non-radiative relaxation of the S_1 state by *trans–cis* isomerisation (see Figure 2.3A). While convenient, this strategy is not applicable to the study of all protein–DNA interactions. For example, binding of λ exonuclease to the dual-labelled substrate shown in Figure 2.11C led to an increase in brightness of both fluorophores, but subsequent degradation steps were only accompanied by changes in E_{FRET}.[127] In some cases, the more difficult task of labelling the protein component must be undertaken to provide the required structural insights. Myong *et al.* engineered a mutant of the Rep helicase bearing a single cysteine residue for labelling with a maleimide-derivative of Cy3 in order to observe repetitive shuttling of the helicase along a ssDNA overhang labelled at the ss/dsDNA junction with Cy5 (Figure 2.11E).[134] SMFS using a fluorescently labelled enzyme has also been used to study how the binding orientation of human immunodeficiency virus (HIV) reverse transcriptase (RT) directs its dual RNaseH and DNA polymerase activities (Figure 2.11F), showing an unusual dynamic flipping of the orientation upon binding to a DNA–RNA/DNA duplex.[135]

2.4   Conclusions

This chapter highlights the increasing use of SMFS for the study of DNA sequencing and bioanalysis as well as DNA structure and function. The field has advanced over the last 15 years to the point where studies of single fluorophores in specialist laboratories are fairly routine experiments, so the focus over the next 5–10 years will be on applications to both basic science and diagnostics. Accessibility of SMFS instrumentation to non-physical scientists will play an important role in this process, with cost and ease of use being key factors in the level of uptake. Taking single molecule studies outside the exclusive domain of specialist laboratories and into the hands of biological scientists should open up new areas of investigation in the same way as the commercial availability of tailor-made synthetic oligonucleotides has powered advances in the biological sciences over the last 30 years. Although off-the-shelf SMFS systems are now available, only well-funded groups are able to afford the current price tags of > US$100,000. As with all technological innovations, it is likely that economy of scale and improvements in manufacturing processes will drive the cost down to some extent. In diagnostic applications, particularly at the point of care, it is of paramount importance that any future SMFS devices can be operated by medical personnel without the need for extensive additional training. This requirement is driving the development of 'sample-in–answer-out' molecular diagnostics, which therefore requires integrated modules to be engineered for rapid DNA extraction, purification and detection. Similar principles are beginning to be applied to the study of proteins in fresh cell lysates by SMFS.[136,137]

Increasing the complexity of systems that can be studied using SMFS would be invaluable for further elucidation of the structure and function of DNA *in vivo*. For example, the molecular machinery that comprises the replisome includes helicases, gyrases, primases, DNA polymerases, ligases and SSBs. Carefully designed two-colour smFRET experiments have already provided valuable insight into the coordination of DNA replication,[138] but increasing the number of species whose interactions are simultaneously monitored could provide considerably more information: while two-colour FRET measures one intermolecular separation, three- and four-colour smFRET can report on three and six intermolecular distances respectively. These techniques are made possible by a combination of optical engineering (to separate and detect photons emitted by each fluorophore) and synthetic chemistry (providing robust fluorophores that have sufficient spectral overlap to allow efficient energy transfer without generating excessive crosstalk). Site-specific chemical biology approaches to protein labelling will be necessary for multicolour smFRET experiments to deliver useful data on protein–DNA interactions; this goal can be achieved by using genetically encoded unnatural amino acids,[139] *e.g.* by introducing azide or alkyne side-chains at predetermined residues for coupling to dyes using click chemistry.[140] Super-resolution microscopy might also help to further our understanding of gDNA function in cells, but will probably be better suited to providing snapshots of the processes than dynamic imaging, because the high-density labelling required is likely to be perturbative. The most promising strategy might therefore be incorporation of a small dye 'synthon', such as an alkyne, into gDNA in live cells to provide bioorthogonal functionalisation with a high-performance fluorophore after fixation.[75] Again, developments in optical engineering (*e.g.* by using two microscope objectives to increase photon collection efficiency)[141] and synthetic chemistry (*e.g.* by producing new fluorophores with enhanced photon emission rates, switching kinetics or number of switching cycles before photobleaching), will be of significant benefit. It is clear that collaborative efforts at the interface between engineering, physics, chemistry and biology will be key to future developments in SMFS and will extend its use and application further in the study of DNA sequence, structure and function.

Acknowledgements

Both authors acknowledge the European Commission for funding (Thera-EDGE Network Contract FP7-216027). Single molecule fluorescence research in D. K.'s lab is supported by the BBSRC, MRC, Wellcome Trust and the Frances and Augustus Newman Foundation.

Simulated data shown in Figures 2.1, 2.9 and 2.11 were produced in ImageJ 1.46b (available for free download at http://rsbweb.nih.gov/ij/) using plugins written by Dr. Steven F. Lee (University of Cambridge). We are very grateful to him for this, as well as for his critical reading of the manuscript.

Conflict of Interest Statement

D. K. is a consultant for Illumina, Inc.

References

1. S. Weiss, *Science*, 1999, **283**, 1676–1683.
2. W. P. Ambrose, P. M. Goodwin, J. H. Jett, A. Van Orden, J. H. Werner and R. A. Keller, *Chem. Rev.,* 1999, **99**, 2929–2956.
3. W. E. Moerner and D. P. Fromm, *Rev. Sci. Instrum.*, 2003, **74**, 3597.
4. E. Haustein and P. Schwille, *Curr. Opin. Struct. Biol.*, 2004, **14**, 531–540.
5. P. Tinnefeld and M. Sauer, *Angew. Chem., Int. Ed.*, 2005, **44**, 2642–2671.
6. P. V. Cornish and T. Ha, *ACS Chem. Biol.*, 2007, **2**, 53–61.
7. A. A. Deniz, S. Mukhopadhyay and E. A. Lemke, *J. R. Soc. Interface*, 2008, **5**, 15–45.
8. E. B. Shera, N. K. Seitzinger, L. M. Davis, R. A. Keller and S. A. Soper, *Chem. Phys. Lett.*, 1990, **174**, 553–557.
9. M. B. Schneider and W. W. Webb, *Appl. Opt.*, 1981, **20**, 1382–1388.
10. S. Nie, D. T. Chiu and R. N. Zare, *Science*, 1994, **266**, 1018–1021.
11. J. Widengren, V. Kudryavtsev, M. Antonik, S. Berger, M. Gerken and C. A. M. Seidel, *Anal. Chem.*, 2006, **78**, 2039–2050.
12. D. Magde, E. Elson and W. Webb, *Phys. Rev. Lett.*, 1972, **29**, 705–708.
13. R. Shusterman, S. Alon, T. Gavrinyov and O. Krichevsky, *Phys. Rev. Lett.*, 2004, **92**, 1–4.
14. M. Eigen and R. Rigler, *Proc. Natl. Acad. Sci. USA*, 1994, **91**, 5740–5747.
15. P. Schwille, F. J. Meyer-Almes and R. Rigler, *Biophys. J.*, 1997, **72**, 1878–1886.
16. R. Rigler, Z. Földes-Papp, F. J. Meyer-Almes, C. Sammet, M. Völcker and A. Schnetz, *J. Biotechnol.*, 1998, **63**, 97–109.
17. H. Li, L. Ying, J. J. Green, S. Balasubramanian and D. Klenerman, *Anal. Chem.*, 2003, **75**, 1664–1670.

18. D. Axelrod, *J. Cell Biol.*, 1981, **89**, 141–145.
19. D. Axelrod, *Methods Enzymol.*, 2003, **361**, 1–33.
20. M. J. Levene, J. Korlach, S. W. Turner, M. Foquet, H. G. Craighead and W. W. Webb, *Science*, 2003, **299**, 682–686.
21. M. A. Osborne, S. Balasubramanian, W. S. Furey and D. Klenerman, *J. Phys. Chem. B.*, 1998, **102**, 3160–3167.
22. R. E. Thompson, D. R. Larson and W. W. Webb, *Biophys. J.*, 2002, **82**, 2775–2783.
23. T. Hirschfeld, *Appl. Opt.*, 1976, **15**, 2965–2966.
24. W.-C. Sun, K. R. Gee, D. H. Klaubert and R. P. Haugland, *J. Org. Chem.*, 1997, **62**, 6469–6475.
25. N. Panchuk-Voloshina, R. P. Haugland, J. Bishop-Stewart, M. K. Bhalgat, P. J. Millard, F. Mao and W. Y. Leung, *J. Histochem. Cytochem.*, 1999, **47**, 1179–1188.
26. R. B. Mujumdar, L. A. Ernst, S. R. Mujumdar, C. J. Lewis and A. S. Waggoner, *Bioconjugate Chem.*, 1993, **4**, 105–111.
27. S. R. Mujumdar, R. B. Mujumdar, C. M. Grant and A. S. Waggoner, *Bioconjugate Chem.*, 1996, **7**, 356–362.
28. J. R. Unruh, G. Gokulrangan, G. S. Wilson and C. K. Johnson, *Photochem. Photobiol.*, 2005, **81**, 682–690.
29. R. Philip, A. Penzkofer, W. Bäumler, R. Szeimies and C. Abels, *J. Photochem. Photobiol. A.*, 1996, **96**, 137–148.
30. G. Vamosi, C. Gohlke and R. Clegg, *Biophys. J.*, 1996, **71**, 972–994.
31. A. Glazer and H. Rye, *Nature.*, 1992, **359**, 859–861.
32. M. Bruchez, M. Moronne, P. Gin, S. Weiss and A. P. Alivisatos, *Science*, 1998, **281**, 2013–2016.
33. W. C. Chan and S. Nie, *Science*, 1998, **281**, 2016–2018.
34. M. Nirmal, B. O. Dabbousi, M. G. Bawendi, J. J. Macklin, J. K. Trautman, T. D. Harris and L. E. Brus, *Nature*, 1996, **383**, 802–804.
35. U. Resch-Genger, M. Grabolle, S. Cavaliere-Jaricot, R. Nitschke and T. Nann, *Nat. Methods*, 2008, **5**, 763–775.
36. I. L. Medintz, H. T. Uyeda, E. R. Goldman and H. Mattoussi, *Nat. Mater.*, 2005, **4**, 435–446.
37. W. R. Algar, A. J. Tavares and U. J. Krull, *Anal. Chim. Acta.*, 2010, **673**, 1–25.
38. N. Kaji, M. Tokeshi and Y. Baba, *Chem. Rec.*, 2007, **7**, 295–304.
39. J. E. Whitaker and R. P. Haugland, *Anal. Biochem.*, 1992, **207**, 267–279.
40. R. F. Kubin and A. N. Fletcher, *J. Lumin.*, 1983, **27**, 455–462.
41. J. Tellinghuisen, P. M. Goodwin, W. P. Ambrose, J. C. Martin and R. A. Keiiert, *Anal. Chem.*, 1994, **66**, 64–72.
42. B. K. Nunnally, H. He, L.-C. Li, S. A. Tucker and L. B. McGown, *Anal. Chem.*, 1997, **69**, 2392–2397.
43. D. Magde, G. E. Rojas and P. G. Seybold, *Photochem. Photobiol.*, 1999, **70**, 737–744.
44. M. Cooper, A. Ebner, M. Briggs, M. Burrows, N. Gardner, R. Richardson and R. West, *J. Fluoresc.*, 2004, **14**, 145–150.

45. C. Eggeling, J. Widengren, R. Rigler and C. A. Seidel, *Anal. Chem.*, 1998, **70**, 2651–2659.
46. C. Eggeling, J. Widengren, L. Brand, J. Schaffer, S. Felekyan and C. A. M. Seidel, *J. Phys. Chem. A.*, 2006, **110**, 2979–2995.
47. X. Kong, E. Nir, K. Hamadani and S. Weiss, *J. Am. Chem. Soc.*, 2007, **129**, 4643–4654.
48. J. R. Grunwell, J. L. Glass, T. D. Lacoste, A. A. Deniz, D. S. Chemla and P. G. Schultz, *J. Am. Chem. Soc.*, 2001, **123**, 4295–4303.
49. M. H. Horrocks, H. Li, J.-U. Shim, R. T. Ranasinghe, R. W. Clarke, W. T. S. Huck, C. Abell and D. Klenerman, *Anal. Chem.*, 2012, **84**, 179–185.
50. R. Roy, S. Hohng and T. Ha, *Nat. Methods*, 2008, **5**, 507–516.
51. J. Widengren and P. Schwille, *J. Phys. Chem. A.*, 2000, **104**, 6416–6428.
52. J. Widengren, E. Schweinberger, S. Berger and C. A. M. Seidel, *J. Phys. Chem. A.*, 2001, **105**, 6851–6866.
53. M. Heilemann, E. Margeat, R. Kasper, M. Sauer and P. Tinnefeld, *J. Am. Chem. Soc.*, 2005, **127**, 3801–3806.
54. S. S. White, H. Li, R. J. Marsh, J. D. Piper, N. D. Leonczek, N. Nicolaou, A. J. Bain, L. Ying and D. Klenerman, *J. Am. Chem. Soc.*, 2006, **128**, 11423–11432.
55. K. Jia, Y. Wan, A. Xia, S. Li, F. Gong and G. Yang, *J. Phys. Chem. A.*, 2007, **111**, 1593–1597.
56. Z. Huang, D. Ji, A. Xia, F. Koberling, M. Patting and R. Erdmann, *J. Am. Chem. Soc.*, 2005, **127**, 8064–8066.
57. C. Eggeling, C. Ringemann, R. Medda, G. Schwarzmann, K. Sandhoff, S. Polyakova, V. N. Belov, B. Hein, C. von Middendorff, A. Schönle and S. W. Hell, *Nature*, 2009, **457**, 1159–1162.
58. J. Widengren, A. Chmyrov, C. Eggeling, P.-A. Löfdahl and C. A. M. Seidel, *J. Phys. Chem. A.*, 2007, **111**, 429–440.
59. J. Vogelsang, R. Kasper, C. Steinhauer, B. Person, M. Heilemann, M. Sauer and P. Tinnefeld, *Angew. Chem. Int. Ed.*, 2008, **47**, 5465–5469.
60. R. E. Benesch and R. Benesch, *Science*, 1953, **118**, 447–448.
61. K. D. Weston, P. J. Carson, J. A. DeAro and S. K. Buratto, *Chem. Phys. Lett.*, 1999, **308**, 58–64.
62. I. Rasnik, S. A. McKinney and T. Ha, *Nat. Methods*, 2006, **3**, 891–893.
63. R. B. Altman, D. S. Terry, Z. Zhou, Q. Zheng, P. Geggier, R. A. Kolster, Y. Zhao, J. A. Javitch, J. D. Warren and S. C. Blanchard, *Nat. Methods*, 2011, **9**, 6–11.
64. T. Cordes, J. Vogelsang and P. Tinnefeld, *J. Am. Chem. Soc.*, 2009, **131**, 5018–5019.
65. C. R. Sabanayagam, J. S. Eid and A. Meller, *J. Chem. Phys.*, 2005, **123**, 224708.
66. G. T. Dempsey, M. Bates, W. E. Kowtoniuk, D. R. Liu, R. Y. Tsien and X. Zhuang, *J. Am. Chem. Soc.*, 2009, **131**, 18192–18193.
67. T. Kottke, S. van de Linde, M. Sauer, S. Kakorin and M. Heilemann, *J. Phys. Chem. Lett.*, 2010, **1**, 3156–3159.

68. J. Lee, S. Lee, K. Ragunathan, C. Joo, T. Ha and S. Hohng, *Angew. Chem. Int. Ed.*, 2010, **49**, 9922–9925.

69. S. A. McKinney, A.-C. Déclais, D. M. J. Lilley and T. Ha, *Nat. Struct. Biol.*, 2003, **10**, 93–97.

70. A. H. El-Sagheer and T. Brown, *Chem. Soc. Rev.*, 2010, **39**, 1388–1405.

71. M. H. Lyttle, T. G. Carter, D. J. Dick and R. M. Cook, *J. Org. Chem.*, 2000, **65**, 9033–9038.

72. A. Iqbal, S. Arslan, B. Okumus, T. J. Wilson, G. Giraud, D. G. Norman, T. Ha and D. M. J. Lilley, *Proc. Natl. Acad. Sci. USA*, 2008, **105**, 11176–11181.

73. M. L. Metzker, *Nat. Rev. Genet.* 2010, **11**, 31–46.

74. J. Bowers, J. Mitchell, E. Beer, P. R. Buzby, M. Causey, J. W. Efcavitch, M. Jarosz, E. Krzymanska-Olejnik, L. Kung, D. Lipson, G. M. Lowman, S. Marappan, P. McInerney, A. Platt, A. Roy, S. M. Siddiqi, K. Steinmann and J. F. Thompson, *Nat. Methods*, 2009, **6**, 593–595.

75. P. J. M. Zessin, K. Finan and M. Heilemann, *J. Struct. Biol.*, 2011, **177**, 344–348.

76. S. A. Bustin, ed., *The PCR Revolution: Basic Technologies and Applications*, Cambridge University Press, Cambridge, 2009.

77. D. R. Bentley, S. Balasubramanian, H. P. Swerdlow, G. P. Smith, J. Milton, C. G. Brown, K. P. Hall, D. J. Evers, C. L. Barnes, H. R. Bignell, J. M. Boutell, J. Bryant, R. J. Carter, R. Keira Cheetham, A. J. Cox, D. J. Ellis, M. R. Flatbush, N. A. Gormley, S. J. Humphray, L. J. Irving, M. S. Karbelashvili, S. M. Kirk, H. Li, X. Liu, K. S. Maisinger, L. J. Murray, B. Obradovic, T. Ost, M. L. Parkinson, M. R. Pratt, I. M. J. Rasolonjatovo, M. T. Reed, R. Rigatti, C. Rodighiero, M. T. Ross, A. Sabot, S. V. Sankar, A. Scally, G. P. Schroth, M. E. Smith, V. P. Smith, A. Spiridou, P. E. Torrance, S. S. Tzonev, E. H. Vermaas, K. Walter, X. Wu, L. Zhang, M. D. Alam, C. Anastasi, I. C. Aniebo, D. M. D. Bailey, I. R. Bancarz, S. Banerjee, S. G. Barbour, P. A. Baybayan, V. A. Benoit, K. F. Benson, C. Bevis, P. J. Black, A. Boodhun, J. S. Brennan, J. A. Bridgham, R. C. Brown, A. A. Brown, D. H. Buermann, A. A. Bundu, J. C. Burrows, N. P. Carter, N. Castillo, M. Chiara, E Catenazzi, S. Chang, R. Neil Cooley, N. R. Crake, O. O. Dada, K. D. Diakoumakos, B. Dominguez-Fernandez, D. J. Earnshaw, U. C. Egbujor, D. W. Elmore, S. S. Etchin, M. R. Ewan, M. Fedurco, L. J. Fraser, K. V. Fuentes Fajardo, W. Scott Furey, D. George, K. J. Gietzen, C. P. Goddard, G. S. Golda, P. A. Granieri, D. E. Green, D. L. Gustafson, N. F. Hansen, K. Harnish, C. D. Haudenschild, N. I. Heyer, M. M. Hims, J. T. Ho, A. M. Horgan, K. Hoschler, S. Hurwitz, D. V. Ivanov, M. Q. Johnson, T. James, T. A. Huw Jones, G.-D. Kang, T. H. Kerelska, A. D. Kersey, I. Khrebtukova, A. P. Kindwall, Z. Kingsbury, P. I. Kokko-Gonzales, A. Kumar, M. A. Laurent, C. T. Lawley, S. E. Lee, X. Lee, A. K. Liao, J. A. Loch, M. Lok, S. Luo, R. M. Mammen, J. W. Martin, P. G. McCauley, P. McNitt, P. Mehta, K. W. Moon, J. W. Mullens, T. Newington, Z. Ning, B. Ling, Ng, S. M. Novo, M. J. O'Neill, M. A. Osborne, A. Osnowski, O. Ostadan, L. L. Paraschos, L. Pickering, A. C. Pike, A. C. Pike, D. Chris Pinkard, D. P.

Pliskin, J. Podhasky, V. J. Quijano, C. Raczy, V. H. Rae, S. R. Rawlings, A. Chiva Rodriguez, P. M. Roe, J. Rogers, M. C. Rogert Bacigalupo, N. Romanov, A. Romieu, R. K. Roth, N. J. Rourke, S. T. Ruediger, E. Rusman, R. M. Sanches-Kuiper, M. R. Schenker, J. M. Seoane, R. J. Shaw, M. K. Shiver, S. W. Short, N. L. Sizto, J. P. Sluis, M. A. Smith, J. Ernest Sohna Sohna, E. J. Spence, K. Stevens, N. Sutton, L. Szajkowski, C. L. Tregidgo, G. Turcatti, S. Vandevondele, Y. Verhovsky, S. M. Virk, S. Wakelin, G. C. Walcott, J. Wang, G. J. Worsley, J. Yan, L. Yau, M. Zuerlein, J. Rogers, J. C. Mullikin, M. E. Hurles, N. J. McCooke, J. S. West, F. L. Oaks, P. L. Lundberg, D. Klenerman, R. Durbin and A. J. Smith, *Nature*, 2008, **456**, 53–59.

78. S. Balasubramanian, *Chem. Commun.*, 2011, **47**, 7281–7286.
79. T. D. Harris, P. R. Buzby, H. Babcock, E. Beer, J. Bowers, I. Braslavsky, M. Causey, J. Colonell, J. Dimeo, J. W. Efcavitch, E. Giladi, J. Gill, J. Healy, M. Jarosz, D. Lapen, K. Moulton, S. R. Quake, K. Steinmann, E. Thayer, A. Tyurina, R. Ward, H. Weiss and Z. Xie, *Science*, 2008, **320**, 106–109.
80. J. Eid, A. Fehr, J. Gray, K. Luong, J. Lyle, G. Otto, P. Peluso, D. Rank, P. Baybayan, B. Bettman, A. Bibillo, K. Bjornson, B. Chaudhuri, F. Christians, R. Cicero, S. Clark, R. Dalal, A. Dewinter, J. Dixon, M. Foquet, A. Gaertner, P. Hardenbol, C. Heiner, K. Hester, D. Holden, G. Kearns, X. Kong, R. Kuse, Y. Lacroix, S. Lin, P. Lundquist, C. Ma, P. Marks, M. Maxham, D. Murphy, I. Park, T. Pham, M. Phillips, J. Roy, R. Sebra, G. Shen, J. Sorenson, A. Tomaney, K. Travers, M. Trulson, J. Vieceli, J. Wegener, D. Wu, A. Yang, D. Zaccarin, P. Zhao, F. Zhong, J. Korlach and S. Turner, *Science*, 2009, **323**, 133–138.
81. J. Korlach, P. J. Marks, R. L. Cicero, J. J. Gray, D. L. Murphy, D. B. Roitman, T. T. Pham, G. A. Otto, M. Foquet and S. W. Turner, *Proc. Natl. Acad. Sci. USA*, 2008, **105**, 1176–1181.
82. R. T. Ranasinghe and T. Brown, *Chem. Commun.*, 2005, 5487–5502.
83. R. T. Ranasinghe and T. Brown, *Chem. Commun.*, 2011, 3717–3735.
84. S. Tyagi and F. R. Kramer, *Nat. Biotechnol.*, 1996, **14**, 303–308.
85. C.-Y. Zhang, S.-Y. Chao and T.-H. Wang, *Analyst.* 2005, **130**, 483–488.
86. T.-H. Wang, Y. Peng, C. Zhang, P. K. Wong and C.-M. Ho, *J. Am. Chem. Soc.*, 2005, **127**, 5354–5359.
87. A. Castro, D. A. R. Dalvit and L. Paz-Matos, *Anal. Chem.*, 2004, **76**, 4169–4174.
88. A. Castro and J. G. K. Williams, *Anal. Chem.*, 1997, **69**, 3915–3920.
89. A. Castro and R. T. Okinaka, *Analyst*, 2000, **125**, 9–11.
90. C. M. D'Antoni, M. Fuchs, J. L. Harris, H.-P. Ko, R. E. Meyer, M. E. Nadel, J. D. Randall, J. E. Rooke and E. A. Nalefski, *Anal. Biochem.*, 2006, **352**, 97–109.
91. A. Orte, R. Clarke, S. Balasubramanian and D. Klenerman, *Anal. Chem.*, 2006, **78**, 7707–7715.
92. Z. Földes-Papp, M. Kinjo, M. Tamura, E. Birch-Hirschfeld, U. Demel and G. P. Tilz, *Exp. Mol. Pathol.*, 2005, **78**, 177–189.

93. A. Orte, R. W. Clarke and D. Klenerman, *Anal. Chem.*, 2008, **80**, 8389–8397.

94. C.-Y. Zhang and L. W. Johnson, *Angew. Chem. Int. Ed.*, 2007, **46**, 3482–3485.

95. C.-Y. Zhang, H.-C. Yeh, M. T. Kuroki and T.-H. Wang, *Nat. Mater.* 2005, **4**, 826–831.

96. C.-Y. Zhang and J. Hu, *Anal. Chem.*, 2010, **82**, 1921–1927.

97. L. Chen, A. Manz and P. J. R. Day, *Lab Chip*, 2007, **7**, 1413–1423.

98. P. Yager, T. Edwards, E. Fu, K. Helton, K. Nelson, M. R. Tam and B. H. Weigl, *Nature*, 2006, **442**, 412–418.

99. H. Craighead, *Nature*, 2006, **442**, 387–393.

100. X. Zhuang, *Annu. Rev. Biophys. Biomol. Struct.*, 2005, **34**, 399–414.

101. C. Bustamante, Z. Bryant and S. B. Smith, *Nature*, 2003, **421**, 423–427.

102. S. Hohng, R. Zhou, M. K. Nahas, J. Yu, K. Schulten, D. M. J. Lilley and T. Ha, *Science*, 2007, **318**, 279–283.

103. T. Förster, *Ann. der Physik.*, 1948, **437**, 55–75.

104. L. Stryer and R. P. Haugland, *Proc. Natl. Acad. Sci. USA*, 1967, **58**, 719.

105. T. Ha, T. Enderle, D. F. Ogletree, D. S. Chemla, P. R. Selvin and S. Weiss, *Proc. Natl. Acad. Sci. USA*, 1996, **93**, 6264–6268.

106. N. Di Fiori and A. Meller, *Biophys. J.*, 2010, **98**, 2265–2272.

107. S. A. McKinney, C. Joo and T. Ha, *Biophys. J.*, 2006, **91**, 1941–1951.

108. M. I. Wallace, L. Ying, S. Balasubramanian and D. Klenerman, *Proc. Natl. Acad. Sci. USA*, 2001, **98**, 5584–5589.

109. C. R. Sabanayagam, J. S. Eid and A. Meller, *Appl. Phys. Lett.*, 2004, **84**, 1216.

110. M. C. Murphy, I. Rasnik, W. Cheng, T. M. Lohman and T. Ha, *Biophys. J.*, 2004, **86**, 2530–2537.

111. A. A. Deniz, M. Dahan, J. R. Grunwell, T. Ha, A. E. Faulhaber, D. S. Chemla, S. Weiss and P. G. Schultz, *Proc. Natl. Acad. Sci. USA*, 1999, **96**, 3670–3675.

112. C. Steinhauer, R. Jungmann, T. L. Sobey, F. C. Simmel and P. Tinnefeld, *Angew. Chem. Int. Ed.*, 2009, **48**, 8870–8873.

113. P. D. Simonson, E. Rothenberg and P. R. Selvin, *Nano Lett.*, 2011, **11**, 5090–5096.

114. R. M. Clegg, A. I. H. Murchie, A. Zechel and D. M. J. Lilley, *Proc. Natl. Acad. Sci. USA*, 1993, **90**, 2994.

115. A. K. Wozniak, G. F. Schröder, H. Grubmüller, C. A. M. Seidel and F. Oesterhelt, *Proc. Natl. Acad. Sci. USA*, 2008, **105**, 18337–18342.

116. L. Ying, J. J. Green, H. Li, D. Klenerman and S. Balasubramanian, *Proc. Natl. Acad. Sci. USA*, 2003, **100**, 14629–14634.

117. S. Hohng, C. Joo and T. Ha, *Biophys. J.*, 2004, **87**, 1328–1337.

118. J. Lee, S. Lee, K. Ragunathan, C. Joo, T. Ha and S. Hohng, *Angew. Chem. Int. Ed.*, 2010, **49**, 9922–9925.

119. N. C. Seeman, *Annu. Rev. Biochem.*, 2010, **79**, 65–87.

120. J. Gelles, B. J. Schnapp and M. P. Sheetz, *Nature*, 1988, **331**, 450–453.

121. M. Fernández-Suárez and A. Y. Ting, *Nat. Rev. Mol. Cell Biol.*, 2008, **9**, 929–943.
122. M. Heilemann, P. Dedecker, J. Hofkens and M. Sauer, *Laser Photonics Rev.*, 2009, **3**, 180–202.
123. J. Vogelsang, C. Steinhauer, C. Forthmann, I. H. Stein, B. Person-Skegro, T. Cordes and P. Tinnefeld, *ChemPhysChem*, 2010, **11**, 2475–2490.
124. M. J. Rust, M. Bates and X. Zhuang, *Nat. Methods*, 2006, **3**, 793–795.
125. C. Flors, C. N. J. Ravarani and D. T. F. Dryden, *ChemPhysChem*, 2009, **10**, 2201–2204.
126. F. Persson, P. Bingen, T. Staudt, J. Engelhardt, J. O. Tegenfeldt and S. W. Hell, *Angew. Chem. Int. Ed.*, 2011, **50**, 5581–5583.
127. A. N. Kapanidis, E. Margeat, S. O. Ho, E. Kortkhonjia, S. Weiss and R. H. Ebright, *Science*, 2006, **314**, 1144–1147.
128. T. Ha, I. Rasnik, W. Cheng, H. P. Babcock, G. H. Gauss, T. M. Lohman and S. Chu, *Nature*, 2002, **419**, 638–641.
129. S. Myong, M. M. Bruno, A. M. Pyle and T. Ha, *Science*, 2007, **317**, 513–516.
130. C. Joo, S. A. McKinney, M. Nakamura, I. Rasnik, S. Myong and T. Ha, *Cell*, 2006, **126**, 515–527.
131. R. Roy, A. G. Kozlov, T. M. Lohman and T. Ha, *Nature*, 2009, **461**, 1092–1097.
132. G. Lee, J. Yoo, B. J. Leslie and T. Ha, *Nat. Chem. Biol.* 2011, **7**, 367–374.
133. G. Luo, M. Wang, W. H. Konigsberg and X. S. Xie, *Proc. Natl. Acad. Sci. USA*, 2007, **104**, 12610–12615.
134. S. Myong, I. Rasnik, C. Joo, T. M. Lohman and T. Ha, *Nature*, 2005, **437**, 1321–1325.
135. E. A. Abbondanzieri, G. Bokinsky, J. W. Rausch, J. X. Zhang, S. F. J. Le Grice and X. Zhuang, *Nature*, 2008, **453**, 184–189.
136. B. Huang, H. Wu, D. Bhaya, A. Grossman, S. Granier, B. K. Kobilka and R. N. Zare, *Science*, 2007, **315**, 81–84.
137. A. Jain, R. Liu, B. Ramani, E. Arauz, Y. Ishitsuka, K. Ragunathan, J. Park, J. Chen, Y. K. Xiang and T. Ha, *Nature*, 2011, **473**, 484–488.
138. M. Pandey, S. Syed, I. Donmez, G. Patel, T. Ha and S. S. Patel, *Nature*, 2009, **462**, 940–943.
139. L. Davis and J. W. Chin, *Nat. Rev. Mol. Cell Biol.* 2012, **13**, 168–182.
140. D. P. Nguyen, H. Lusic, H. Neumann, P. B. Kapadnis, A. Deiters and J. W. Chin, *J. Am. Chem. Soc.*, 2009, **131**, 8720–8721.
141. K. Xu, H. P. Babcock and X. Zhuang, *Nat. Methods*, 2012, **9**, 185–188.

CHAPTER 3

Small Molecule–Oligonucleotide Conjugates

DAVID A. RUSLING AND KEITH R. FOX*

Centre for Biological Sciences, Life Sciences Building 85, University of
Southampton, Highfield, Southampton SO17 1BJ, UK
*Email: k.r.fox@soton.ac.uk

3.1 Introduction

The recognition of nucleic acids by short synthetic oligonucleotides has been an
active area of research for over 30 years. The hybridisation of exogenous oligonucleotides to cellular RNA and DNA offers a means to modulate the
expression of specific genes, with applications in the treatment of viral infections, cancer and other diseases.[1] More than one hundred oligonucleotides are
currently undergoing human clinical trials and there is a high level of optimism
that some of these will make it to the market.[2] Continued success in this area
will depend on synthetic efforts that improve the pharmacokinetic, pharmacodynamic and hybridization properties of these molecules.[1,3–9] The covalent
attachment (conjugation) of various low molecular weight compounds to oligonucleotides has proved particularly useful in this respect.[10] These conjugates
may be designed to improve an already existing feature of the oligonucleotide,
such as the strength and specificity of hybridisation, or may endow the oligonucleotide with entirely new properties, such as the ability to react chemically
with its target. The aim of this review is to summarise the properties of some of
these small molecule–oligonucleotide conjugates, with an emphasis on their
biological applications.

RSC Biomolecular Sciences No. 26
DNA Conjugates and Sensors
Edited by Keith R Fox and Tom Brown
© The Royal Society of Chemistry 2012
Published by the Royal Society of Chemistry, www.rsc.org

3.2 The Recognition of Nucleic Acids by Synthetic Oligonucleotides

Short synthetic oligonucleotides can be designed to hybridise to single- or double-stranded nucleic acids by exploiting Watson–Crick (W-C) or Hoogsteen hydrogen bonding, generating either double- or triple-helical structures. Hydrogen bonding drives the specificity of these interactions whilst base stacking is the predominant stabilising factor. By exploiting the hybridisation of oligonucleotides to either cellular mRNA or DNA it is possible to disrupt the flow of genetic information at different levels (Figure 3.1).

3.2.1 Duplex Formation and the Antisense Strategy

In 1978, Zamecnik and Stephenson first reported that exogenous oligonucleotides could bind to cognate single-stranded RNA by W-C hybridisation and inhibit translation and replication of the Rous sarcoma virus in cell culture.[11,12] This strategy was termed the antisense approach because the oligonucleotides are designed to bind to the coding 'sense' mRNA strand of the targeted gene (Figure 3.1B). Upon hybridisation, the oligonucleotide acts sterically to block the action of the ribosome on the mRNA and inhibit protein synthesis. In addition, the binding of an oligodeoxynucleotide to mRNA generates an RNA-DNA duplex that is a target for RNase H, a cellular enzyme that selectively degrades the mRNA partner of a RNA–DNA hybrid. Once degraded, the oligodeoxynucleotide is free to go on to hybridise to further mRNA and the same process is repeated. One of the positive aspects of this approach is the fact that exogenous oligonucleotides can access mRNA found in the cytosol much more easily than genomic DNA found in the cell nucleus. However targeting of mRNA is subject to up-regulation of the target gene and permanent modification of gene activity is not possible using this approach.

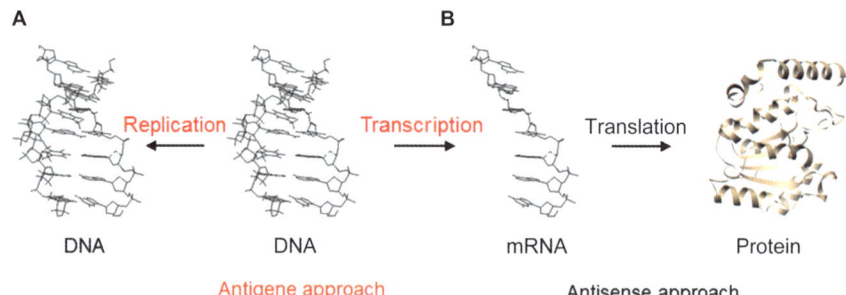

A   **B**

Replication   Transcription   Translation

DNA   DNA   mRNA   Protein

Antigene approach   Antisense approach

Figure 3.1   Repressing the central dogma of molecular biology using oligonucleotides and their conjugates. (A) Hybridisation to double-stranded DNA can repress the replication and/or expression of specific genes; a strategy termed the antigene approach. (B) Hybridisation to single-stranded RNA can inhibit mRNA translation and processing; a strategy termed the antisense approach.

3.2.2 Triplex Formation and the Antigene Strategy

In 1988, the Hélène and Dervan groups simultaneously proposed that exogenous oligonucleotides could be exploited to bind to genomic DNA by Hoogsteen hybridisation to form a triple helix and this was later used to inhibit the transcription of the *c-myc* gene *in vitro*.[13–15] By analogy with the above strategy this was later termed the antigene approach (Figure 3.1A). A triplex is generated by the binding of a triplex-forming oligonucleotide (TFO) within the major groove of a double-stranded DNA or RNA duplex by forming Hoogsteen hydrogen bonds between the TFO and exposed groups on the base pairs, generating base triplets (Figure 3.2A). Two triplex motifs have been described which differ in the orientation of the third strand relative to the central strand of the duplex. Third strands composed of pyrimidine bases bind in a parallel orientation to the central strand of the duplex, generating C^+.GC and T.AT triplets, while those containing purines bind in an antiparallel orientation, forming A.AT and G.GC triplets.[13–17] (The notation X.YZ refers to a triplet in which the third-strand base, X, interacts with the duplex base pair YZ, forming hydrogen bonds to base Y.) Triplex formation with pyrimidine-containing oligonucleotides requires conditions of low pH, necessary for the protonation of cytosine at N3 for generating the C^+.GC triplet. Both motifs are also hampered by a requirement for oligopurine–oligopyrimidine duplex target sites. To address these limitations a variety of base and nucleoside analogues have been developed and triplex formation at mixed sequence targets under physiological conditions is now possible.[6–10] Upon hybridisation the TFO can be designed to block transcription initiation or elongation, as well as DNA replication.[18–20] One of the positive aspects of this approach is that, unlike the targeting of mRNA, there are only two copies of the duplex target per diploid cell. Moreover, oligonucleotide-directed modifications to the duplex

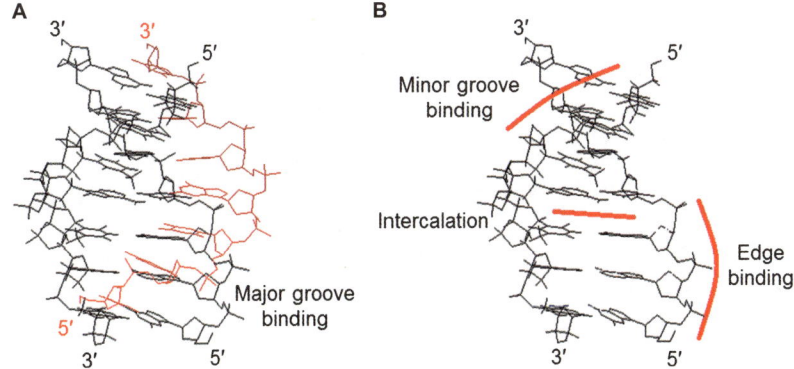

Figure 3.2 Nucleic acid and small molecule recognition of double-stranded DNA and RNA. (A) Formation of a triplex by the specific binding of a triplex-forming oligonucleotide (shown in red) within the major groove of a double-stranded duplex (shown in black). (B) Possible sites of small molecule interactions with a double-stranded duplex.

target sequence can invoke permanent alterations to gene activity and lead to heritable genetic change.

3.3 Small Molecule–Oligonucleotide Conjugation

Developments in oligonucleotide synthesis have allowed the covalent attachment of small molecules and other compounds to a variety of positions within an oligonucleotide structure.[10,21] The easiest sites of attachment are at the 5'- and 3'-ends of the oligonucleotide because these can be added either pre- or post-oligonucleotide synthesis [Figure 3.3(i) and (ii)]. Modifications at the termini have the additional advantage that they impart greater resistance of the oligonucleotide to exonucleases *in vivo*. Small molecules have also been attached to internal positions within the oligonucleotide at internucleoside phosphates [Figure 3.3(iii)], the 2'-position of the ribose sugar [Figure 3.3(iv)] and at various positions within the base [Figure 3.3(v)]. In general, modifications to the 2'-position of deoxyribose convert it from a C2'-*endo* (S-type) to a C3'-*endo* (N-type) configuration which favours oligonucleotide hybridisation to single-stranded RNA as well as triplex formation on double-stranded DNA. In each case, the appendage of the small molecule is done in such a fashion so as to minimise disruption of W-C and Hoogsteen pairing between strands.

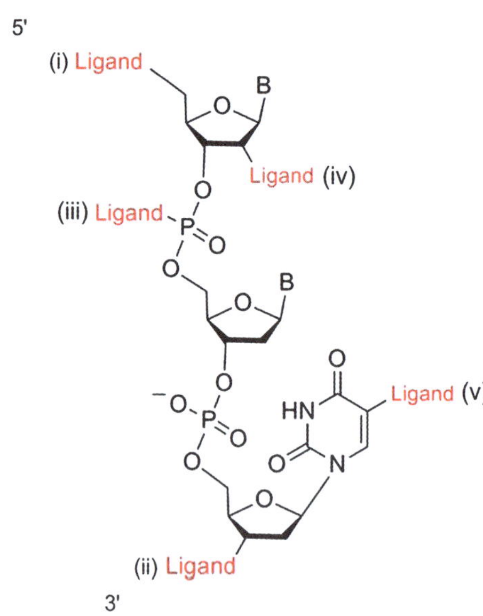

Figure 3.3 Small molecule–oligonucleotide conjugation. Small molecules have been attached to a variety of positions within an oligonucleotide, including the 5' (i) and 3' temini (ii), the internucleoside phosphate (iii), the 2' position on the ribose (iv), and at various positions of the base [the 5-position of dU is shown (v)]. The linker attaching the ligand to the oligonucleotide has varied depending on the study and is omitted for clarity.

3.4 Improving the Hybridisation Properties of Oligonucleotides

A variety of small molecules have been isolated or chemically synthesised that bind reversibly to double-stranded nucleic acids,[22] and many of these have been conjugated to oligonucleotides. The impetus for attaching these molecules is to improve the strength of oligonucleotide hybridisation with their single- or double-stranded targets. In this way sequence selectivity is achieved by the oligonucleotide while the ligand acts as a specific/non-specific anchor to increase affinity. There are three principle ways in which small molecules can reversibly interact with the features present within a double helix, including outside edge binding, binding by intercalation and groove binding (see Figure 3.2B). Often binding occurs by one or more of these modes so the small molecules described in the following section will be classified by their primary mode of interaction.

3.4.1 Stabilisation by Edge Binders

The sugar phosphate backbone of both DNA and RNA is negatively charged and compounds that contain positive charges can associate with the outside edge of the helix through non-specific electrostatic interactions. Polycations that interact in this fashion include the polyamines spermine and spermidine. As well as improving hybridisation, the addition of positively charged groups to an oligonucleotide can increase cellular permeation, a factor that currently limits the bioavailability of unmodified oligonucleotides.

3.4.1.1 Polyamines

Spermidine and spermine [Figure 3.4A(i) and (ii)] are found at millimolar concentrations in the nuclei of eukaryotic cells where they help to stabilise double-stranded DNA against denaturation. They are largely protonated at physiological pH and exhibit net positive charges close to $+3$ and $+4$, respectively. Both polycations have been attached to the 3′ and 5′ end of oligonucleotides and shown to generate duplexes with a higher stability than their unmodified counterparts.[23,24] The most stable duplexes have been generated with multiple residues attached to the oligonucleotide termini with an increase in duplex melting temperature (T_m) of around 6 °C per additional residue in a low salt buffer.[24] Furthermore, the base sequence and conjugation site (3′ or 5′ end) hardly influenced the effect of spermine on T_m. Spermine has been appended to different positions on a base with varying degrees of success. Attachment at the N4 position of ^{Me}C was shown to decrease duplex stability owing to the modification disrupting the ability of C to form a W-C base pair with G.[25] Conversely, attachment at the C5 position of dU,[26] and N2 or C8 positions of dG,[27–29] does not disrupt W-C base pairing and results in duplexes of higher stability. In fact, the conjugation of spermine to two or more bases within an oligonucleotide led to strand displacement of a complementary

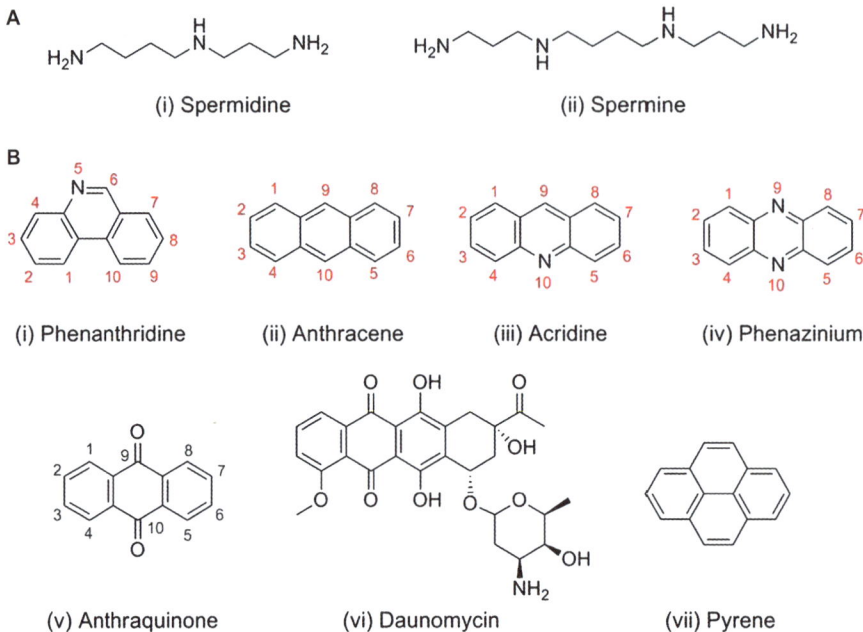

Figure 3.4   Small molecule edge binders and intercalators. (A) Polyamines that interact with DNA via edge binding. (B) Compounds composed of various fused ring systems that interact with DNA via intercalation.

double-stranded target.[27,29] It has been suggested that the increase in affinity of spermine-conjugated oligonucleotides stems from charge neutralisation and faster association kinetics.[27]

DNA and RNA triplexes are particularly stabilised by polycations that help to alleviate the charge repulsion between the three negatively charged strands.[30,31] Consequently, spermine has been attached to the 5′-end of the TFO,[23] 2′-position of the ribose,[32,33] 5-position of dU,[26] and N4-position of dC.[34–36] In all cases the introduction of spermine increased the affinity of the TFO for its duplex target, except with ribose modifications that were located in the centre of the TFO which inhibited triplex formation.[33] In fact, the addition of spermine to the base actually extended triplex formation with pyrimidine-rich TFOs to physiological conditions, even though the N3 of cytosine should not be protonated at this pH.[35] In most cases the appended spermine moiety was found only to affect the T_m of the triplex, but not the subsequent melting of the underlying duplex. Oligonucleotides containing tetraethoxyleneoxyamine, which contains a single protonation site, produced similar effects on triplex stability.[36]

3.4.2   Stabilisation by Intercalators

Compounds that contain extended planar aromatic ring systems bind to DNA by inserting between adjacent base pairs, perpendicular to the helix axis. Most intercalators tend to favour binding between 5′-pyrimdine-purine-3′ steps (YpR) because of stacking requirements but can also bind between other base pairs with a lower affinity. Increasing hybridisation by the addition of an intercalator to the end of an oligonucleotide has the advantage over modifications to the internal structure of an oligonucleotide in that the resulting RNA–DNA hybrid remains a substrate for RNase H.[4]

3.4.2.1 Phenanthridine

The archetypal DNA intercalators are based on the phenathridine ring system [see Figure 3.4B(i)], and 3,8-diamino-5-ethyl-6-phenylphenanthridinium (ethidium) is the most well characterised. Ethidium was one of the first intercalators to be attached to an oligonucleotide and the conjugate examined for its ability to bind to single-stranded DNA.[37] It was first appended to the backbone of a thymidine dinucleotide *via* a position on its 6-phenyl ring and shown to interact with poly(dA) under conditions where the unmodified dinucleotide did not bind. This approach was later extended to longer oligonucleotides and resulted in a T_m enhancement of around 8 °C for a single internal substitution.[38] Free ethidium exhibits a preference to bind between adjacent GC base pairs but surprisingly the environment surrounding the intercalation position seemed to have little influence on the degree of duplex stabilisation.[39]

3.4.2.2 Acridine

Acridines are perhaps the most studied of the DNA intercalators that have been attached to oligonucleotides. Acridine is an isomer of phenanthridine and is structurally related to anthracene [see Figure 3.4B(ii)] with one of the central CH groups replaced by nitrogen [Figure 3.4B(iii)]. Anthracene itself has been appended to an oligonucleotide at either its 9 or 10 position and shown to stabilise the interaction of the oligonucleotide with single-stranded DNA.[40] Acridine is most often employed as its 2-methoxy,6-chloro,9-amino derivative and is usually conjugated to an oligonucleotide *via* its 9-amino group. The first oligonucleotide to be synthesised was a tetranucleotide that contained acridine at its 3′ terminus.[41] The presence of this intercalator strongly stabilised its interaction with poly(rA) but the strength of this stabilisation was dependent on the linker length; a compound with a methylene chain of six carbon atoms gave a more stable complex than a compound with three. Subsequent studies showed that conjugation to the 5′-end, both 5′- and 3′-ends, and to the internucleosidic phosphate was also stabilising, however attachment at the ends was favoured over attachment at the centre.[42] The interactions of acridine with double-stranded DNA have been proposed to occur either through end-stacking, with the acridine moiety stacked on the terminal duplex base pair, or through intercalation between the last two adjacent base pairs of the helix.[43] Interestingly, attachment of acridine to oligodeoxynucleotides containing [α]-anomers of nucleosides allowed the formation of parallel stranded helices, and binding with RNA was favoured over DNA.[43–46] Addition of the acridine at the 3'-end was also shown to protect the oligonucleotide against nuclease digestion, an effect that was enhanced when attached to [α]-deoxynucleotides. Acridine–oligonucleotide conjugates have been successfully exploited to inhibit both eukaryotic and prokaryotic mRNA translation in cell extracts,[47–49] viral reverse transcription and replication in cell culture,[50] as well as mRNA transcripts generated from oncogenes.[51,52] In each of these studies, unmodified oligonucleotides required much higher concentrations to elicit the same effect as the acridine conjugates.

Various acridines have also been attached to TFOs to promote triplex formation.[53–63] In this way the TFO positions the acridine to bind to the duplex region adjacent to the triplex site to enhance binding affinity. This is particularly pronounced with the intercalator attached to the 5'-end of the TFO because intercalation can occur at a 5'-YpR-3' step located at the triplex–duplex junction, a site that is particularly susceptible to intercalation owing to perturbations in the duplex structure upon binding of the third strand.[54] In general, attachment of acridines increases triplex stability by around 100-fold but also increases the interaction of the TFO with non-cognate sequences.[57] Most of the studies have been done with pyrimidine-containing TFOs but the same effect can be seen in purine-containing TFOs.[58–61] Triplex stabilisation has been shown to be due to a decrease in the dissociation rate constant of the TFO but also leads to a slight increase in the association rate constant, though the latter is dependent on the flanking base sequence.[62] Acridine-TFOs have also been shown to exhibit biological activity via a triplex-directed mechanism.[55,63,64]

Another intercalator structurally related to both anthracene and acridine is phenazinium [see Figure 3.4B(iv)], and this has been appended to various positions within an oligonucleotide and shown to enhance duplex stability.[65,66] The phenazinium derivative used in these studies was attached *via* its 2-position to the oligonucleotide and stabilisation was greatest at the 5'-end of the oligonucleotide with a change in free energy of the complex that was approximately equivalent to extending the duplex by one GC or two AT base pairs.

3.4.2.3 Anthraquinone

Anthraquinones have also received a great deal of attention and are similar in structure to anthracene with two of the central CH groups replaced by carbonyls [see Figure 3.4B(v)]. Anthraquinones were first appended to the 5'-end of oligonucleotides *via* piperazinyl or methylene linkers and in both cases the duplex T_m values generated by these oligonucleotides were increased by 6–12 °C relative to the appropriate oligonucleotide controls.[67] Interestingly, the anthraquinones were also compatible with oligonucleotides synthesised with a phosphorothioate backbone in which one of the non-bridging oxygens is replaced by sulfur. Phosphorothioate oligonucleotides have greater resistance to nuclease digestion than unmodified oligonucleotides and have been used for antisense applications for a number of years. Anthraquinones have also been appended to oligonucleotides at either their 3' or 5' ends,[68] the 2'-position of the ribose sugar,[69,70] and at the 5-position of dU.[69] In most cases, introduction of the anthraquinone was more stabilising at the oligonucleotide ends than in the centre. Further stabilisation is afforded by the inclusion of more than one anthraquinone and was again more stabilising at the ends of the duplex.[69–71] Positioning of a pair of anthraquinones opposite one another within a duplex is also possible but only when the anthraquinone is attached *via* its 2,6 positions.[72] Anthraquinones have themselves been modified with spermine and the resultant spermine–anthraquinone conjugate generated stable duplexes that were less dependent on ionic conditions.[73] It has also been reported that positioning an anthraquinone at the duplex termini can increase base pairing fidelity.[68,74] Association of the anthraquinone oligonucleotide with its cognate strand exhibited a ΔT_m of $+18\,°C$ relative to an unmodified oligonucleotide, whilst association with a sequence containing a single mismatch at the 3′-end resulted in a ΔT_m of -24 and $-30\,°C$ when the complement was DNA or RNA, respectively.[74]

Anthraquinones are particularly effective at stabilising triplexes, and TFOs modified at both ends generate parallel triplexes with T_ms of over $40\,°C$ even at physiological pH.[75,76] Moreover, stabilisation occurred irrespective of the base step at the triplex–duplex junction.[76] Interestingly, two adjacent polypurine sequences located on different strands can be targeted using a single anthraquinone–TFO conjugate.[77,78] These oligonucleotides were designed to contain both α and β anomers so that half of the oligonucleotide recognised one of the purine sequences (5′-3′) and the other half recognises the 'inverted' second sequence (3′-5′). By positioning the anthraquinone between the two purine sequences it acted as a stabilising linker that traversed the DNA helix.

The anthraquinone-like intercalator daunomycin [see Figure 3.4B(vi)] has been conjugated to TFOs and used to increase the strength of hybridisation.[79] Daunomycin is composed of an aglycone moiety, consisting of the intercalating anthraquinone, and an amino sugar that interacts in the minor groove. Appendage at the 5′-end of a dodecamer increased the affinity of the TFO but the degree of stabilisation was dependent on the site of attachment and no gain was observed when it was attached to the amino sugar. Daunomycin conjugates have been targeted to the c-*myc* promoter and reduced activity *in vitro*, as well as in prostate and breast cancer cells.[80] Further studies revealed that binding of these conjugates also inhibited growth and induced apoptosis in prostate cancer cells.[81]

3.4.2.4   Pyrene

Pyrene contains four fused benzene rings with a surface area more similar to that of a W-C base pair than either acridine or anthraquinone [see Figure 3.4B(vii)]. It has been conjugated to several positions within an oligonucleotide, including the internucleotide phosphorous,[82] ribose sugar,[83] the 3′ and 5′ termini,[76,84] as well as at the 5-position of dU.[85] In all of these cases the appendage of pyrene to these positions leads to an increase in duplex stabilisation. Its incorporation into a TFO also stabilises triplex formation but not as much as incorporation of an anthraquinone.[76] Pedersen and co-workers have shown that the choice of the linker attaching pyrene to an oligonucleotide has profound effects on both selectivity and extent of stabilisation. Short linkers such as glycerol favour duplex over triplex stabilisation and are more effective for stabilisation of single-stranded DNA over RNA targets.[86–89] It was suggested that linker length must be sufficiently long to insert pyrene into the duplex region within a triplex. Consequently, a longer linker containing a phenyl group attached to pyrene via an acetylene bridge has been employed for triplex stabilisation (termed twisted intercalating nucleic acid; TINA).[90–92] In this design the phenyl mimics a nucleobase in the TFO part of the triple helix and the acetylene bridge provides structural rigidity and twisting ability that helps the intercalator to adjust itself to an appropriate position inside the dsDNA. A pyrimidine-containing TFO containing two bulged substitutions formed a stable triplex with a T_m of 43 °C at physiological pH, whereas the native oligonucleotide was unable to bind to the target duplex. Further studies revealed that the most potent stabilisation is afforded with at least three nucleotides between insertions.[91] The incorporation of pyrene (and presumably other intercalators) into G-rich oligonucleotides has the additional advantage that it acts to inhibit intramolecular quadruplex formation, the formation of which can decrease the efficacy of oligonucleotides.[92]

3.4.2.5   Triplex Specific Ligands

Several compounds have been rationally designed to intercalate between consecutive base triplets in a DNA triplex. These usually contain extended aromatic ring systems and exhibit a higher preference for binding to triplex over duplex DNA. The first triplex binders to be developed were benzopyridoindole [BePI; Figure 3.5(i)] and benzopyridoquinoline [BQQ; Figure 3.5(ii)] and both were subsequently conjugated to TFOs *via* the terminal nitrogens on each of the side chains.[93–95] Both binders increased the stability of DNA triplexes when attached at the ends and internal positions within a TFO. In the latter case, these intercalators were added as an additional nucleotide rather than replacing an existing one. Both compounds contain positive charges and therefore exhibit a greater preference for binding within triplexes containing contiguous runs of T.AT triplets on account of charge repulsion with C^+.GC triplets. BQQ was more effective when positioned at the TFO termini whilst BePI was more effective when added at an internal position. The addition of both ligands did not affect the stringency of triplex formation and conjugated TFOs targeted to duplexes containing a single mismatch were less stable than their perfectly matched counterparts. Naphthylquinoline [Figure 3.5(iii)] and a naphthalene diimide [Figure 3.5(iv)] have also been attached to TFOs and have been shown to increase third strand affinity to a greater extent than an equivalent concentration of the free ligand.[96,97]

3.4.3   Stabilisation by Groove Binders

Small molecules that contain appropriately positioned hydrogen bond donor and acceptor moieties are capable of interacting within the major and minor

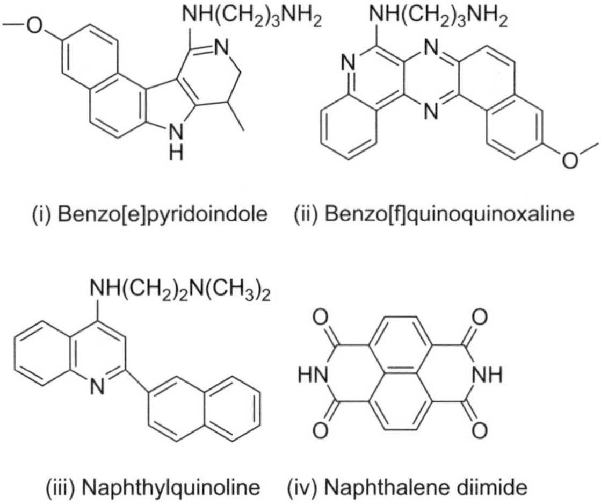

(i) Benzo[e]pyridoindole   (ii) Benzo[f]quinoquinoxaline

(iii) Naphthylquinoline   (iv) Naphthalene diimide

Figure 3.5   Triplex-specific intercalators. Compounds composed of extended ring systems that exhibit a greater preference for intercalating within triplex over duplex DNA.

grooves of a DNA helix by making contacts with functional groups that protrude from the bases. These molecules are often crescent shaped and composed of simple aromatic ring systems connected by bonds with torsional freedom that allow them to twist and become isohelical upon binding within the duplex groove. Unlike intercalation, binding within the duplex groove is more sequence dependent, and simple minor groove binders generally prefer to bind to sequences containing adjacent AT base pairs. These molecules are therefore limited to improving oligonucleotide hybridisation to targets containing appropriate ligand binding sites.

3.4.3.1   Polyamides

Netropsin [Figure 3.6(i)] and distamycin A [Figure 3.6 (ii)] are classic minor groove binding ligands that exhibit a preference for AT-rich minor grooves. Both ligands have been conjugated to the termini of oligonucleotides and the thermal stability of DNA duplexes containing appropriately positioned 5'-TTAAA and 5'-TATA binding sites studied.[98] In both cases, the attachment of the ligand at both ends of the oligonucleotide resulted in a dramatic T_m increase of $>25\,^\circ\mathrm{C}$ and had a stronger impact on the T_m values than did two free molecules of each ligand per duplex. Surprisingly, the addition of a single distamycin A residue did not result in stabilisation, whilst the addition of a single netropsin did. Furthermore, the conjugation of both ligands was shown to be compatible with phosphorothioate oligonucleotide modifications.

Given that both netropsin and distamycin are amide-containing di- and tri-peptides of *N*-methylpyrrolocarboxyamide [MPC; Figure 3.6(iii)], peptides containing different numbers of MPC units have also been conjugated to

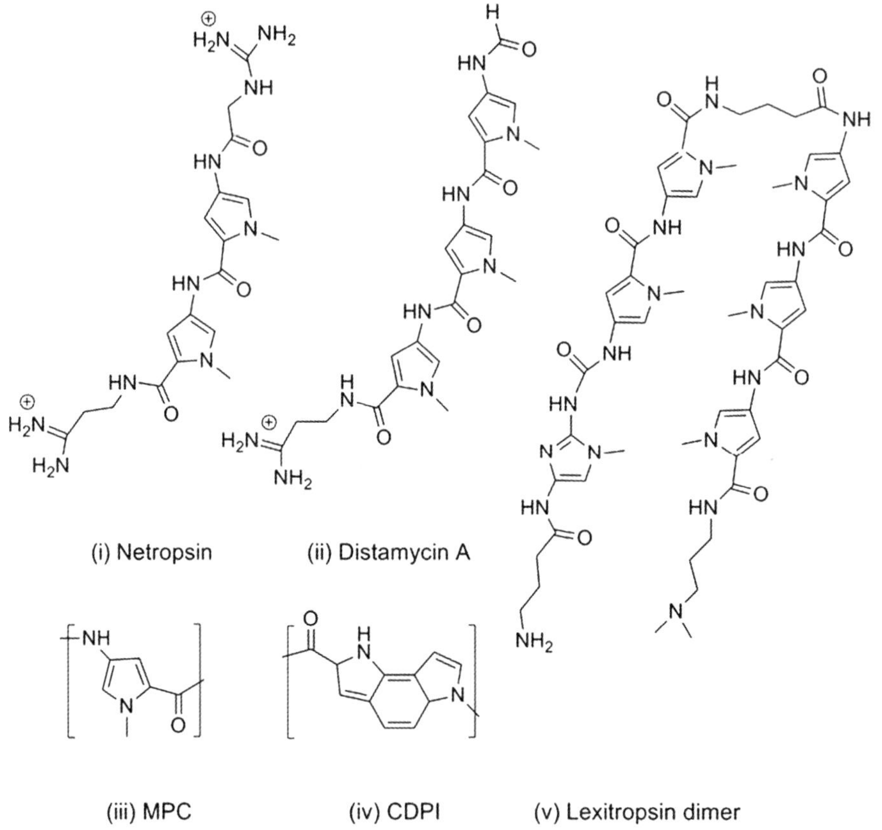

Figure 3.6 Minor groove-binding polyamides composed of repeating ring systems connected by peptide bonds.

oligonucleotides. As expected these were also stabilising and the degree of this stabilisation was dependent on the number of MPC residues incorporated within the attached peptide.[99] Again these studies were undertaken with oligonucleotides that generate duplexes containing multiple AT base pairs. This approach was later extended to the addition of peptides containing 1,2-dihydro-3*H*-pyrrolo[3,2-*e*]indole-7-carboxylate subunits [CDPI; Figure 3.6(iv)] and resulted in a larger increase in stability (ΔT_m of $>40\,^\circ$C).[100] The addition of CDPI peptides to the oligonucleotide was more stabilising with DNA than RNA, exhibited an enhanced specificity within the minor groove binding region of the duplex and was also compatible with phosphorothioate oligonucleotides.[100–103]

Peptides containing MPC units have also been attached to TFOs *via* sufficiently long ethylene glycol linkers and targeted to duplexes containing runs of AT base pairs.[104] The triplexes generated by these conjugates exhibit much higher T_m values, 15-fold faster association kinetics, and dissociation kinetics that were an order of magnitude slower than the non-conjugated individual components. As several studies have shown that polyamides can bind to the minor groove in a 2 : 1 fashion the same polyamide was synthesised as a hairpin dimer and upon conjugation afforded an even greater increase in duplex stability compared with the addition of a single polyamide. In fact, temperature gradient electrophoresis revealed that, above a certain temperature, the TFO dissociated from its binding site while the hairpin polyamide remained bound, generating a complex with a slower mobility than in the presence of either free ligand, oligonucleotide or bound conjugate.

Polyamides based on lexitropsins have been conjugated to TFOs to enhance triplex formation.[105] Initially a tetrapyrrole was covalently attached to the end of the TFO *via* a hexamethylenediamine linker and targeted to a duplex containing a polypurine tract with five contiguous AT base pairs, but no increase in triplex stability was observed. It was therefore synthesised as a hairpin dimer [Figure 3.6(v)] and attached via a longer triethleneglycol linker, and this time the stability of both the duplex and the triplex was markedly increased. The level of stabilisation was in fact similar to that of an intercalator conjugate containing a triplex-specific ligand.

3.4.3.2 Other Groove Binders

The conjugation of the bicyclic bis-benzimidazole Hoechst 33258 [Figure 3.7(i)] to an oligonucleotide has also been shown to increase the thermal stability of DNA duplexes by 10–16 °C.[106] A variety of tether lengths were examined to allow the ligand to reach beyond the terminus of the DNA duplex and bind to internal AT-rich target sequences as far as four base pairs from the site of attachment. Effective stabilisation required a four base pair binding site that could also include a GC base pair, albeit with a lower degree of stabilisation ($\Delta T_\mathrm{m} = +4$ °C). Conjugation of the ligand was also shown to be compatible with an oligonucleotide containing a deoxynucleic guanidine (DNG) backbone, in which positively charged guanidine residues replaced phosphodiester linkages.[107] This is a particularly useful advance because oligonucleotides containing DNG linkages increase the cellular permeability and nuclease resistance of oligonucleotides. Hoechst 33258 has also been attached to the termini of TFOs and shown to increase triplex stability dramatically, by up to 28 °C.[107,108] The oligonucleotide was capable of binding to a variety of triplex sequences and could even be used to to stabilise the interaction of the TFO with a 5′-AATT sequence located within the oligopurine tract, in which the two T pyrimidine inversions were targeted by guanine and generated stable G.TA triplets.[108]

Unfortunately there has been little success in exploiting conjugated groove binders to improve hybridisation to TFO target duplex sequences containing adjacent GC base pairs. However, a TFO conjugated to a pyrrolobenzodiazepine derivative [PBD; Figure 3.7(ii)] has been shown to interact with a 5′-GAGGG sequence located adjacent to the triplex site but the conjugate did not increase triplex stability.[109] The interaction of the conjugated TFO was demonstrated by targeting it to an identical sequence containing inosine in place of guanine, which cannot interact with the PBD, and showed no TFO

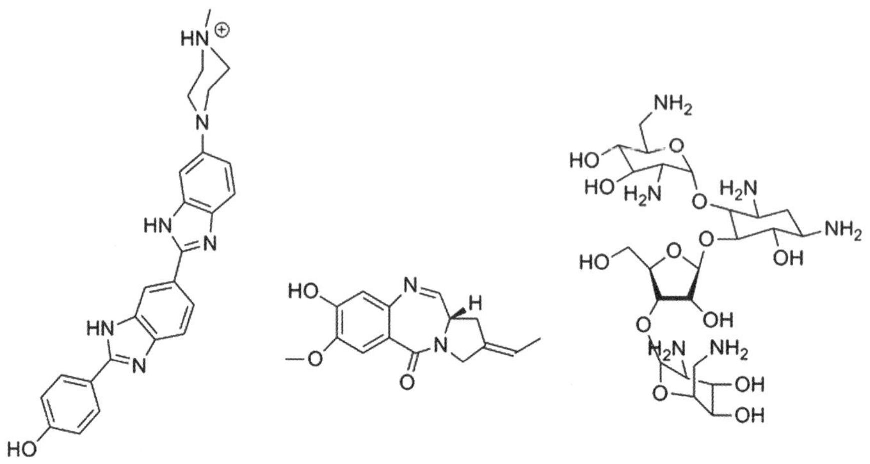

(i) Hoechst 33258 (ii) Pyrrolobenzodiazepine (iii) Neomycin

Figure 3.7 Other groove-binding ligands.

binding. It was suggested that the lower stability of the complex was due to the slower association of the conjugate with its target.

Increasing hybridisation of an oligonucleotide to single-stranded RNA by attachment of a major groove binding ligand has also been demonstrated. The aminoglycoside antibiotic neomycin [Figure 3.7(iii)] has been shown to interact with RNA duplexes with high affinity and has therefore been conjugated to an oligonucleotide.[110] It was covalently attached at the 5-position of dU and substituted into any given site in an oligonucleotide where a thymidine was present. The increase in T_m for a duplex generated with RNA using this conjugate was 6 °C whilst stabilisation was not observed for a triplex, attributed to the relatively short linker employed in the study.

3.5 Cross-linking Oligonucleotides to Nucleic Acid Targets

The oligonucleotide conjugates discussed so far have contained small molecule binders that improve the hybridisation properties of an oligonucleotide. These molecules interact reversibly with their single- or double-stranded targets and may therefore be removed by cellular factors. Another approach is to conjugate small molecules containing cross-linking group(s) so that upon hybridisation, and under appropriate conditions, the oligonucleotide becomes covalently attached to its target. Such derivatised oligonucleotides have the advantage that they can act at lower concentrations than unmodified ones. Moreover, covalent modifications to double-stranded DNA may also be recognised by cellular repair pathways and used to introduce permanent mutations through inefficient repair mechanisms.

3.5.1 Photoactive Cross-linkers

Cross-linkers that are activated upon exposure to ultraviolet (UV) light have been appended to the termini of oligonucleotides. The simplicity of this approach was first demonstrated using an oligonucleotide containing an attached azidophenacyl group [Figure 3.8A(i)].[111] Upon excitation of the azidophenacyl group with UV light, an oligo-[α]-thymidylate was cross-linked specifically to a complementary oligo-[β]-adenylate sequence. These cross-links were then converted to chain breaks under alkaline conditions and the parallel orientation of the two chains assigned unambiguously. An azido derivative of proflavine (3,6 diaminoacridine) [Figure 3.8A(ii)] has also been used to cross-link double-stranded and triple-stranded structures.[14] In fact, the targeting of a TFO containing this photoreactive group was one of the first demonstrations of

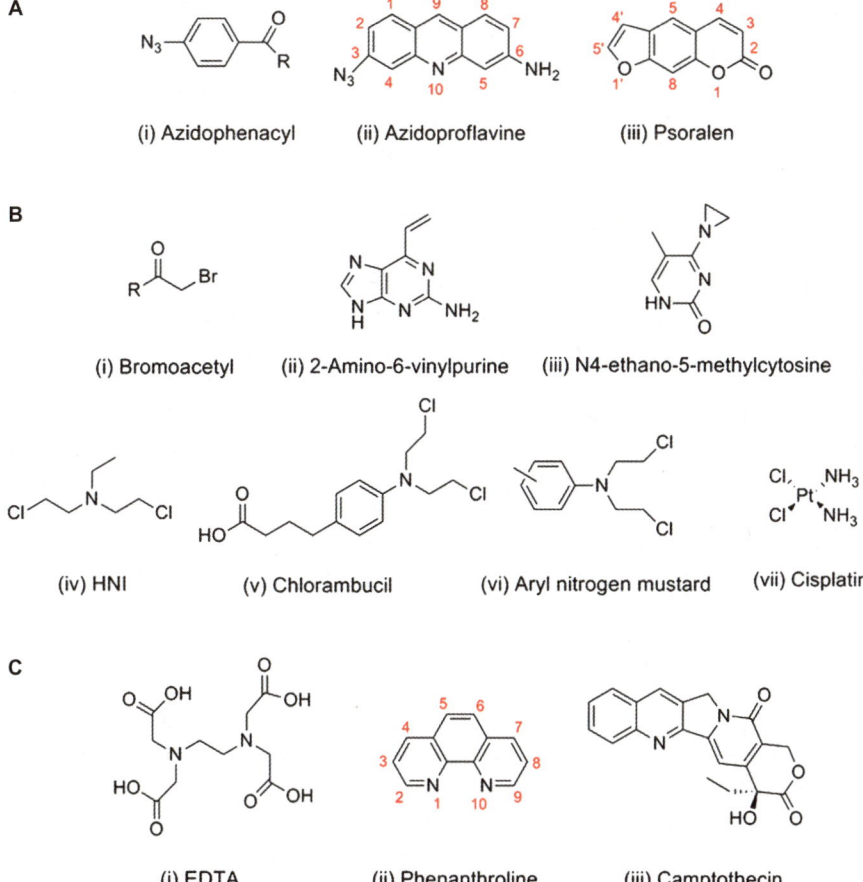

Figure 3.8 Small molecules that can invoke chemical reactions with nucleic acids. (A) Cross-linking agents. (B) Alkylating agents. (C) Agents that can direct cleavage of nucleic acids.

oligonucleotide-directed recognition of double-stranded DNA. The derivative was appended to the 3'-end of the oligonucleotide *via* a linker attached to its 6 amino group. In each of these cases mono-adducts between the TFO and a single duplex strand are generated upon cross-linking.

3.5.1.1 Psoralen

Psoralen [Figure 3.8A(iii)] interacts with double-stranded DNA by intercalation and upon irradiation with long wavelength UV light leads to a $2 + 2$ cycloaddition with adjacent thymidines cross-linking the two duplex strands at TpA steps.[112] Psoralen is most often employed as its 4,5',8-trimethyl derivative and conjugated to oligonucleotides *via* either its 4 or 5' positions using a suitable linker. Psoralen oligonucleotides have been used to bind to single-stranded DNA and induce specific cross-links when the psoralen is positioned opposite thymidine at the end of an oligonucleotide.[113] These conjugates have been exploited in antisense applications and have been used to inhibit collagenase expression by binding to a region spanning the initiation codon on collagenase mRNA.[114] Psoralen-oligonucleotides have also been targeted to a TATA box region within single-stranded DNA.[115]

Psoralen-linked oligonucleotides have had more success in the triplex targeting of double-stranded DNA because the TFO can position the psoralen to bind to a TpA step adjacent to the triplex target site, thereby cross-linking the oligonucleotide to one or both strands, generating mono- and bis-adducts respectively. The first experiments of this kind were undertaken by the Hélène laboratory and demonstrated the feasibility of this approach by targeting a TFO to a proviral sequence *in vitro*.[116,117] This was later extended to targeting of the aromatase gene within intact cells.[118,119] These studies revealed that the attached psoralen intercalated at the triplex–duplex junction with a strong preference for the orientation that positions the furan-side of the psoralen close to the purine-rich strand. Oligonucleotides containing psoralen at both ends, or at positions located at the centre of the TFO, can be exploited to cross-link the TFO to two appropriately positioned TpA steps simultaneously, generating triplex 'staples'.[120] Importantly, although psoralen requires mild deprotection conditions during oligonucleotide synthesis, it is still compatible with conjugation to TFOs containing stabilising base and nucleoside modifications.[121]

As well as increasing the residence time of a TFO with its target, cross-linking can invoke a biological response that leads to mutagenesis. Upon cross-linking, the DNA helix becomes distorted and is recognised by cellular repair pathways including nucleotide excision repair (NER) and homology-directed repair (HDR). It is likely that processing of TFO-mediated lesions is error prone and therefore results in mutagenesis. Cross-links can also enhance the frequency of homologous recombination in mammalian cells and psoralen-TFOs have been shown to induce intramolecular recombination in both episomal and chromosomal targets.[122–126] It is important to emphasise the specificity that the conjugation of psoralen to a TFO imparts; plasmids treated with free psoralen have a higher general mutation rate and a lower survival frequency than those treated with pso-TFOs. A variety of factors are known to influence the ability of pso-TFOs to stimulate repair, and longer TFOs and doubly modified TFOs are less mutagenic than shorter TFOs containing a single modification.[127,128]

3.5.2 Alkylating and Alkylating-like Cross-linkers

By attaching electrophilic alkylating or alkylating-like agents to oligonucleotides it is possible to cross-link these conjugates to both single- and double-stranded DNA targets. Unlike photoactive cross-linkers these agents offer the advantage that they do not require activation by external stimuli and can react with DNA under physiological conditions. Most alkylating agents exhibit a preference to react with the N7 of guanine and are therefore most effective when positioned at DNA sequences containing this base. However, alkylation of the N7 of adenine as well as alkylation at various positions on pyrimidine bases is also possible.

3.5.2.1 *Electrophilic Alkylators*

One of the first conjugates to be exploited in this manner contained an *N*-bromoacetyl [see Figure 3.8B(i)] electrophile attached to the 5-position of a thymine at the 5′-end of a pyrimidine-containing TFO.[129] Upon binding to its duplex target the oligonucleotide alkylated a single guanine residue located two base pairs to the 5′-side of this sequence with an 87% efficiency under physiologically relevant conditions. Furthermore, the reaction was specific and generated only mono-adducts with the TFO covalently linked to a single strand within the target duplex. *N*-bromoacetyl has also been conjugated to purine-containing TFOs and shown to generate site-specific mono-adducts.[130] The alkylation of both duplex strands has also been achieved by targeting two TFO-conjugates to adjacent triplex sites located on opposite Watson–Crick strands.[131] Subsequent depurination of the alkylated DNA led to the generation of a double-strand break between the two triplex sites.

In a similar fashion other electrophilic alkylating groups have been appended to bases and incorporated into oligonucleotides. For example, N_4,N_4-ethano-5-methyldeoxycytidine [Figure 3.8B(iii)], which contains an electrophilic methylene, was incorporated into a TFO and was shown to cross-link a duplex target with 95% efficiency.[132] It was positioned in the third strand in such a manner so as to generate a base triplet with a GC base pair, positioning the methylene group in close proximity to the N7 and O6 sites located on the guanine Hoogsteen partner. 2-Amino-6-vinylpurine [AVP; see Figure 3.8B(ii)] has also been incorporated into oligonucleotides and both antisense and antigene effects examined.[133,134] Oligonucleotides containing AVP were shown to inhibit translation of mRNA in a luciferase translation assay in cell lysate to a greater extent that the unmodified oligonucleotide.[133] AVP was conjugated to 2'-OMe ribose in place of the base and shown to cross-link specifically to uridine. Triplex-directed cross-linking has also been achieved with a high selectivity towards the C of the GC target site. The TFO conjugates produced adducts at the complementary position within a *supF* reporter gene, leading to mutations at the site during replication and repair.[134] AVP introduced at an internal position within the TFO reacted with cytosine whilst those introduced at the terminal position of the TFO reacted with the adenine adjacent to the triplex site.

Several alkylating nitrogen mustards have been conjugated to the termini of oligonucleotides, the first of which was 2-chloroethylamine [HNI; Figure 3.8B(iv)].[135] TFO–HNI conjugates were capable of site-specific cross-linking reactions with single- and double-stranded DNA targets but with a relatively low yield. Greater success has been achieved by conjugating chlorambucil [Figure 3.8B(v)] to both pyrimidine- and purine-containing TFOs.[136,137] The attachment of chlorambucil to each end of the TFO allowed cross-linking at separate sites on each strand of the duplex, generating bis-adducts with the TFO covalently attached to both strands. Cross-links were formed preferentially at 5'-GNC-3' rather than 5'-GC-3' sites but the efficiency was dependent on the flanking sequence, solution conditions and rate of triplex formation relative to the rate of the chlorambucil reaction. The ability to repair adducts generated by the cross-linking of a TFO–chlorambucil conjugate to its duplex target has been examined in an *in vitro* repair assay.[138] Surprisingly, this study demonstrated that the NER pathway was capable of repairing more than 25% of the adducts generated after 24 h. It was suggested that greater success may be achieved by using TFOs containing modifications at both ends of the oligonucleotide. Indeed, a chlorambucil bis-conjugate was later shown to suppress promoter activity in cancer cells by 60–70%.[139] TFOs containing different 2,3,5,6-tetrafluorophenyl ester derivatives of aryl nitrogen mustards [Figure 3.8B(vi)] have also been prepared and shown to exhibit different reaction rates.[140] The ability to vary the reactivity of mustard-TFOs is likely to improve the pharmacodynamics of these TFOs, as well as to prevent self-reactions between the mustard and the oligonucleotide carrier.

3.5.2.2 *Platinum-containing Agents*

One of the problems of using alkylating agents to cross-link DNA is that alkylated guanine is prone to depurination and release of the bound oligonucleotide. An alternative is to use alkylating-like platinum compounds that generate highly stable coordination complexes with DNA. The primary target of these platinum compounds is guanine and they can generate both intrastrand and interstrand cross-links. Colombier *et al.* were one of the first groups to show that triplex formation could be applied to deliver a platinated complex to a specific DNA site.[141,142] *Trans*-diamminedichloroplatinum (II) (transplatin) was first reacted with a pyrimidine containing TFO and subsequently shown to generate interstrand cross-links upon binding to its target duplex. The yield and rate of cross-linking depended on the location and nature of the appended group, with transplatin-modified adducts of cytosine more reactive than guanine. Cisplatin [see Figure 3.8B(vii)] has also been conjugated to the 5′-end of TFOs and exploited to cross-link to guanines located adjacent to the TFO target site.[143,144] In these studies oligo dT was used to minimise the reactivity of cisplatin to the TFO and the highest reactivity was achieved when the dinucleotide d(GpG) was present in the target strand. In an alternative approach both *trans* and *cis*platin have been appended to *N*4-(2-aminoethyl)cytosine, a base analogue designed to recognise a CG inversion within a polypurine–polypyrimidine triplex target site.[145,146] In this manner the platinum complex is positioned so as to react with the guanine of the CG base pair. Although the platinated base resulted in a triplex with decreased stability, specific cross-linking occurred and was most effective with a tether containing five methylene groups between the platinated compound and the base. Lastly, bis-platinated TFOs containing transplatin at both ends of the oligonucleotide have also been shown to generate interstrand cross-links with both strands of the duplex target.[147] TFOs of this type were shown to bind to plasmid DNA and inhibit transcription *in vitro*, as well as to inhibit plasmid replication in *E. coli* cells.

3.6 Oligonucleotide-directed Cleavage of Nucleic Acid Targets

Oligonucleotide conjugates can be exploited to induce sequence-specific cleavage of DNA and RNA targets upon hybridisation. The small molecules attached to these oligonucleotides are either capable of cleaving a nucleic acid by themselves, or they can recruit a cellular enzyme to perform the same action. The recruitment of a cellular enzyme that cleaves double-stranded DNA is similar to the process of the recruitment of RNase H by RNA–DNA hybrids.

3.6.1 Metal Complexes

Metal complexes attached to oligonucleotides can be exploited to cleave single- and double-stranded nucleic acids in the presence of molecular oxygen and a reducing agent by generating free hydroxyl radicals. Hybridisation-dependent cleavage of single-stranded DNA was first shown by attaching ethylenediaminetetraacetic acid [EDTA; see Figure 3.8C(i)] to an oligonucleotide in the presence of Fe^{2+} and dithiothreitol.[148] The attachment of Fe^{2+}.EDTA *via* an ethylenediamine spacer to a 16-mer oligonucleotide allowed the targeting and cleavage of a 37-mer single-stranded target containing a complementary region, with cleavage occurring up to four residues either side of the duplex termini. Similar results have also been observed with attachment of EDTA to the 5-position of dU.[149] EDTA has also been exploited to cleave double-stranded DNA when attached to a TFO.[13,150]

Phenanthroline [see Figure 3.8C(ii)] has also been used to direct cleavage to single- and double-stranded targets upon addition of Cu^+ and mercaptopropionic acid.[151–153] Unlike EDTA, the addition of phenathroline, a DNA intercalator, to an oligonucleotide offers the additional advantage of enhancing oligonucleotide hybridisation. It has been attached to oligonucleotides at its 4 or 5 positions using varying linker lengths. A variety of porphyrins have also been used to cleave poly(dA) and poly(rA) using complementary oligonucleotide poly(dT).[154–157] Although this work is promising, several factors have limited the applications of these conjugates under biological conditions. A particular concern is the possibility of non-specific reactions with other nucleic acids, as well as self-cleavage.

The glycopeptide antibiotic bleomycin has also been appended to oligonucleotides and exploited to invoke strand cleavage of single- and double-stranded targets.[158–160] Free bleomycin interacts with the DNA minor groove and induces oxidative cleavage in the presence of certain cofactors – molecular oxygen, Fe(II) and a reducing agent. Attachment at the 5′-end of an oligonucleotide resulted in selective degradation of its duplex complement at a site adjacent to the oligonucleotide binding site,[159] whilst attachment at the centre allowed cleavage within the oligonucleotide binding site.[159] In the latter case, the conjugate was then capable of cleaving further targets (at least three further molecules). Bleomycin has also been attached to the 3′ or 5′ terminus of a hexadecathymidylate TFO and shown to cleave site-specifically both strands of a duplex target with a yield of about 61%.[160]

3.6.2 Topoisomerase Inhibitors

TFOs covalently linked to topoisomerase I inhibitors can be used to trigger topoisomerase I-mediated DNA cleavage at specific locations within double-stranded DNA.[161] Perhaps the most studied of these inhibitors is camptothecin, a quinolone alkaloid [see Figure 3.8C(iii)]. Topoisomerase I acts to relax DNA supercoiling by nicking one of the two duplex strands, rotating the DNA about this cleaved position and subsequently religating the two DNA strands. Camptothecin traps a covalent intermediate during this process, preventing the enzyme from undertaking the religation step. Camptothecin-TFOs have been shown to induce cleavage in the nanomolar concentration range, as compared with micromolar for the free drug.[162] The specificity and efficacy of this cleavage have been shown to depend on both the length of the TFO and the linker between the oligonucleotide and camptothecin.[163] Surprisingly, DNA cleavage only occurs with TFOs bearing the inhibitor at the 3′-end of the oligonucleotide, and is attributed to the structural differences of the 3′ and 5′ triplex–duplex junctions.[164,165] As well as these *in vitro* studies camptothecin conjugates have been shown to act in cells using a luciferase reporter gene system; 50% of luciferase reporter activity was inhibited in the presence of 0.5 μM TFO.[166] Furthermore, the formation of covalent topoisomerase–conjugate–DNA complexes was detected in cell nuclei. Another topoisomerase I inhibitor, rebeccamycin, has also been conjugated to a TFO and exhibited similar results.[167] The activity of topoisomerase II, which simultaneously cuts both strands of the DNA helix, has also been targeted using inhibitor–oligonucleotide conjugates.[168]

3.7 Summary and Future Directions

The successful application of oligonucleotides as drug molecules will require that they possess certain properties, including high affinity binding to their intended nucleic acid targets, stability against cellular and serum nucleases, as well as the ability to penetrate cell and nuclear membranes. It is clear from this review that the conjugation of small molecules to oligonucleotides has tremendous potential in helping to endow oligonucleotides with these desired characteristics. The appendage of edge binders, intercalators and groove binders has been shown to improve the strength of oligonucleotide hybridisation to nucleic acids without compromising on selectivity. Conjugating molecules to the ends of oligonucleotides and at the 2′-position of the ribose has the additional advantage that it helps protect against nuclease digestion. Oligonucleotides bearing positively charged groups are also less susceptible to nuclease digestion and increase permeation through cellular and nuclear membranes. It is likely that the development of oligonucleotides containing a combination of these small molecules will prove the most successful, as well as those that are compatible with existing oligonucleotide chemistries, such as phosphorothioate and methylphosphonate modifications that have shown promise in clinical trials.

As well as improving the existing properties of oligonucleotides, small molecules containing reactive groups have been shown to bestow novel properties on oligonucleotides and allow directed cross-linking and cleavage of nucleic acids. This is a promising advance, however many of these studies have been undertaken *in vitro* and less is known about their efficacy *in vivo*. Some of these groups require activation by external stimuli and these conditions might not be possible within intact cells or organisms. There are also concerns about non-specific reactions of these conjugates with other nucleic acids and cellular components. Despite this, the ability to modify gene activity permanently has several advantages over hybridisation-dependent modulation and on-going efforts in this area could be very rewarding.

Oligonucleotide-directed recognition of nucleic acids has primarily been driven by a desire to use these molecules for therapeutic applications but other research areas will benefit from these efforts, such as diagnostics, functional genomics and molecular biology. Oligonucleotide conjugates are also likely to have a big impact in the burgeoning field of bionanotechnology which relies on oligonucleotide hybridisation to generate complex architectures and devices with nanoscale dimensions.

References

1. C. Hélène and J. J. Toulmé, *Biochim. Biophys. Acta*, 1990, **1049**, 99–125.
2. D. Jones, *Nat. Rev. Drug Discov.*, 2011, **10**, 401–402.
3. J. Goodchild, *Curr. Opin. Mol. Ther.*, 2004, **6**, 120–128.
4. M. Manoharan, *Antisense Nucleic Acid Drug Dev.*, 2002, **12**, 103–128.
5. J. Kurreck, *Eur. J. Biochem.*, 2003, **270**, 1628–1644.

6. D. M. Gowers and K. R. Fox, *Nucleic Acids Res.*, 1999, **27**, 1569–1577.

7. K. R. Fox, *Curr. Med. Chem.*, 2000, **7**, 17–37.

8. J. Robles, A. Grandas, E. Pedroso, F. J. Luque, R. Eritja and M. Orozco, *Curr. Org. Chem.*, 2002, **6**, 1333–1368.

9. S. Buchini and C. J. Leumann, *Curr. Opin. Chem. Biol.*, 2003, **7**, 717–726.

10. J. Goodchild, *Bionconjug. Chem.*, 1990, **1**, 165–187.

11. P. C. Zamecnik and M. L. Stephenson, *Proc. Natl. Acad. Sci. USA*, 1978, **75**, 280–284.

12. M. L. Stephenson and P. C. Zamecnik, *Proc. Natl. Acad. Sci. USA*, 1978, **75**, 285–288.

13. H. E. Moser and P. B. Dervan, *Science*, 1987, **238**, 645–650.

14. T. Le Doan, L. Perrouault, D. Praseuth, N. Habhoub, J-L. Decout, N. T. Thuong, J. Lhomme and C. Hélène., *Nucleic Acids Res.*, 1987, **15**, 7749–7760.

15. M. Cooney, G. Czernuszewicz, E. H. Postel, S. J. Flint and M. E. Hogan, *Science*, 1988, **241**, 456–459.

16. P. A. Beal and P. B. Dervan, *Science*, 1991, **251**, 1360–1363.

17. R. H. Durland, D. J. Kessler, S. Gunnell, M. Duvic, B. M. Pettitt and M. E. Hogan, *Biochemistry*, 1991, **30**, 9246–9255.

18. D. A. Rusling, V. J. Broughton-Head, T. Brown and K. R. Fox, *Curr. Chem. Biol.*, 2008, **2**, 2–11.

19. M. M. Seidman and P. M. Glazer, *J. Clin. Invest.*, 2003, **112**, 487–494.

20. A. Jain, G. Wang and K. M. Vasquez, *Biochimie.*, 2008, **90**, 1117–1130.

21. Y. Singh, P. Murat and E. Defrancq, *Chem. Soc. Rev.*, 2010, **39**, 2054–2070.

22. P. B. Dervan, *Bioorg. Med Chem.*, 2001, **9**, 2215–2235.

23. C. Sund, N. Puri and J. Chattopadhyaya, *Nucleosides Nucleotides*, 1997, **16**, 755–760.

24. R. Noir, M. Kotera, B. Pons, J.-S. Remy and J.-P. Behr, *J. Am. Chem. Soc.*, 2008, **130**, 13500–13505.

25. T. P. Prakash, D. A. Barawkar, K. Vaijayanti and K. N. Ganesh, *Biorg. Med. Chem. Lett.*, 1994, **4**, 1733–1738.

26. H. Nara, A. Ono and A. Matsuda, *Bioconjug. Chem.*, 1995, **6**, 54–61.

27. N. Schmid and J.-P. Behr, *Tetrahedron Lett.*, 1995, **36**, 1447–1450.

28. A. R. Diaz, R. Eritja and R. G. Garcia, *Nucleosides Nucleotides*, 1997, **16**, 2035–2051.

29. P. F. Potier and J.-P. Behr, *Nucleosides Nucleotides Nucleic Acids*, 2001, **20**, 809–813.

30. K. J. Hampel, P. Crosson and J. S. Lee, *Biochemistry*, 1991, **7**, 4455–4459.

31. T. Thomas and T. J. Thomas, *Biochemistry*, 1993, **32**, 14068–14074.

32. C. H. Tung, K. J. Breslauer and S. Stein, *Nucleic Acids Res.*, 1993, **21**, 5489–5494.

33. C. Sund, N. Puri and J. Chattopadhyaya, *Tetrahedron*, 1996, **52**, 12275–12279.

34. D. A. Barawkar, K. G. Rajeev, V. A. Kumar and K. N. Ganesh, *Biochem. Biophys. Res. Commun.*, 1994, **205**, 1665–1670.

35. D. A. Barawkar, K. G. Rajeev, V. A. Kumar and K. N. Ganesh, *Nucleic Acids Res.*, 1996, **24**, 1229–1237.
36. K. G. Rajeev, V. R. Jadhav and K. N. Ganesh, *Nucleic Acids Res.*, 1997, **25**, 4187–4193.
37. R. L. Letsinger and M. E. Schott, *J. Am. Chem. Soc.*, 1981, **103**, 7394–7396.
38. E. N. Timofeev, I. P. Smirnov, L. A. Haff, E. I. Tishchenko, A. D. Mirzabekov and V. L. Florentiev, *Tetrahedron Lett.*, 1996, **37**, 8467–8470.
39. R. Huber, N. Amann and H.-A. Wagenknecht, *J. Org. Chem.*, 2004, **69**, 744–751.
40. K. Yamana, R. Aota and H. Nakano, *Tetrahedron Lett.*, 1995, **36**, 8427–8430.
41. U. Asseline, M. Delarue, G. Lancelot, F. Toulmé, N. T. Thuong, T. Montenay-Garestier and C. Hélène, *Proc. Natl. Acad. Sci. USA*, 1984, **81**, 3297–3301.
42. U. Asseline, F. Toulmé, N. T. Thuong, M. Dealrue, T. Montenay-Garestier and C. Hélène, *EMBO J.*, 1984, **3**, 795–800.
43. G. Lancelot, U. Asseline, N. T. Thuong and C. Hélène, *Biochemistry.*, 1985, **24**, 2521–2529.
44. N. T. Thuong, U. Asseline, V. Roig, M. Takasugi and C. Hélène, *Proc. Natl. Acad. Sci. USA*, 1987, **84**, 5129–5133.
45. U. Asseline, T. T. Nguyen and C. Hélène, *J. Biol. Chem.*, 1985, **260**, 8936–8941.
46. J.-S. Sun, U. Asseline, D. Rouzaud, T. Montenay-Garestier, N. T. Thoung and C. Hélène, *Nucleic Acids Res.*, 1987, **15**, 6149–6158.
47. J. J. Toulmé, H. M. Krisch, N. Loreau, N. T. Thuong and C. Hélène, *Proc. Natl. Acad. Sci. USA*, 1986, **83**, 1227–1231.
48. C. Cazenave, N. Loreau, N. T. Thoung, J.-J. Toulmé and C. Hélène, *Nucleic Acids Res.*, 1987, **15**, 4717–4735.
49. C. Cazenave, M. Chevrier, T. T. Nguyen and C. Hélène, *Nucleic Acids Res.*, 1987, **15**, 10507–10521.
50. A. Zerial, N. T. Thuong and C. Hélène, *Nucleic Acids Res.*, 1987, **15**, 9909–9919.
51. T. Saison-Behmoaras, B. Tocqué, I. Rey, M. Chassignol, N. T. Thuong and C. Hélène, *EMBO J.*, 1991, **10**, 1111–1118.
52. T. Saison-Behmoaras, I. Duroux, N. T. Thoung, U. Asseline and C. Hélène, *Antisense Nucleic Acid Drug Dev.*, 1997, **7**, 361–368.
53. J.-S. Sun, C. Giovannangeli, J. C. Francois, K. Kurfurst, T. Monteny-Garestier, U. Asseline, T. Saison-Behmoaras, N. T. Thuong and C. Hélène, *Proc. Natl. Acad. Sci. USA*, 1991, **88**, 6023–6027.
54. J.-S. Sun, J.-C. François, T. Montenay-Garestier, T. Saison-Behmoaras, V. Roig, N. T. Thuong and C. Hélène, *Proc. Natl. Acad. Sci. USA*, 1989, **86**, 9198–9202.
55. F. Birg, D. Praseuth, A. Zerial, N. T. Thuong, U. Asseline, T. Le Doan and C. Hélène, *Nucleic Acids Res.*, 1990, **18**, 2901–2908.

56. M. Grigoriev, D. Praseuth, P. Robin, A. Hemar, T. Saison-Behmoaras, A. Dautry-Varsat, N. T. Thuong, C. Hélène and A. Harel-Bellan, *J. Biol. Chem.*, 1992, **267**, 3389–3395.
57. T. J. Stonehouse and K. R. Fox, *Biochim. Biophys. Acta*, 1994, **1218**, 322–330.
58. S. A. Cassidy, L. Strekowski, W. D. Wilson and K. R. Fox, *Biochemistry*, 1994, **33**, 15338–15347.
59. H. B. Gamper, I. V. Kutyavin, R. L. Rhinehart, S. G. Lokhov, M. W. Reed and R. B. Meyer, *Biochemistry*, 1997, **36**, 14816–14826.
60. F. M. Orson, J. Klysik, D. E. Bergstrom, B. Ward, G. A. Glass, P. Hua and B. M. Kinsey, *Nucleic Acids Res.*, 1999, **27**, 810–816.
61. J. Klysik, B. M. Kinsey, P. Hua, G. A. Glass and F. M. Orson, *Bioconjug. Chem.*, 1997, **8**, 318–326.
62. K. R. Fox, *FEBS Lett.*, 1995, **357**, 312–316.
63. E. Brunet, M. Corgnali, L. Perrouault, V. Roig, U. Asseline, M. D. Sørensen, B. R. Babu, J. Wengel and C. Giovannangeli, *Nucleic Acids. Res.*, 2005, **33**, 4223–4234.
64. D. A. Stewart, X. Xu, S. D. Thomas and D. M. Miller, *Nucleic Acids Res.*, 2002, **30**, 2565–2574.
65. I. V. Kutyavin, M. A. Podyminogin, Y. N. Bazhina, O. S. Fedorova, D. G. Knorre, A. S. Levina, S. V. Mamayev and V. F. Zarytova, *FEBS Lett.*, 1988, **238**, 35–38.
66. S. G. Lokhov, M. A. Podyminogin, D. S. Sergeev, V. N. Silnikov, I. V. Kutyavin, G. V. Shishkin and V. P. Zarytova, *Bioconjug. Chem.*, 1992, **3**, 414–419.
67. K. Mori, C. Subasinghe and J. S. Cohen, *FEBS Lett.*, 1989, **249**, 213–218.
68. K.-Y. Lin and M. Matteucci, *Nucleic Acids Res.*, 1991, **19**, 3111–3114.
69. T. H. Keller and R. Häner, *Nucleic Acids Res.*, 1993, **21**, 4499–4505.
70. K. Yamana, T. Mitsui, J. Yoshioka, T. Isuno and H. Nakano, *Bioconjug. Chem.*, 1996, **7**, 715–720.
71. J. P. May, L. J. Brown, I. van Delft, N. Thelwell, K. Harley and T. Brown, *Org. Biomol. Chem.*, 2005, **3**, 2534–2542.
72. N. Bouquin, V. L. Malinovskii and R. Häner, *Eur. J. Org. Chem.*, 2008, 2213–2219.
73. T. Moriguchi, H. Sekiguchi, M. Tachibana and K. Shinozuka, *Nucleosides Nucleotides Nucleic Acids.*, 2006, **25**, 601–612.
74. A. Patra and C. Richert, *J. Am. Chem. Soc.*, 2009, **131**, 12671–12681.
75. Z. Zhao, G. Peng, J. Michels, K. R. Fox and T. Brown, *Nucleosides Nucleotides Nucleic Acids.*, 2007, **26**, 921–925.
76. N. B. Gaied, Z. Zhao, S. R. Gerrard, K. R. Fox and T. Brown, *Chembiochem.*, 2009, **10**, 1839–1851.
77. T. Moriguchi, A. T. Azam and K. Shinozuka, *Nucleic Acids Symp. Ser.*, 2005, **49**, 7–8.
78. T. Miyashita, N. Matsumoto, T. Moriguchi and K. Shinozuka, *Tetrahedron Lett.*, 2003, **44**, 7399–7402.

79. A. Garbesi, S. Bonazzi, S. Zanella, M. L. Capobianco, G. Giannini and F. Arcamone, *Nucleic Acids Res.*, 1997, **25**, 2121–2128.
80. G. M. Carbone, E. McGuffie, S. Napoli, C. E. Flanagan, C. Dembech, U. Negri, F. Arcamone, M. L. Capobianco and C. V. Catapano, *Nucleic Acids Res.*, 2004, **32**, 2396–2410.
81. S. Napoli, U. Negri, F. Arcamone, M. L. Capobianco, G. M. Carbone and C. V. Catapano, *Nucleic Acids Res.*, 2006, **34**, 734–744.
82. K. Yamana and R. L. Letsinger, *Nucleic Acids Symp. Ser.*, 1985, **16**, 169–172.
83. K. Yamana, Y. Ohashi, K. Nunota, M. Kitamura, H. Nakano, O. Sangen and T. Shimidzu, *Tetrahedron Lett.*, 1991, **32**, 6347–635083.
84. J. S. Mann, Y. Shibata and T. Meehan, *Bioconjug. Chem.*, 1992, **3**, 554–558.
85. J. Telser, K. A. Cruickshank, L. E. Morrison, T. L. Netzel and C.-K. Chan, *J. Am. Chem. Soc.*, 1989, **111**, 7226–7232.
86. U. B. Christensen and E. B. Pedersen, *Nucleic Acids Res.*, 2002, **30**, 4918–4925.
87. U. B. Chrisensen, M. Wamberg, F. A. G. El-Essawy, A. E-H. Ismail, C. B. Nielsen, V. V. Filichev, C. H. Jessen, M. Petersen and E. B. Pedersen, *Nucleosides Nucelotides Nucleic Acids.*, 2004, **23**, 207–225.
88. V. V. Filichev, B. Vester, L. H. Hansen and E. B. Pedersen, *Nucleic Acids Res.*, 2005, **33**, 7129–7137.
89. V. V. Filichev, B. Vester, L. H. Hansen, A. M. T. Abdel, B. R. Babu, J. Wengel and E. B. Pedersen, *Chembiochem.*, 2005, **6**, 1181–1184.
90. V. V. Filichev and E. B. Pedersen, *J. Am. Chem. Soc.*, 2005, **127**, 14849–14858.
91. U. V. Schneider, N. D. Mikkelsen, N. Jøhnk, L. M. Okkels, H. Westh and G. Lisby, *Nucleic Acids Res.*, 2010, **38**, 4394–4403.
92. M. Paramasivam, S. Cogoi, V. V. Filichev, N. Bomholt, E. B. Pedersen and L. E. Xodo, *Nucleic Acids Res.*, 2008, **36**, 3494–3507.
93. G. C. Silver, C. H. Nguyen, A. S. Boutorine, E. Bisagni, T. Garestier and C. Hélène, *Bioconjug. Chem.*, 1997, **8**, 15–22.
94. G. C. Silver, J.-S. Sun, C. H. Nguyen, A. S. Boutorine, E. Bisagni and C. Hélène, *J. Am. Chem. Soc.*, 1997, **119**, 263–268.
95. S. Vinogradov, V. Roig, Z. Sergueeva, C. H. Nguyen, P. Arimondo, N. T. Thuong, E. Bisagni, J.-S. Sun, C. Hélène and U. Asseline, *Bioconjug. Chem.*, 2003, **14**, 120–135.
96. M. D. Keppler, C. M. McKeen, O. Zegrocka, L. Strekowski, T. Brown and K. R. Fox, *Biochim. Biophys. Acta.*, 1999, **1447**, 137–142.
97. D. A. Gianolio, J. M. Segismundo and L. W. McLaughlin, *Nucleic Acids Res.*, 2000, **28**, 2128–2134.
98. A. S. Levina, V. G. Metelev, A. S. Cohen and P. C. Zamecnik, *Antisense Nucleic Acid Drug Dev.*, 1996, **6**, 75–85.
99. N. Alexander, S. G. Sinyakov, I. V. Kutyavin, H. B. Gamper and R. B. Meyer, *J. Am. Chem. Soc.*, 1995, **117**, 4995–4996.
100. E. A. Lukhtanov, I. V. Kutyavin, H. B. Gamper and R. B. Meyer, *Bioconjug. Chem.*, 1995, **6**, 418–426.

101. I. V. Kutyavin, I. A. Afonina, A. Mills, V. V. Gorn, E. A. Lukhtanov, E. S. Belousov, M. J. Singer, D. K. Walburger, S. G. Lokhov, A. A. Gall, R. Dempcy, M. W. Reed, R. B. Meyer and J. Hedgpeth, *Nucleic Acids Res.*, 2000, **28**, 655–661.

102. I. V. Kutyavin, E. A. Lukhtanov, H. B. Gamper and R. B. Meyer, *Nucleic Acids Res.*, 1997, **25**, 3718–3723.

103. I. Afonina, I. Kutyavin, E. Luktanov, R. B. Meyer and H. Gamper, *Proc. Natl. Acad. Sci.*, 1996, **93**, 3199–3204.

104. D. S. Novopashina, A. N. Sinyakov, V. A. Ryabinin, A. G. Venyaminova, L. Halby, J.-S. Sun and A. S. Boutorine, *Chem. Biodiversity.*, 2005, **2**, 936–952.

105. A. N. Sinyakov, V. A. Ryabinin, G. N. Grimm and A. S. Boutorine, *Mol. Biol.*, 2001, **35**, 251–260.

106. S. B. Rajur, J. Robles, K. Wiederholt, R. G. Kuimelis and L. W. McLaughlin., *J. Org. Chem.*, 1997, **62**, 523–529.

107. P. M. Reddy and T. C. Bruice, *J. Am. Chem. Soc.*, 2004, **126**, 3736–3747.

108. J. Robles and L. W. McLaughlin, *J. Am. Chem. Soc.*, 1997, **119**, 6014–6021.

109. Z. V. Zhilina, A. J. Ziemba, J. O. Trent, M. W. Reed, V. Gorn, Q. Zhou, W. Duan, L. Hurley and S. Ebbinghaus, *Bioconjug. Chem.*, 2004, **15**, 1182–1192.

110. I. Charles, H. Xi and D. P. Arya, *Bioconjug. Chem.*, 2007, **18**, 160–169.

111. D. Praseuth, M. Chassignol, M. Takasugi, T. Le Doan, N. T. Thuong and C. Hélène, *J. Mol. Biol.*, 1987, **196**, 939–942.

112. E. Sage and E. Moustacchi, *Biochemistry.*, 1987, **26**, 3307–3314.

113. B. L. Lee, A. Murakami, K. R. Blake, S. B. Lin and P. S. Miller, *Biochemistry.*, 1988, **27**, 3197–3203.

114. M. Lin, K. L. Hultquist, D. H. Oh, E. A. Bauer and W.K. Hoeffler, *FASEB J.*, 1995, **9**, 1371–1377.

115. W. R. Kobertz and J. M. Essigmann, *J. Am. Chem. Soc.*, 1997, **119**, 5960–5961.

116. C. Giovannangeli, N. T. Thuong and C. Hélène, *Nucleic Acids Res.*, 1992, **20**, 4275–4281.

117. M. Takasugi, A. Guendouz, M. Chassignol, J. L. Decout, J. Lhomme, N. T. Thuong and C. Hélène, *Proc. Natl. Acad. Sci. USA*, 1991, **88**, 5602–5606.

118. V. M. Macaulay, P. J. Bates, M. J. McLean, M. G. Rowlands, T. C. Jenkins, A. Ashworth and S. Neidle, *FEBS Lett.*, 1995, **372**, 222–228.

119. P. J. Bates, V. M. Macaulay, M. J. McLean, T. C. Jenkins, A. P. Reszka, C. A. Laughton and S. Neidle, *Nucleic Acids Res.*, 1995, **23**, 4283–4289.

120. H. Li, V. J. Broughton-Head, G. Peng, V. E. Powers, M. J. Ovens, K. R. Fox and T. Brown, *Bioconjug. Chem.*, 2006, **17**, 1561–1567.

121. H. Li, V. J. Broughton-Head, K. R. Fox and T. Brown, *Nucleosides Nucleotides Nucleic Acids.*, 2007, **26**, 1005–1009.

122. K. M Vasquez, K. Marburger, Z. Intody and J. H. Wilson, *Proc. Natl. Acad. Sci. USA*, 2001, **98**, 8403–8410.

123. Z. Sandor and A. Bredberg, *Biochim. Biophys. Acta*, 1995, **1263**, 235–240.
124. Z. Luo, M. A. Macris, A. F. Faruqi and P. M. Glazer, *Proc. Natl. Acad. Sci. USA*, 2000, **97**, 9003–9008.
125. A. F. Faruqi, M. M. Seidman, D. J. Segal, D. Carroll and P. M. Glazer, *Mol. Cell Biol.*, 1996, **16**, 6820–6828.
126. A. F. Faruqi, H. J. Datta, D. Carroll, M. M. Seidman and P. M. Glazer, *Mol. Cell Biol.*, 2000, **20**, 990–1000.
127. Z. Sandor and A. Bredberg, *FEBS Lett.*, 1995, **374**, 287–291.
128. G. Wang and P. M. Glazer., *J. Biol. Chem.*, 1995, **270**, 22595–22601.
129. T. J. Povsic and P. B. Dervan, *J. Am. Chem. Soc.*, 1990, **112**, 9428–9430.
130. K. B. Grant and P. B. Dervan, *Biochemistry*, 1996, **35**, 12313–12319.
131. T. J. Povsic, S. A. Strobel and P. B. Dervan, *J. Am. Chem. Soc.*, 1992, **114**, 5934–5941.
132. J-P. Shaw, J. F. Milligan, S. H. Krawczyk and M. Matteucci, *J. Am. Chem. Soc.*, 1991, **113**, 7765–7766.
133. S. Imoto, T. Hori, S. Hagihara, Y. Taniguchi, S. Sasaki and F. Nagatsugi, *Bioorg. Med. Chem. Lett.*, 2010, **20**, 6121–6124.
134. F. Nagatsugi, S. Sasaki, P. S. Miller and M. M. Seidman, *Nucleic Acids Res.*, 2003, **31**, e31.
135. O. S. Fedorova, D. G. Knorre, L. M. Podust and V. F. Zarytova, *FEBS Lett.*, 1988, **228**, 273–276.
136. I. G. Kutyavin, H. B. Gamper, A. A. Gall and R. B. Meyer, *J. Am. Chem. Soc.*, 1991, **113**, 7765–7766.
137. J. N. Lampe, I. V. Kutyavin, R. Rhinehart, M. W. Reed, R. B. Meyer and H. B. Gamper, *Nucleic Acids Res.*, 1997, **25**, 4123–4131.
138. A. Ziemba, L. C. Derosier, R. Methvin, C. Y. Song, E. Clary, W. Kahn, D. Milesi, V. Gorn, M. Reed and S. Ebbinghaus, *Nucleic Acids Res.*, 2001, **29**, 4257–4263.
139. A. J. Ziemba, M. W. Reed, K. D. Raney, A. B. Byrd and S. Ebbinghaus, *Biochemistry*, 2003, **42**, 5013–5024.
140. M. W. Reed, E. A. Lukhtanov, V. Gorn, I. Kutyavin, A. Gall, A. Wald and R. B. Meyer, *Bioconjug. Chem.*, 1998, **9**, 64–71.
141. C. Colombier, B. Lippert and M. Leng, *Nucleic Acids Res.*, 1996, **24**, 4519–4524.
142. E. Bernal-Mendez, J.-S. Sun, F. Gonzalez-Vilchez and M. Leng, *New J. Chem.*, 1998, 1479–1483.
143. S. K. Sharma and L. W. McLaughlin, *J. Am. Chem. Soc.*, 2002, **124**, 9658–9659.
144. S. K. Sharma and L. W. McLaughlin, *J. Inorg. Biochem.*, 2004, **98**, 1570–1577.
145. M. A. Campbell and P. S. Miller, *J. Biol. Inorg. Chem.*, 2009, **14**, 873–881.
146. M. A. Campbell, T. M. Mason and P. S. Miller, *Can. J. Chem.*, 2007, **85**, 241–248.
147. M. A. Campbell and P. S. Miller, *Bioconjug. Chem.*, 2009, **20**, 2222–2230.
148. B. C. Chu and L. E. Orgel, *Proc. Natl. Acad. Sci. USA*, 1985, **82**, 963–967.
149. G. B. Dreyer and P. B. Dervan, *Proc. Natl. Acad. Sci. USA*, 1985, **82**, 968–972.

150. M. Boidot-Forget, M. Chassignol, M. Takasugi, N. T. Thuong and C. Hélène, *Gene.*, 1988, **72**, 361–371.
151. C. H. Chen and D. S. Sigman, *Proc. Natl. Acad. Sci. USA*, 1986, **83**, 7147–7151.
152. J. C. Francois, T. Saison-Behmoaras, M. Chassignol, N. T. Thuong, J.-S. Sun and C. Hélène, *Biochemistry*, 1998, **27**, 2272–2274.
153. J. S. Sun, J. C. François, R. Lavery, T. Saison-Behmoaras, T. Montenay-Garestier, N. T. Thuong and C. Hélène, *Biochemistry.*, 1988, **27**, 6039–6045.
154. T. Le Doan, L. Perrouault, C. Hélène, M. Chassignol and N. T. Thuong, *Biochemistry*, 1986, **25**, 6736–6739.
155. T. Le Doan, T, D. Praseuth, L. Perrouault, M. Chassignol, N. T. Thuong and C. Hélène, *Bioconjug Chem.*, 1990, **1**, 108–113.
156. H. Seliger, H. Knoller, A. Ruck, K. Heckelsmiller and R. Steiner, *Nucleosides Nucleotides*, 1998, **17**, 2053–2061.
157. D. Magda, M. Wright, S. Crofts, A. Lin and J. L. Sessler, *J. Am. Chem. Soc.*, 1997, **119**, 6947–6948.
158. D. S. Sergeev, V. F. Zarytova, S. V. Mamaev, T. S. Godovikova and V. V. Vlassov, *Antisense Res. Dev.*, 1992, **2**, 235–241.
159. D. S. Sergeyev, T. S. Godovikova and V. F. Zarytova, *Nucleic Acids Res.*, 1995, **23**, 4400–4406.
160. P. Vorobjev, O. Tchaika and V. Zarytova, *Nucleosides Nucleotides Nucleic Acids*, 2004, **23**, 1047–1051.
161. M. Matteucci, K.-Y. Lin, T. Huang, R. Wagner, D. D. Sternbach, M. Mehrotra and J. M. Besterman, *J. Am. Chem. Soc.*, 1997, **119**, 6939–6940.
162. P. B. Arimondo, C. Bailly, A. S. Boutorine, P. Moreau, M. Prudhomee, J.-S. Sun, T. Garestier and C. Hélène, *Bioconjug. Chem.*, 2001, **12**, 501–509.
163. P. B. Arimondo, A. S. Boutorine, B. Baldeyrou, C. Bailly, M. Kuwahara, S. M. Hecht, J.-S. Sun, T. Garestier and C. Hélène, *J. Biol. Chem.*, 2002, **277**, 3132–3140.
164. P. B. Arimondo, S. Angenault, L. Halby, A. Boutorine, F. Schmidt, C. Monneret, T. Garestier, J-S. Sun, C. Bailly and C. Hélène, *Nucleic Acids Res.*, 2003, **31**, 4041–4040.
165. P. B. Arimondo, G. S. Laco, C. J. Thomas, L. Halby, D. Pez, P. Schmitt, A. Boutorine, T. Garestier, Y. Pommier, S. M. Hecht, J.-S. Sun and C. Bailly, *Biochemistry*, 2005, **44**, 4171–4180.
166. P. B. Arimondo, C. J. Thomas, K. Oussedik, B. Baldeyrou, C. Mahieu, L. Halby, D. Guianvarc'h, A. Lansiaux, S. M. Hecht, C. Bailly and C. Giovannangeli, *Mol. Cell Biol.*, 2006, **26**, 324–333.
167. P. B. Arimondo, P. Moreau, A. Boutorine, C. Bailly, M. Prudhomme, J.-S. Sun, T. Garestier and C. Hélène, *Bioorg. Med. Chem.*, 2000, **8**, 777–784.
168. M. Duca, K. Oussedik, A. Ceccaldi, L. Halby, D. Guianvarc'h, D. Dauzonne, C. Monneret, J.-S. Sun and P. B. Arimondo, *Bioconjug. Chem.*, 2005, **16**, 873–884.

CHAPTER 4

Small Molecule–RNA Conjugates

SANJUKTA MUHURI,[a] GOPAL GUNANATHAN JAYARAJ[b] AND SOUVIK MAITI*[b,c]

[a] Department of Chemistry, Barasat Government College, West Bengal – 700124, India; [b] Institute of Genomics and Integrative Biology, (CSIR-IGIB), CSIR, Delhi – 110007, India; [c] National Chemical Laboratory, (CSIR-NCL), Pune – 411008, India
*Email: Souvik@igib.res.in

4.1 Introduction

RNA is composed of all four common ribonucleosides, A, G, C and U, coupled together by $5'$ to $3'$ phosphodiester linkages. The chemical composition of an RNA molecule allows it to play important roles within biological systems. Classically, RNA was considered as the information transducer of the genetic code (in the form of mRNA), though it was later discovered that RNA could play more versatile roles in regulation of the central dogma of biology by functioning as carrier molecules (tRNA) or as functional components of ribosomes (rRNA). The centrality of RNA is further elucidated in its role as a key regulator of gene expression, in the form of spliceosomal RNPs. More recently, the discovery and description of small and long non-coding RNA has added to the growing repertoire of RNA functions.[1,2]

The last few decades have witnessed a tremendous paradigm shift in RNA biology. As well as being a simple biochemical transducer of genetic information, the discovery of riboswitches,[3] RNA quadruplexes[4] and other functional

RSC Biomolecular Sciences No. 26
DNA Conjugates and Sensors
Edited by Keith R Fox and Tom Brown
© The Royal Society of Chemistry 2012
Published by the Royal Society of Chemistry, www.rsc.org

secondary and tertiary structures has established that RNA structure and its dynamics play an important role in biological regulation. Having noted its biological importance, it is also important to understand that RNA possesses a chemical diversity that far exceeds that of many other biomolecules.[5] It has been established that RNA can be chemically modified in over one hundred different ways,[6] which impart a range of properties that contribute to different aspects of its structural and regulatory roles within the cell. These modifications have several roles within the cell, which include tRNA function, nuclease resistance, stress sensors, *etc.*[6] What is perhaps less well understood is the existence of other small molecule–RNA conjugates. The best described example of an RNA–small molecule conjugate is the charged tRNA, in which an amino acid is enzymatically coupled to a tRNA by the amino-acyl tRNA synthetase. Only recently has a considerable amount of attention also focused on the discovery of natural small molecule–RNA conjugates, which have implications in the explanation of the previously hypothesised 'RNA world scenario', where such template directed chemistries could have given rise to life and the course of evolution as we know it.[7]

Unlike DNA, RNA exhibits a high level of diversity in terms of its secondary and tertiary folding and it therefore has greater potential for selective recognition based on structure rather than sequence *per se*. Therefore, it seems reasonable to investigate whether changes in RNA structure, through the incorporation of small molecules, might play a key role in the selective recognition of the biological macromolecules that are essential for many vital biological processes. In parallel, understanding of small molecule–RNA conjugates also forms a prerequisite for the development of novel RNA/small molecule based therapeutic agents.

In the following sections, we discuss some recently discovered naturally occurring modifications and conjugations of RNA. We also highlight some of the important synthetic small molecule–RNA conjugates that have been recently described, with a separate note on modifications of therapeutic importance.

4.2 Small Molecule–RNA Conjugates

4.2.1 Examples of Naturally Occurring Small Molecule–RNA Conjugates

The peptidyl transferase centre (PTC) is contained within the central part of domain V of the 23S rRNA, which also contains binding sites for various classes of antibiotics. Alterations in the PTC site affect drug binding and thereby cause antibiotic resistance. A recent report showed that methylation at the PTC mediated by cfr methyltransferase confers combined resistance to five different classes of antibiotics that bind to the PTC.[8] It was proposed that cfr causes methylation at the 8-position of A2503 in 23S rRNA to generate C8-methyl adenosine at that position, leading to resistance to several ribosomal-targeted antibiotics. This was the first example of a naturally occurring methyl attachment at the purine 8-position of RNA (Figure 4.1). The antibiotic resistance conferred by the methylation at this position is purely steric in nature; the methyl group at the C-8 position of A2503 points directly into the drug binding site and prevents the binding of drugs targeted to the PTC, leading to the loss of recognition of the RNA by those drugs. This modification is an example of why it is important to understand the purpose of RNA modifications in living systems and how they can be utilised for therapeutic purposes. Interestingly, the presence of dimethylated A2503 (m^2m^8A) has also been reported,[8] however this modification has yet to be fully characterised or structurally confirmed.

Agmatidine (2-agmatinylecytidine) (Figure 4.2) is another naturally occurring nucleotide present in the anticodon of archaeal tRNA[Ile] that decodes AUA codons specifically instead of AUG.[9,10] Before the discovery of Agmatidine, the mechanism by which archaea translated AUA codons was unknown. Agmatidine, a 2-position modified cytidine, was synthesised from agmatine and ATP by the enzyme tRNA [Ile] –agm^{2}C synthetase and performs a function similar to lysidine in bacteria. In *Escherichia coli*, the presence of lysidine at the first position of the anticodon allows for AUA decoding by tRNA [Ile] to recognise A and not G in the third position of the codon.[11] Agmatidine has a similar structure to lysidine, in which agmatine replaces the C2-oxo- group of cytidine and the first position of the anticodon performs the same function of differentiating between A and G at the third position of the codon in *Haloarcula marismortui*.[9] Conjugation of an agmatine moiety at the C2 position of cytosine induces a tautomeric conversion, with protonation of the N3 position and imino group formation at C4. This modification completely alters the hydrogen bond donor–acceptor patterns of cytosine, enabling agm^{2}C to base pair with adenine instead of guanine so that recognition of the anticodon changes from AUG to AUA. Considering the structural similarity between agm^{2}C and

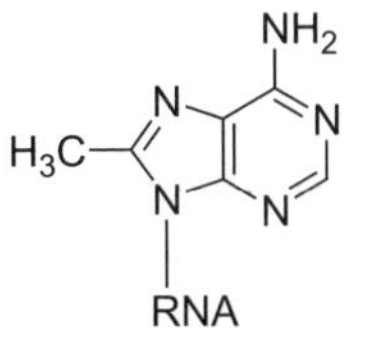

Figure 4.1 C-8 methyl modification of adenosine.

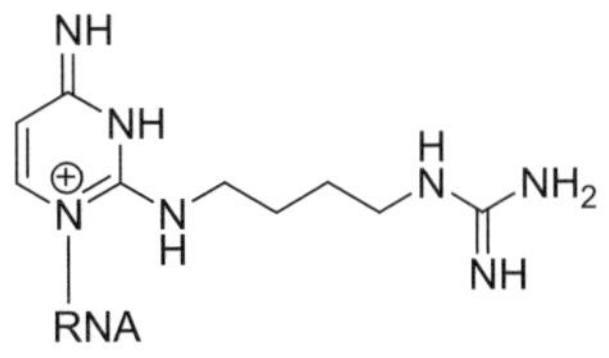

Figure 4.2 2-Agmatinyl cytidine.

lysidine, it seems likely that agm^{2}C–adenosine pairing occurs in the same manner as the lysidine–adenosine pairing in bacterial tRNA. The discrimination between A and G that results from the presence of lysidine or agmatidine is hypothesised to play an important role in stabilising the potential positively charged tautomeric structures at neutral pH, which are able to base pair selectively with A.[9]

Very recently Liu and coworkers reported the discovery of a surprising modification at the 5′ end of the small RNA molecule found in bacteria.[12] They developed a general approach to the discovery of small molecule–RNA conjugates that does not depend on the specific structure or the biological function of the conjugate. Assuming that small molecule-like chemical modifications of RNAs are labile under certain conditions, they treated fractions of short RNA molecules, isolated from bacteria, with either a base or a nucleophile. This treatment removed any labile molecules associated with the RNA and the products were subsequently analysed by mass spectrometry. They applied this method to RNA extracts obtained from *E. coli* and *Streptomyces venezuelae* and found several unknown molecules associated with RNA. One has been further characterised and identified as hydroxyfuranone, or rather its tautomer succinic anhydride. However, because no nucleotide connection of the succinyl group was detectable under these conditions, it was hypothesised that this succinyl group came from some other larger molecule instead of being directly associated with RNA. To address this hypothesis and to investigate the intact small molecule–RNA conjugate, a second method was introduced. A nuclease P1 treatment step was included in the protocol prior to base or nucleophile treatment. This method gave rise to the discovery of a new species of covalent RNA modification, namely 3′-dephospho-CoA and its succinyl, acetyl and methyl malonyl esters. These observations suggest that CoA attachment to RNA occurs after a thioesterification step. The liberation of 3′-dephospho-CoA derivatives (Figure 4.3) from cellular RNA by nuclease P1 digestion, together with their presence on the 5′ terminus of RNA, strongly suggests that the CoA–RNA conjugate has a phosphodiester bond linking the 3′-phosphate of CoA to the 5′-end of the RNA. The discovery of covalent 5′ modifications of small RNAs [≤200 nucleotides (nt)] with 3′-dephospho-CoA might open the door to a new area in RNA biology and in the chemical diversity of naturally occurring RNA molecules.

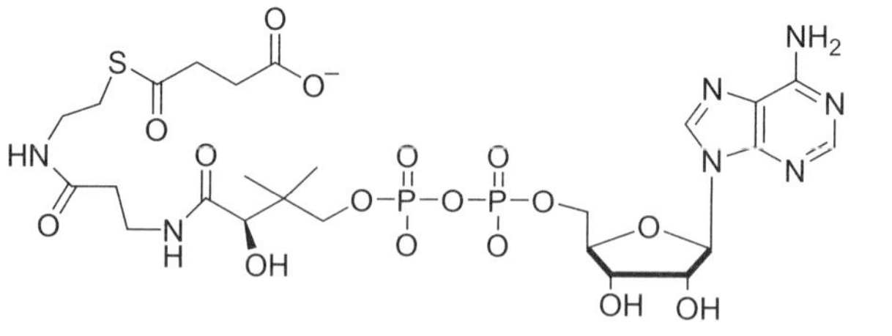

Figure 4.3   3′-Dephospho-CoA.

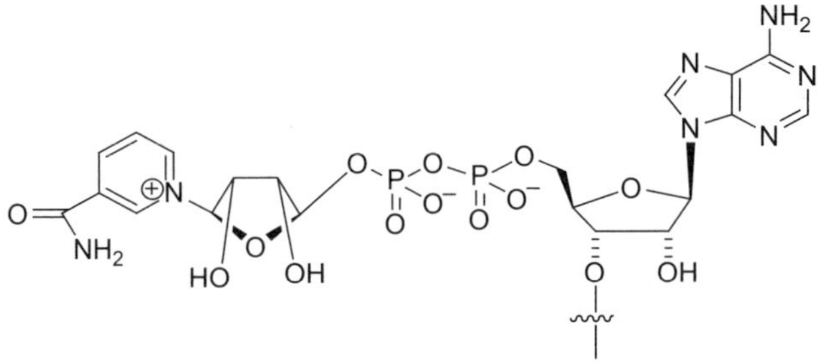

Figure 4.4 NAD-linked RNA.

In addition to the above method for detecting small molecule–RNA conjugates, the same group developed a more general method for detecting small molecule–RNA conjugates that can be applied to any such conjugates, regardless of their chemical reactivity towards bases and nucleophiles.[13] They discovered another naturally occurring modification of RNA in which nicotinamide adenine dinucleotide (NAD) is bound at the 5′ terminus of small RNAs (Figure 4.4). To identify the covalently attached small molecule–RNA conjugates in bacterial RNA, these investigators used a combination of size exclusion chromatography, nuclease catalysed fragmentation and mass spectrometry. They also determined that NAD is mostly attached to the 5′ terminus of RNAs below 200 nucleotides in length, and aberrant transcriptional initiation is not responsible for incorporation of NAD into RNAs. Although the specific biological functions of these NAD-linked RNAs are not yet known, it can be speculated that these NAD groups may cause the localisation of the NAD-linked RNAs to NAD binding proteins and enable new redox chemistries. It may also be possible that NAD-linked RNAs play a vital role in RNA stability or even in gene expression.

4.2.2 Synthetic Small Molecule–RNA Conjugates

The production of synthetic small molecule–RNA conjugates has had a strong impact on RNA research. This rapidly growing field of research has three main applications: i) Structural studies for controlling RNA structures that generate specialised architectures and functional materials. ii) Imaging; the fluorescent labelling of RNA is useful for monitoring RNA production, processing and relocation inside the cell and for following the interaction with other intracellular components and the degradation of RNA molecules. iii) Nanotechnology; interest in RNA nanotechnology[14,15] has increased because of its potential application in nanomedicine, including the treatment of cancer, viral infections and other genetic diseases. The advances in synthetic RNA modifications will depend on the ability to manipulate the composition of RNA and to probe the structure, enabling a broader understanding of RNA structure and function. Synthetically modified RNA nucleotides play an important role in cellular functions and the progression of disease. Chemical tools that can modify RNA nucleotides with specificity and efficiency may alter the specific function of the target RNA and thereby enable manipulation of gene expression.

There are many methods for site-specific incorporation of useful labels or reactive groups into RNA that can be used for subsequent labelling and covalent attachment of other biomolecules. These reactive groups or modified residues can be attached by chemical synthesis or by enzymatic incorporation.[16] Chemical methods involve either the direct incorporation of reactive residues into the RNA via solid-phase synthesis using phosphoramidite chemistry or through post-synthetic labelling with a reactive group. On the other hand, enzymatic methods are especially useful for labelling of RNA at the 5′ or 3′ terminus *in vitro* or *in vivo*.

Among the chemical methods, post-synthetic modification has been more widely used than direct solid phase synthesis owing to the high cost of directly labelled synthetic RNA or the inaccessibility of appropriate phosphoramidites. The post-synthetic modification of RNA can be achieved through four general chemical methods based on the reacting species: i) periodate chemistry, ii) amine chemistry (using activated phosphates or activated succinimide esters), iii) thiol chemistry and iv) 'click' chemistry.

Periodate reactions depend on the ribose 2′, 3′-diol and are therefore specific to RNA. The 3′-ribose of RNA molecules can be oxidised by sodium periodate to a dialdehyde. The resulting dialdehyde can couple with amine-containing molecules such as hydrazine derivatives to produce morpholino linked RNA conjugates (Figure 4.5).[17] This methodology has been successfully used for the quantitative introduction of one amine functionalised fluorescent dye per RNA molecule in solution.[18]

The amine chemistry involves the reaction between a primary amine and N-hydroxysuccinimide (NHS) esters (Figure 4.6). In this strategy, nucleophilic attack by the primary amine moiety takes place at the ester carbonyl to form an imido-linkage. Generally this method has been adapted to modify the 5′- terminus of RNA. This conjugation of a site-specific amino group within an RNA molecule to succinimide-derivatised molecules is specific and fairly efficient. However buffers that contain amine-functionalities, such as TRIS, cannot be used.

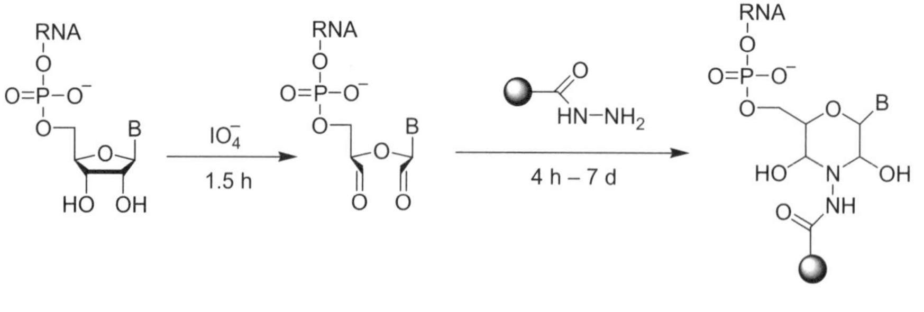

Figure 4.5   Morpholino-linked RNA conjugates.

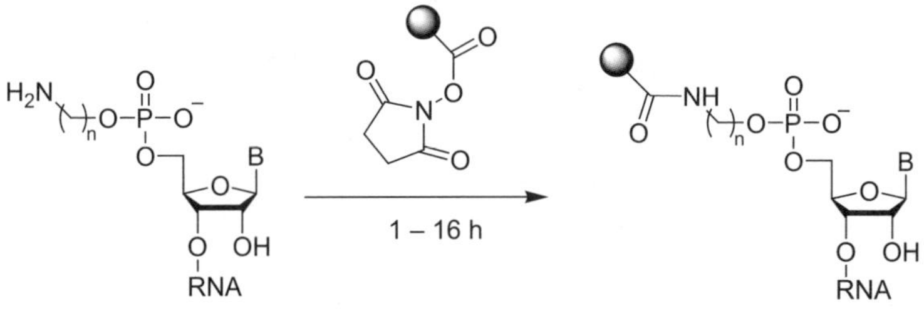

Figure 4.6 Modification of 5′-terminus of the RNA molecule using the coupling reaction between a primary amine and the N-hydroxy succinimide (NHS) esters.

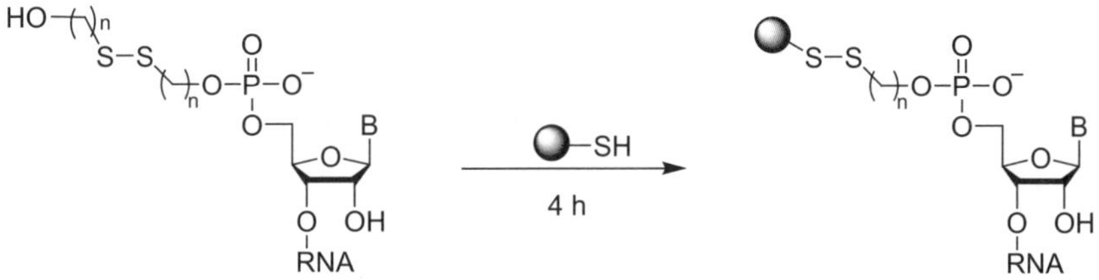

Figure 4.7 Conjugation through disulfide bond formation.

A further extension of this methodology involves the activation of a 5′-terminal phosphate with an imidazole moiety to react with amine functionalised molecules such as fluorophores, to achieve the efficient synthesis of fluorescently labelled RNA.[17]

Thiol chemistry involves the formation of a disulfide bridge covalently linking two thiol-containing molecules (Figure 4.7). This methodology is extensively used for obtaining information about the proximity of functional groups within highly structured RNAs. It has also been used for the attachment of biotin or other reactive cross-linkers for structural probing.

Very recently, Gothelf and Kjems developed[19] a highly flexible strategy for functionalising long RNA at internal positions by attaching a short linker to the target RNA (Figure 4.8).

For this purpose, a tetrameric DNA–RNA complex was designed to enable the transfer of a chemical group from a labelled DNA strand onto a non-labelled RNA molecule in a site-specific manner. This method involves the 4-(4,6-dimethoxy-1,3,5-triazin-2-yl)-4-methyl-morpholinium chloride (DMTMM) catalysed formation of a bond between the carboxylic acid containing DNA-donor molecule and the 2′-OH group of target RNA, by activation of the carboxylic acid moiety. Close proximity of the carboxylic acid group and the non-labelled RNA molecule in the four-way junction plays an important role, favouring the esterification reaction. The donor molecule was generated by treating an amino-modified DNA strand with the commercially

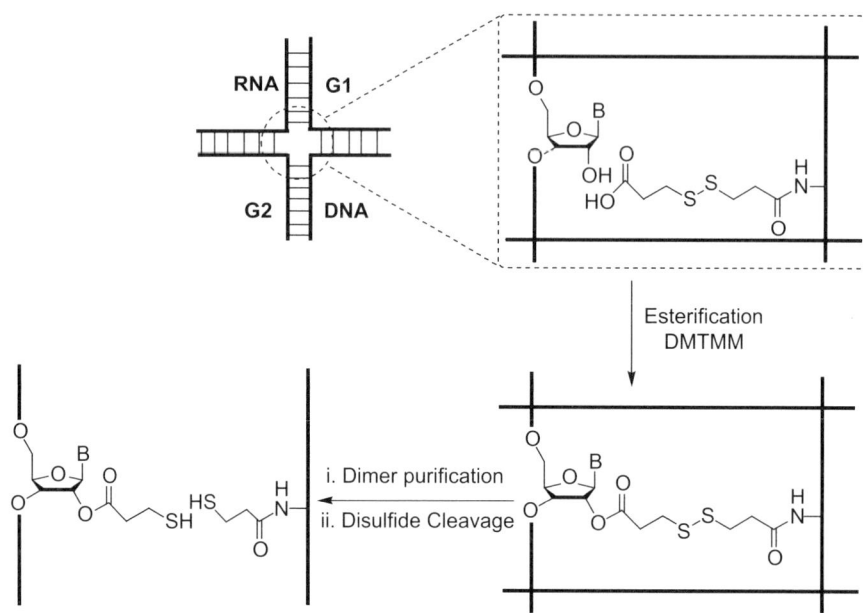

Figure 4.8   Strategy for the functionalisation of long RNA at internal positions by attaching a short linker to the target RNA.

available DSP linker containing a cleavable disulfide spacer with *N*-hydroxysuccinimide (NHS) esters at each end. It involved the formation of an amide linkage between the amino group on the DNA and one of the NHS esters. Finally, hydrolysis of the other NHS ester resulted in the formation of a free carboxylic acid at the end of the linker for further esterification with the 2′-OH group of target RNA. Importantly this method allows the incorporation of a free thiol moiety onto RNA that enables efficient conjugation with various functional groups.

In another report Famulok and co-workers[20] used this thiol chemistry to introduce different functional groups onto the 5′-end of RNA. Their straightforward method involved the introduction of a suitable functionality via derivatisation with guanosine monophosphorothioate (GMPS) at the 5′-end and the subsequent coupling of the RNA with a 2-thiopyridine activated substrate by the nucleophilic sulfur at the phosphorothioate, to form a disulfide linkage (Figure 4.9). Such modifications at the 5′-end of the RNA molecule allow the generation of highly functionalized RNA libraries for *in vitro* selection of novel catalytic RNAs and also the attachment of photoaffinity tags for mapping of RNA–protein interactions in large RNA–protein complexes.

'Click' chemistry is among the most popular and efficient strategies used for RNA functionalisation. The term click chemistry encompasses various reactions that are selective and proceed with high yields with few or no side products under simple reaction conditions and solvents. Over the past few years several research groups have applied the powerful copper-catalyzed

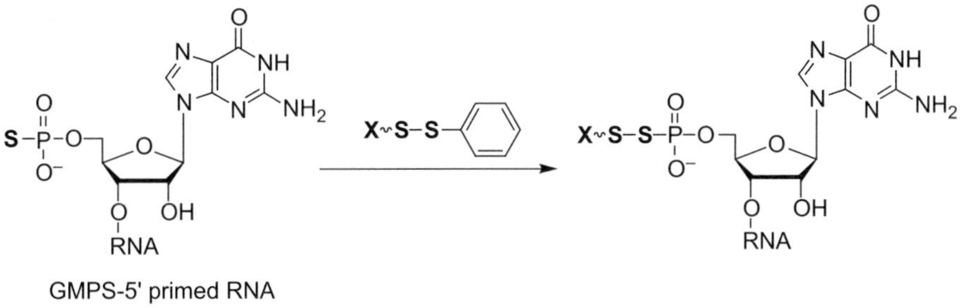

GMPS-5' primed RNA

Figure 4.9 Conjugation through derivatization with guanosine monophosphorothioate (GMPS) at the 5'-end.

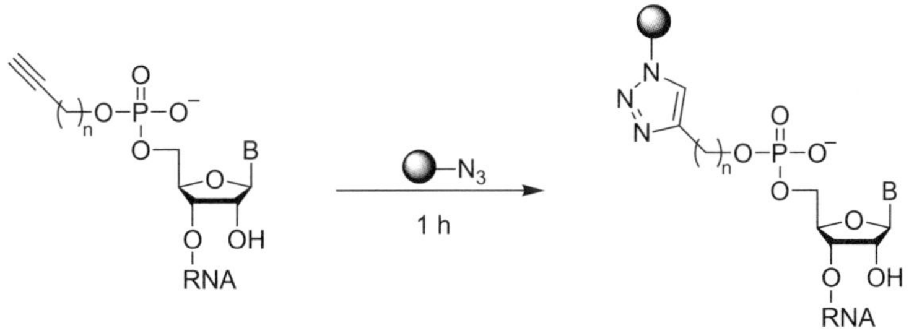

Figure 4.10 Conjugation through 'click chemistry'.

azide–alkyne cycloaddition reaction (CuAAC) for the introduction of a variety of useful modifications at specific sites in RNA molecules (Figure 4.10). Jao and Salic were the first to report click chemistry on propyne-containing RNA that had been generated by the biosynthetic incorporation of the uridine analogue 5-ethynyluridine (EU).[21] This pioneering work provided a sensitive and efficient alternative for mapping cellular transcription.

Beal and coworkers[22] also used click chemistry for the post-synthetic labelling of 2-amino purine-containing RNA with mannose derivatives or azidothymidines. However this was hampered by the low yield and significant degradation of the RNA. Although certainly beneficial to the RNA research community, the CuAAC reaction requires millimolar concentrations of Cu(I) salts. This restricts its use with inherently labile RNA molecules. Recently Das and co-workers investigated the conditions in which Cu(I)-catalysed chemistry can be used efficiently to label RNA (with free 2'-OH groups) that was obtained synthetically or enzymatically while keeping the RNA intact.[23] The degradation of RNA can be prevented by degassing and the use of acetonitrile as a co-solvent rather than a commercially available ligand to stabilise Cu(I). RNA can be readily conjugated with alkyne groups at the 5'-, 3'- or internal 2' position by synthetic or enzymatic incorporation. Once the clickable groups have been installed onto the RNA, click labelling is opened up subsequently for further functionalisation.

Azides, being relatively inert under physiological conditions, have recently became very popular functional groups in bioconjugate chemistry, largely owing to their efficient functionalisation by click chemistry. However, unlike alkynes, azides are unstable under solid phase synthesis conditions. Therefore azido groups can only be introduced into the 5′-terminus of the synthetic RNA in the last step of solid phase synthesis prior to deprotection.[24] Recently Micura and co-workers have described the solid phase chemical synthesis of 2′-azido modified RNA (Figure 4.11) for the first time.[25] They introduced a phosphodiester, rather than a phosphoramidite, which can be used to incorporate 2′-azides during regular solid phase synthesis of RNA. The azido group installed into RNA by this method may provide a reliable foundation for a wide range of applications in bioconjugation strategies. In another report Helm and co-workers[26] presented N3BC [7-azido-4-(bromomethyl)-2H-chromen-2-one] as a versatile multifunctional reagent, which allows the post-synthetic derivatisation of RNA. N3BC selectively alkylates uridine residues (Figure 4.12) thereby introducing an azido moiety for further functionalization by click chemistry. Interestingly, it was found that N3BC offered a much wider range of applications because it could be used on native RNA, synthetic RNA and *in vitro* transcripts alike.

Another interesting example of highly selective terminal labelling of RNA was reported by Fujimoto and co-workers,[27] who used template-directed

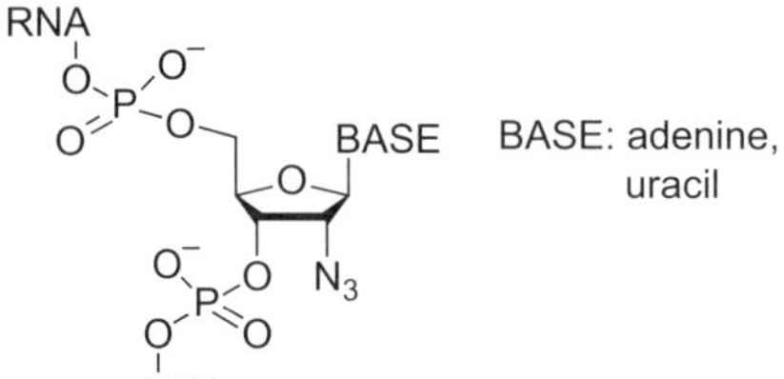

Figure 4.11   2′-Azido modified RNA.

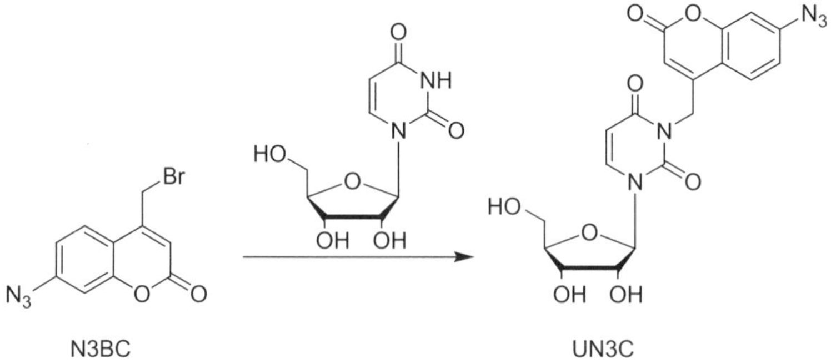

Figure 4.12   Selective alkylation of uridine residue with N3BC to produce UN3C.

photoligation through ^{cv}U (5-carboxyvinyl-2'-deoxyuridine). When an ODN containing ^{cv}U at the 5'-terminus was photoirradiated with an ORN containing a pyrimidine base at the 3'-terminus in the presence of template DNA, efficient terminal labelling was observed with high degree of single nucleotide specificity.

In addition to these chemical methods several enzymatic methods have been used to incorporate labels or small molecules into RNA, permitting the conjugation of RNA strands that cannot be obtained by chemical synthesis.[28] Several enzymes have been used for such RNA labelling. T4 polynucleotide kinase transfers a radioactive phosphate from ATP to the 5' end of RNA and is the standard method for the 5' radiolabelling of RNA. On the other hand, T4 RNA ligase is able to incorporate a radiolabel at the 3' end. Szostak and coworkers reported[29] a unique approach for selective labelling and detection of a specific RNA in a mixture of RNAs using DNA polymerase and a short synthetic DNA template complementary to the 3'-terminus of the RNA. They took advantage of the template-directed selectivity principle, *i.e.* the natural function of DNA polymerase to elongate the RNA primers on DNA templates. Sometimes RNA may form 3'-terminal secondary structures or intramolecular duplexes; in those cases the RNA 3'-terminus may be not be accessible for conventional labelling at the 3'-end. The DNA template used in this method competes with these intermolecular and/or intramolecular duplexes to expose the RNA 3'-terminus, providing an alternative strategy for labelling.

RNA is generated by *in vitro* transcription from DNA template predominantly with T7 RNA polymerase (T7 RNA pol). Guo and co-workers described[30] a simple procedure for 5'-biotin labelling of RNA by one-step *in vitro* transcription. For this purpose a number of functional group AMP conjugates were prepared as transcription initiators. For this the transcription initiator biotin-HDAAMP (Figure 4.13) was chemically synthesised by a two-step pathway. First, the condensation of HDA with AMP produces the amino derivative HDAAMP which then undergoes conjugation with biotin-NHS, generating the desired nucleotide initiator with good coupling efficiency. Following a similar methodology, guanosine monophosphate (GMP) derivatives can be selectively incorporated at the 5'-end of RNA transcripts. Jäschke and co-workers developed[31] a concise synthetic strategy for the preparation of GMP-derivatised transcription initiators bearing a range of potential substrates attached to the decaethylene glycol spacer (Figure 4.13).

This incorporation of GMP derivatives at the 5'-end of RNA transcripts provides another reliable method for producing RNA conjugates bearing organic substrates such as benzylallyl ether, anthracene, benzyl carbamate and a primary amino group for *in vitro* selection of novel ribozymes. The T7 RNA pol has been shown to incorporate various 5' labels useful for RNA detection when primed with the respective GMP or AMP analogues. Recently, Das and co-workers exploited the versatility of T7 RNA polymerase's transcriptional priming to use 5'-azido-5'-deoxyguanosine for incorporating a 5'-azide reactive group.[23] This 5'-azide provides a good platform for subsequent click chemistry labelling and conjugation. In an enzymatic approach, Silverman and co-workers[33] developed a deoxyribozyme-catalysed labelling strategy

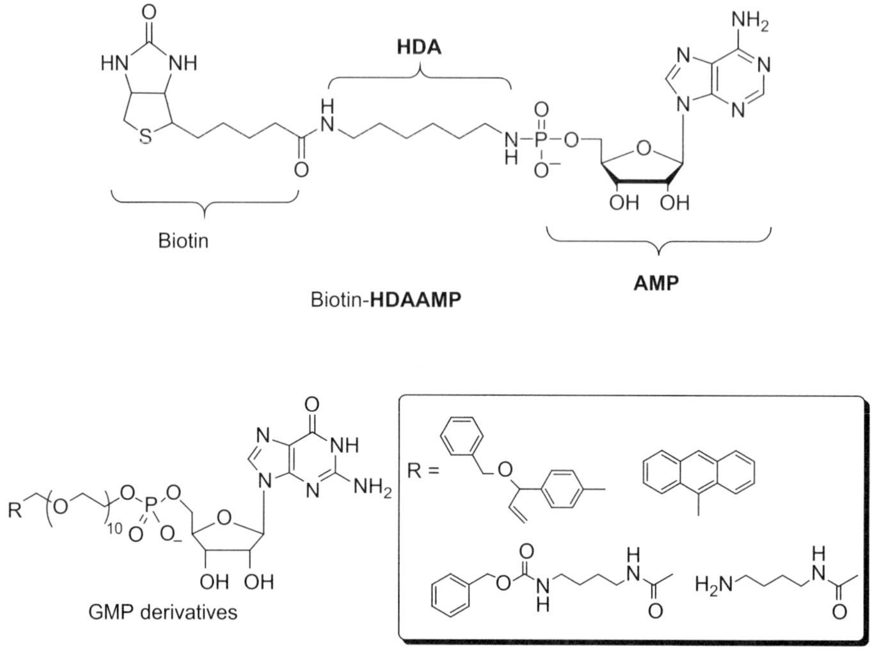

Figure 4.13 Two-step pathway for the synthesis of the transcription initiator biotin-HDAAMP.

(DECAL) for direct site-specific internal RNA modification. They used the 10DM24 deoxyribozyme, a catalytic DNA molecule, for such modification. First, a single 5-aminoallylcytidine nucleotide was incorporated at the second position of a short 'tagging RNA' by *in vitro* transcription. The aminoallyl modified 'tagging RNA' was then labelled with the desired biophysical probe using amine chemistry. Finally the tagging RNA was attached by the deoxyribozyme to an internal 2'-OH group of the target RNA. For large RNA targets, preparation of mutants is relatively cumbersome. However, the 10DM24 deoxyribozyme tolerates different biophysical labels on the tagging RNA which suggests the versatility of the DECAL strategy for applications that require covalent RNA modifications.

4.2.3   Other Synthetic Small Molecule–RNA Conjugates

As previously mentioned, RNA can be conjugated to various small molecules using the phosphoramidite chemistry. Here, we mention specific examples of therapeutically important small molecule conjugates with short interfering (si)RNA and micro- (mi)RNA. RNAi (interference) based therapeutics requires the siRNA of interest to be efficiently delivered into the cell and to be stable. One way to do this is to conjugate lipid moieties to the RNA.[32] An interesting example was created by Beal and co-workers[34] in which a new base modification in siRNA produced effective blocking of immune stimulation and provided an additional route for the development of new liver cancer therapeutics. Human microRNA 122 (miR-122) is abundantly expressed in healthy liver cells, but substantially down-regulated in hepatocellular carcinomas (HCC);[35] restoration of miR-122 levels to normal in HCC cells can reverse their tumorigenic properties.[36,37] Thus, providing the liver with a source of miR-122 in the form of a siRNA guide strand may represent an attractive therapy for the treatment of HCC. However, the miR-122 sequence contains an alternating U/G-rich motif which is recognised by immune receptors, which are primarily toll-like receptors 7 and 8 (TLR 7/8). Therefore, the authors used nucleobase analogues of adenosine and guanosine containing cyclopentyl and propyl minor groove projections (Figure 4.14) that alter the shape while maintaining the Watson–Crick base pairing effectively to decrease the interaction between the miRNA mimic and the immune receptors (TLRs). They thus synthesised a novel guanosine analogue containing an N^2-cyclopentyl group (cPent-G) and incorporated it in the guide strand of the miR-122 mimic, replacing guanosine within the putative immunostimulatory motif (U/G rich motif) of the guide strand. Similarly, 2-aminopurine containing N^2-propyl and N^2-cyclopentyl substituents (Pr-AP and cPent-AP) was incorporated in the passenger strand, replacing the adenosines at two positions opposite to the 5′ immunostimulatory motif in the guide strand. It was found that this miR-122 mimic, when administered in human peripheral blood mononuclear cells, is capable of blocking cytokine production, which is an undesirable off-target effect of RNA-based therapeutics, while maintaining the native miRNA activity.

Another interesting example of siRNA modification is the incorporation of a *North-* locked form of a nucleoside analogue, based on a carbocyclic bicyclo[3.1.0]hexane system (2′-deoxy-methanocarba nucleosides).[36] Like locked nucleic acids (LNA), the conformationally restricted bicyclo[3.1.0]hexne pseudosugars also adopt an A-type helical structure, necessary for effective gene silencing. Moreover, the 2′-deoxy-methanocarba nucleoside has an open 2′ position, allowing exploration of the chemical diversity at this site to give new *North*-MC modified siRNAs, which is not possible for LNAs. Eritja and colleagues[38] were the first to explore the replacement of the natural sugar rings in siRNAs by a *North*-2-deoxyMC thymidine derivative (T^N) (Figure 4.15),[38,39]

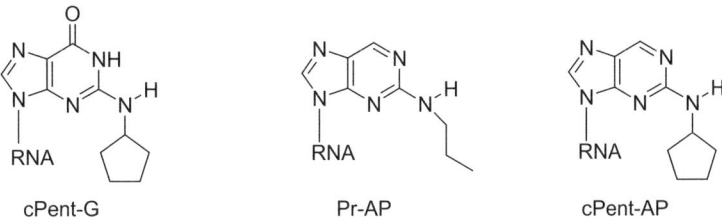

Figure 4.14 Base modification with cyclopentyl and propyl substituents.

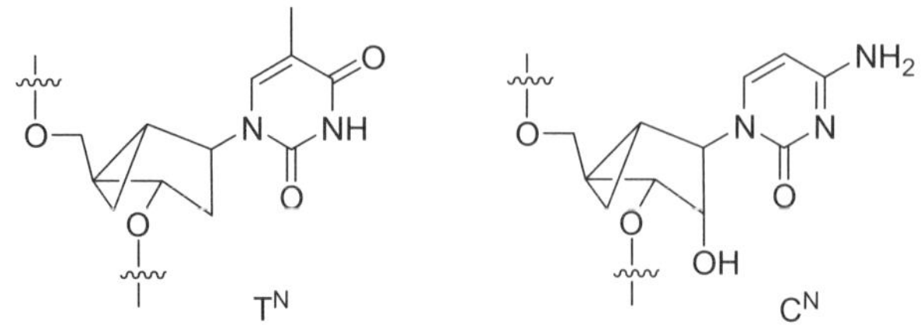

Figure 4.15   *North* methanocarba-thymidine monomer (T^N) and *North* ribo-metha-
nocarba-cytidine monomer (C^N).

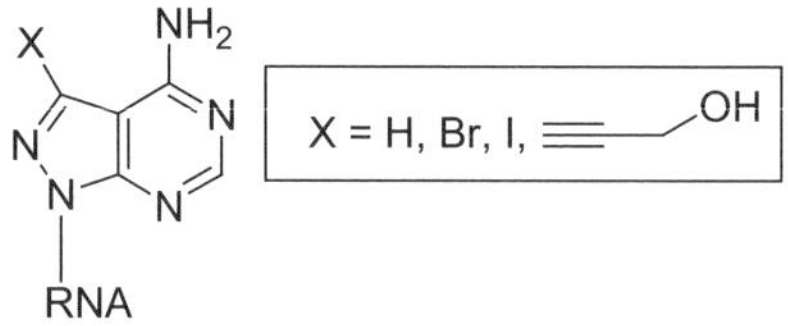

Figure 4.16   8-Aza-7-deazaadenosine derivatives.

and showed that the RNAi machinery was not impaired by this modification.
Unlike T^N, C^N has a 2′-hydroxyl group which allows further structural
alterations for potential therapeutic applications.

   In another example nucleoside modification was used to establish the
structure–activity relationship of the RNA editing by adenosine deaminase
(ADAR2) in order to understand the mechanism of conversion of adenosine to
inosine in RNA substrates.[40] To date, no RNA substrates have been obtained
in crystalline form with ADAR2. Beal and colleagues[40] used RNA substrates
containing synthetic nucleoside analogues to probe substrate recognition by
this deaminase. They introduced RNA editing substrates containing 8-aza-7-
deazaadenosine derivatives (Figure 4.16) bearing different C7 groups and
described the role of the active site structure in controlling ADAR2's reaction
with these compounds. Interestingly, it was found that the rate of deamination
of these three bulky 7-substituted analogues was much lower than that
observed for adenosine. Analysis of the crystal structure of the domain
ADAR2 with AMP modelled into the active site suggested a possible steric
clash between the bulky substituent at the purine 7-position and the side chain
of R455. Such an unfavourable interaction may cause poor fitting of the
enzyme/substrate within the active site.

4.3   Concluding Remarks

To summarise, we have discussed examples of both natural and non-natural/
synthetic RNA modifications and small molecule–RNA conjugates, which
represent the current state of research progress in the field. These types of
modification include those used to restrict certain geometries, including

backbone alteration, to enable unusual base pairing and to act as tools for probing structure–activity relationships. We further highlight examples that enhance properties of certain therapeutically important RNA-based modalities such as siRNA and miRNA mimics. Although the studies described here are introductory and explain little about the clinical efficacy, they pave the way for improvement in design of these modalities. Most importantly, the role of modified RNA nucleosides in mediating ligand-binding processes also needs to be assessed. Such studies will be necessary for the future use of such synthetic conjugates for both basic research and therapy.

Further, we briefly explored the spectrum of modifications which are highly abundant in naturally occurring RNAs and that have been known to modulate biological function. We highlight recent studies in which the repertoires of such naturally occurring RNA modifications and small molecule conjugates have been described. These reports demonstrate that such modifications can potentially alter or regulate the RNA tertiary structure and thus regulate its *in vivo* activity in different conditions. From these studies it is evident that a new paradigm in the understanding of RNA function is necessary in terms of chemical modifications. The future development of RNA-targeting drugs will rely on a deeper understanding of these modifications and conjugates. We believe that an increased understanding of RNA–ligand interactions will emerge in the near future as more structural information is obtained, which may eventually lead to the discovery of improved antibiotics, antiviral agents or other RNA-specific drugs with increased efficiency and lower toxicity.

References

1. N. Ban, P. Nissen, J. Hansen, P. B. Moore and T. A. Steitz, *Science*, 2000, **289**, 905.
2. P. A. Sharp, *Cell*, 2009, **136**, 577.
3. M. Mandal and R. Breaker, *Nature Rev. Mol. Cell Biol.*, **5**, 451.
4. X. Ji, H. Sun, H. Zhou, J. Xiang, Y. Tang and C. Zhao, *Nucleic Acid Ther.*, 2011, **3**, 185.
5. D. A. Hiller and S. A. Strobel, *Phil. Trans. R. Soc B.*, 2011, **366**, 2929.
6. C. Yi and T. Pan, *Acc. Chem. Res.*, 2011, **44**, 1380.
7. J. W. Szoztak, D. P. Bartel and L. Luisi, *Nature*, 2001, **409**, 387.
8. A. M. B. Giessing, S. S. Jensen, A. Rasmussen, L. H. Hansen, A. Gondela, K. Long, B. Vester and F. Kirpeker, *RNA.*, 2009, **15**, 327.
9. D. Mandal, C. Köhrer, D. Su, S. P. Russell, K. Krivos, C. M. Castleberry, P. Blum, P. A. Limbach, D. Söll and U. L. Rajnbhandary, *Proc. Natl. Acad. Sci. USA*, 2010, **107**, 2872.
10. Y. Ikeuchi, S. Kimura, T. Numata, D. Nakamura, T. Yokogawa, T. Ogata, T. Wada, T. Suzuki and T. Suzuki, *Nat. Chem. Biol.*, 2010, **6**, 277.
11. T. Muramatsu, K. Nishikawa, F. Memoto, Y. Kuchino, S. Nishimura, T. Miyajawa and S. Yokoyama, *Nature*, 1988, **336**, 179.
12. W. E. Kowtoniuk, Y. Shen, J. M. Heemstra, I. Agarwal and D. R. Liu, *Proc. Natl. Acad. Sci. USA*, 2009, **106**, 7768.

13. Y. G. Chen, W. E. Kowtoniuk, I. Agarwal, Y. Shen and D. R. Liu, *Nat. Chem. Biol.*, 2009, **5**, 879.
14. P. Guo, *Nature Nanotechnology*, 2010, **5**, 833.
15. E. Paredes, M. Evans and S. R. Das, *Methods*, 2011, **54**, 251.
16. S. Sasaki, K. Onizuka and Y. Taniguchi, *Chem. Soc. Rev.*, 2011, **40**, 5698.
17. P. Z. Qin and A. M. Pyle, *Methods*, 1999, **18**, 60.
18. D. Proudnikov and A. Mirzabekov, *Nucleic Acids Res.*, 1996, **24**, 4535.
19. K. Jahn, E. M. Olsen, M. M. Nielsen, T. Tørring, R. MohammadZadegan, E. S. Andersen, K. V. Gothelf and J. Kjems, *Bioconjugate Chem.*, 2011, **22**, 95.
20. G. Sengle, A. Jenne, P. S. Arora, B. Seelig, J. S. Nowick, A. Jäschke and M. Famulok, *Biorg. Med. Chem.*, 2000, **8**, 1317.
21. C. Y. Jao and A. Salic, *Proc. Natl. Acad. Sci. USA*, 2008, **105**, 15779.
22. H. Peacock, O. Maydanovych and P. A. Beal, *Org. Lett.*, 2010, **12**, 1044.
23. E. Paredes and S. R. Das, *Chem. Bio. Chem.*, 2011, **12**, 125.
24. A. H. El-Sagheer and T. Brown, *Proc. Natl. Acad. Sci. USA*, 2010, **107**, 15329.
25. M. Aigner, M. Hartl, K. Fauster, J. Steger, K. Bister and R. Micura, *Chem. Bio. Chem.*, 2011, **12**, 47.
26. S. Kellner, S. Seidu-Larry, J. Burhenne, Y. Motorin and M. Helm, *Nucleic Acids Res.*, 2011, **39**, 7348.
27. Y. Yoshimura, Y. Noguchi and K. Fujimoto, *Org. Biomol. Chem.*, 2007, **5**, 139.
28. E. Hilario, *Mol. Biotech.* 2004, **28**, 77; J. Temsamani and S. Agrawal, *Mol. Biotech.*, 1996, **5**, 223.
29. Z. Huang and J. W. Szostak, *Anal. Biochem.*, 2003, **315**, 129.
30. F. Huang, J. He, Y. Zhang and Y. Guo, *Nature Protocols*, 2008, **3**, 1848.
31. R. Fiammengo, K. Musílek and A. Jäschke, *J. Am. Chem. Soc.*, 2005, **127**, 9271.
32. C. Lorenz, P. Hadwiger, M. John, H. P. Vornlocher and C. Unverzagt, *Bioorg. Med. Chem.*, 2004, **14**, 4975.
33. D. A. Baum and S. K. Silverman, *Angew. Chem. Int. Ed.*, 2007, **46**, 3502.
34. H. Peacock, R. V. Fucini, P. Jayalath, J. M. Ibarra-Soza, H. J. Haringsma, W. M. Flanagan, A. Willingham and P. A. Beal, *J. Am Chem. Soc.*, 2011, **133**, 9200.
35. C. Coulouarn, V. Factor, J. Anderson, M. Durkin and S. Thorgeirsson, *Oncogene*, 2009, **28**, 3526.
36. S. Bai, M. W. Nasser, B. Wang, S. H. Hsu, J. Datta, H. Kutay, A. Yadav, G. Nuovo, P. Kumar and K. Ghoshal, *J. Biol. Chem.*, 2009, **284**, 32015.
37. W. C. Tsai, P. W. C. Hsu, T.-C. Lai, G. Y. Chau, C. W. Lin, C. M. Chen, C. D. Lin, Y. L. Liao, J. L. Wang, Y. P. Chau, M. T. Hsu, M. Hsiao, H. D. Huang and A.-P. Tsou, *Hepatology*, 2009, **49**, 1571.
38. M. Terrazas, S. M. Ocampo, J. C. Perales, V. E. Marquez and R. Eritja, *Chem. Bio. Chem.*, 2011, **12**, 1056.
39. M. Terrazas, A. Aviñó, M. A. Siddiqui, V. E. Marquez and R. Eritja, *Org. Lett.*, 2011, **13**, 2888.
40. S. Pokharel, P. Jayalath, O. Maybanovych, R. A. Goodman, S. C. Wang, D. J. Tantillo and P. A. Beal, *J. Am. Chem. Soc.*, 2009, **131**, 11882.

CHAPTER 5

Click Chemistry – a Versatile Method for Nucleic Acid Labelling, Cyclisation and Ligation

AFAF H. EL-SAGHEER[a,b] AND TOM BROWN*[a]

[a] School of Chemistry, University of Southampton, Highfield, Southampton SO17 1BJ, UK; [b] Chemistry Branch, Dept. of Science and Mathematics, Faculty of Petroleum and Mining Engineering, Suez Canal University, Suez, 43721, Egypt
*Email: tb2@soton.ac.uk

5.1 Introduction to Click Chemistry

Click chemistry is a new concept that was developed to provide an atom-efficient method to join together organic molecules in high yield under mild conditions in the presence of a diverse range of functional groups.[1] The best example of click chemistry is the Cu(I) catalysed [3 + 2] azide–alkyne cycloaddition reaction, known as the CuAAC reaction.[2,3] This new reaction is a highly efficient variant of the original Huisgen [3 + 2] cycloaddition which has been used in organic chemistry for more than 50 years.[4] The CuAAC reaction has found great popularity in the nucleic acids field and has been the subject of recent reviews.[5,6] It has been utilised as a method of joining together single strands of DNA,[7] cross-linking complementary strands,[8] cyclising single and

RSC Biomolecular Sciences No. 26
DNA Conjugates and Sensors
Edited by Keith R Fox and Tom Brown
© The Royal Society of Chemistry 2012
Published by the Royal Society of Chemistry, www.rsc.org

double strands,[7,9] labelling oligonucleotides with reporter groups,[10] immobilising DNA on surfaces,[11] constructing nanowires from DNA templates,[12] building DNA nanostructures,[13] producing analogues of DNA with modified nucleobases[14,15] and backbones,[16–21] synthesising chemically modified RNA constructs,[22,23] monitoring DNA synthesis *in vivo*[24] and creating biologically active PCR templates.[25] Several of the above applications will be discussed in this chapter, and other click reactions that have been applied to nucleic acids will also be described. This is a large and growing field, so this chapter cannot constitute a comprehensive review. It is a description of various areas of nucleic acid chemistry in which click reactions have made a significant impact.

5.2 The CuAAC Reaction for Oligonucleotide Labelling

Research on new methods to produce chemically labelled nucleic acids has been an intense field of activity ever since techniques to synthesise oligonucleotides were first invented.[26,27] The most important examples of oligonucleotide labels are fluorescent dyes, which continue to be essential to the development of platform technologies such as DNA sequencing and genetic analysis.[28–30] Historically, the most common method of introducing chemically sensitive labels into synthetic oligonucleotides has been amide bond formation between amino-modified oligonucleotides and active ester derivatives of the labels. Although this reaction remains a very important source of fluorescent oligonucleotides it has some drawbacks. Active esters are not stable in the buffers that are used to maintain the pH of the labelling reaction above 8.0. In addition they are sensitive to moisture and are susceptible to decomposition on storage. Alternative labelling methods such as the reaction of thiol-labelled DNA with maleimide or iodoacetamide derivatives of dyes (Michael-type reactions) suffer from similar problems due to the instability of the electrophile and the tendency of thiol-modified oligonucleotides to dimerise. Recently the field received a boost when the copper catalysed alkyne–azide cycloaddition reaction (CuAAC reaction) was introduced as a method of oligonucleotide labelling.[10,31] The reaction between alkyne-labelled oligonucleotides and azide derivatives of various reporter groups (Figure 5.1) proceeds in high yield in aqueous buffer in conditions under which the azide and alkyne are otherwise completely stable. In addition to its valuable contribution to DNA labelling, the CuAAC reaction has also been used to add fluorescent tags to alkyne-functionalised RNA.[23] Alkyne-modified oligonucleotides are simple to synthesise using phosphoramidite monomers derived from simple terminal alkynes, and such alkynes are stable under oligonucleotide synthesis and deprotection conditions. Moreover the alkyne moiety can be attached to nucleobases without perturbing base pairing or duplex stability, offering the possibility of internally labelling oligonucleotides.[32,33] The common points of attachment to the nucleobases are the 5-position of pyrimidines and the 7-position of 7-deazapurines.

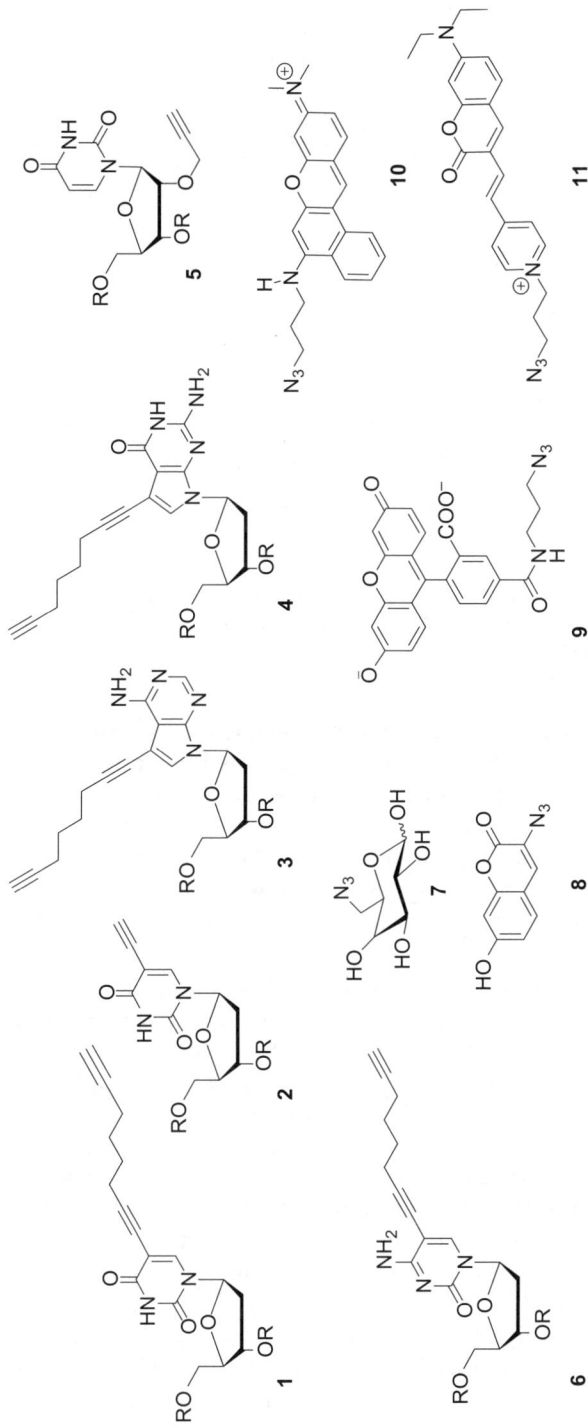

Figure 5.1 Examples of alkyne-modified nucleosides that can be incorporated into oligonucleotides by solid-phase synthesis and azide labels that have been used in click labelling of oligonucleotides by the CuAAC reaction.[10,32–37]

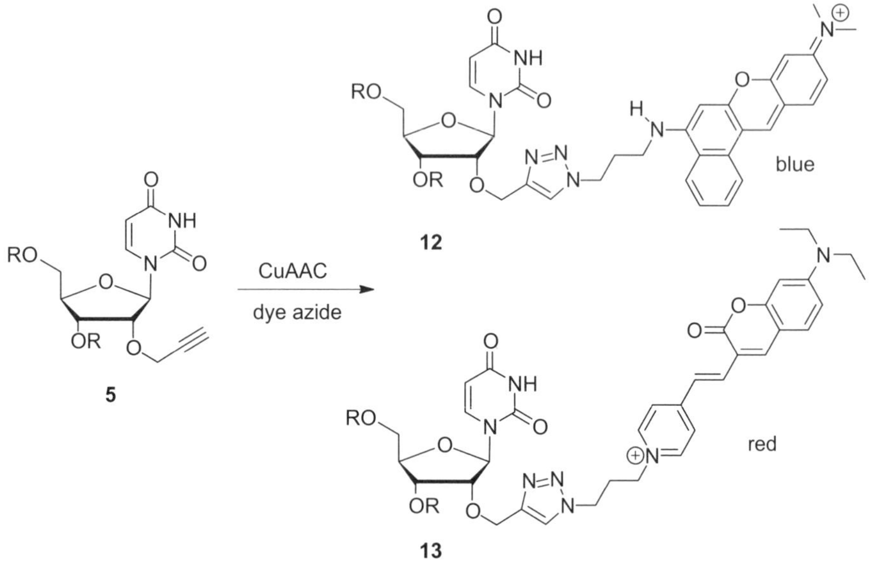

Figure 5.2 Labelling oligonucleotides at the 2′-position of the ribose sugar with fluorescent dyes that are unstable to the conditions of oligonucleotide deprotection.[34]

Post-synthetic labelling is particularly valuable for dyes that are difficult to synthesise and/or are expensive, because very small quantities are required in the labelling reaction (typically <1 mg). Moreover, this is the only viable method to label synthetic DNA with dyes that are not stable to the conditions of oligonucleotide deprotection, and cannot therefore be added as phosphoramidite monomers during solid-phase synthesis. Click chemistry has been used to advantage in this context by a number of workers, including Wagenknecht (Figure 5.2).[34]

One drawback of any post-synthetic DNA labelling method is the lack of control, *i.e.* it is difficult to introduce different labels at specific predefined loci within a particular oligonucleotide strand. This problem has been solved for click labelling by incorporating an unprotected and two differentially protected terminal alkynes into the DNA sequence (Figure 5.3) during solid-phase synthesis (a procedure named 'click-click-click'). After oligonucleotide assembly the unprotected alkyne (octadiynyl dU, **1**) is labelled with the first dye (as a suitable azide-derivative) either on-resin or post-synthetically, then the second alkyne (trimethylsilyl octadiynyl dC, **14**) is deprotected on the resin by treatment with acetic acid, or post-synthetically during the ammonia cleavage step, after which the second label is added (again as an azide-derivative). The third alkyne (tri-isopropylsilyl dC, **15**) is then deprotected post-synthetically with tetrabutylammonium fluoride before click labelling. The procedure is high yielding and suitable for labels that are unstable to the oligonucleotide synthesis conditions. In addition to the 5-octadiynyl derivatives of dU and dC, two

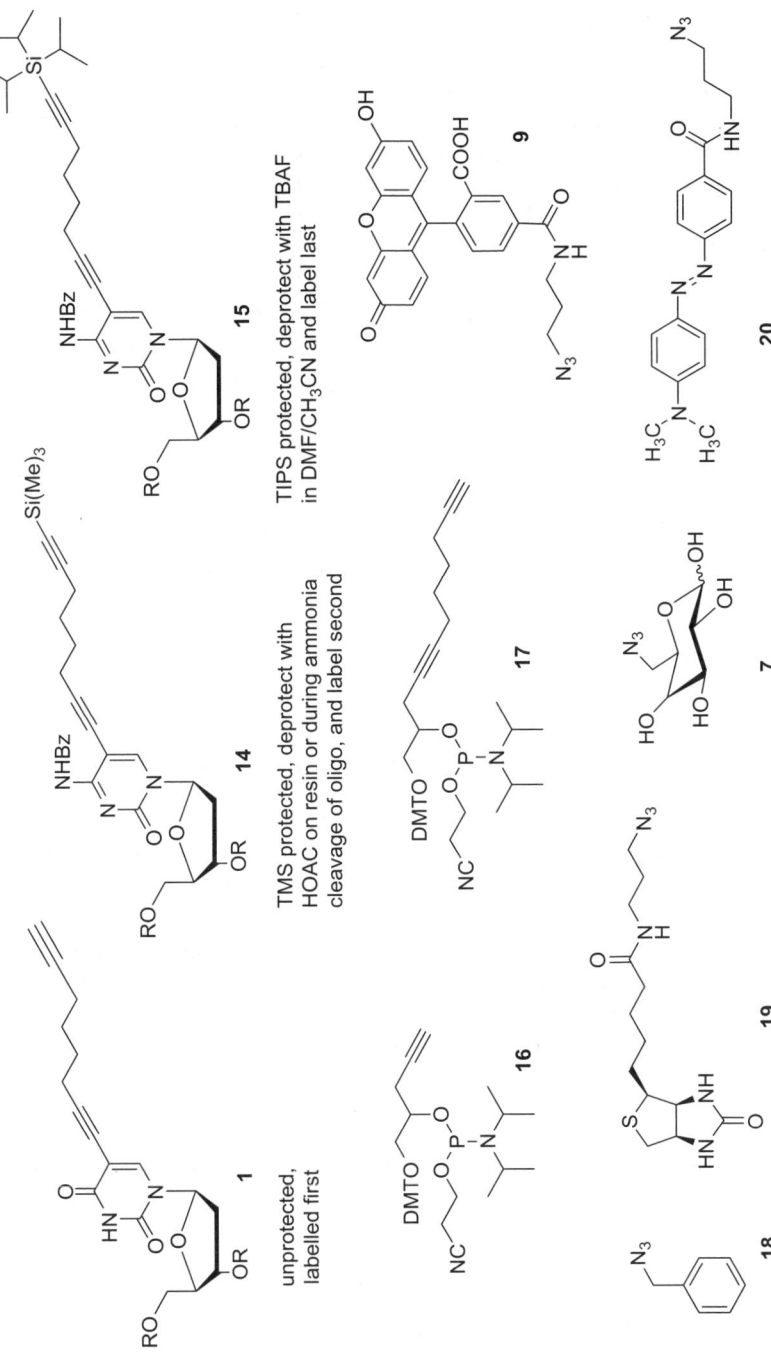

Figure 5.3 'Click-click-click' labelling of oligonucleotides using differential protection of alkynes. Benzyl, biotin, dabcyl and galactose azides were used amongst others.[38]

new non-nucleosidic alkyne phosphoramidite monomers were also in used in this study (**16** and **17**). Examples of azides that have been added by this procedure are shown in Figure 5.3.[35–38]

Click labelling of DNA is not restricted to applications involving synthetic oligonucleotides. Alkynes can also be introduced into long biochemically synthesised DNA strands during PCR via alkyne-modified deoxynucleoside triphosphates (dNTPs). The resultant amplicons can be labelled with azide derivatives of various reporter groups,[36,39] and in this way large numbers of fluorescent dyes can be introduced throughout the sequences of long DNA constructs. The efficiency of incorporation of the modified dNTPs depends upon several factors including the precise PCR conditions, the mode of attachment of the alkyne to the nucleobase and the nature of the polymerase. This approach to DNA labelling is successful because alkyne-labelled dNTPs are good substrates for DNA polymerases. This is because terminal alkynes are small and sterically undemanding, so they do not greatly inhibit the binding of the dNTPs to DNA polymerases. In contrast, the direct use of bulky dye-labelled dNTPs in PCR is not a good strategy for labelling amplicons. It results in low levels of dye incorporation, and if high ratios of dye-labelled dNTPs to unlabelled dNTPs are used to compensate for this shortcoming, the PCR amplification becomes very inefficient. The above labelling method is not restricted to fluorescent dyes; it can be used to functionalise DNA with many reporter groups. In principle the procedure can also be carried out using azide derivatives of dNTPs and alkyne-functionalised dyes (see following section).

5.3   Reverse Click Labelling of Oligonucleotides by the CuAAC Reaction

The click labelling reaction can be carried out in reverse, *i.e.* alkyne-modified dyes can be conjugated to azide-labelled oligonucleotides. However, this is more complicated because stable azide-containing phosphoramidite monomers for use in oligonucleotide synthesis cannot be prepared. This is because azides react with P(III) (Staudinger reaction); consequently the azide monomers decompose as they are being synthesised or purified. However, an alternative strategy exists. Functional groups such as alkyl iodides, bromides or mesylates can be introduced into oligonucleotides during solid-phase synthesis and subsequently converted to azides by reaction with NaN_3 which is a good nucleophile (Figure 5.4).[40] In addition, it is possible to introduce azides to the 3′-end of oligonucleotides during solid-phase synthesis by means of an azide-modified nucleoside attached to the solid support.[41] Azide-functionalised nucleosides can also be added to oligonucleotides using H-phosphonate[42] or phosphotriester chemistry.[43,44] These strategies utilise phosphoramidite chemistry for the addition of all other monomers during oligonucleotide synthesis, and therefore demonstrate that azides are stable in the presence of the large excess of phosphoramidite used in the repeated coupling steps. This is because the phosphoramidite group is protonated by the mildly acidic coupling reagent

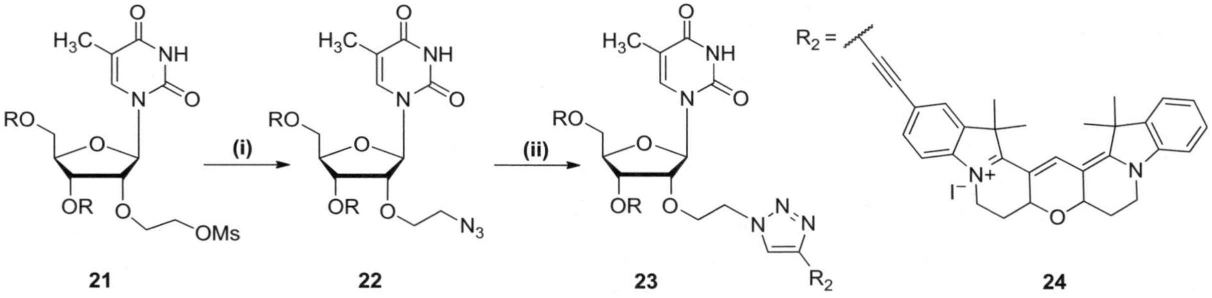

Figure 5.4 Reverse click labelling using oligonucleotide containing 2′-mesyl dT and an alkyne Cy-Dye, in this example Cy3B. Conversion of 2′-mesyloxyethyl ribothymidine (**21**) to 2′-azidoethyl ribothymidine (**22**). Conditions: i) NaN₃, DMF, 65 °C, 20 h, ii) alkyne Cy-Dye (**24**): CuSO₄, sodium ascorbate, tris-hydroxypropyl triazole ligand,[45] DMSO, 55 °C, 2 h to give dye-labelled oligonucleotide R = DNA strand.[40]

(*e.g.* tetrazole) and behaves as an electrophile, not a nucleophile. The reverse click labelling approach is useful in cases when the alkyne derivatives of fluorescent dyes or other reporter groups are more readily accessible than the azide derivatives, as in the case of some Cy-dyes (*e.g.* **24**, Figure 5.4).[40] In this study a range of Cy-dyes were prepared as alkyne derivatives and conjugated by click chemistry to 2′-azidoethyl groups in oligonucleotides, both on the synthesis resin and in solution.

5.4 DNA Labelling *in vivo*

A procedure has been developed to detect DNA synthesis in proliferating cells (Figure 5.5). It is based on the incorporation of 5-ethynyl-2′-deoxyuridine (**25a**) into genomic DNA and its subsequent derivatisation with a fluorescent azide [fluorescein (**26**), tetramethylrhodamine or Alexa dyes] to give fluorescently labelled DNA (**27a**). The method, which utilises the CuAAC reaction, is highly sensitive because the small fluorescent azides used for detection give a high degree of specimen penetration. In contrast to existing methods this new approach does not require sample fixation or DNA denaturation, and can be used in cultured cells or in whole animals.[24] The same approach has been adapted to detect RNA synthesis in cells by incorporating 5-ethynyluridine into newly transcribed RNA (**25b**, Figure 5.5). An incorporation rate of around one label every 35 uridine residues was achieved.[46] Transcription rates in whole animals were measured and found to vary greatly among different tissues and among different cell types, giving insights into this important biological process. The methodology probably cannot be used in living cells for extended periods because of the toxicity of copper, but it is a quick procedure, so this does not appear to be a major drawback.

An alternative procedure for labelling RNA has been developed using 5-azidopropyl UTP (**28**), a DNA template and T7 RNA polymerase, allowing

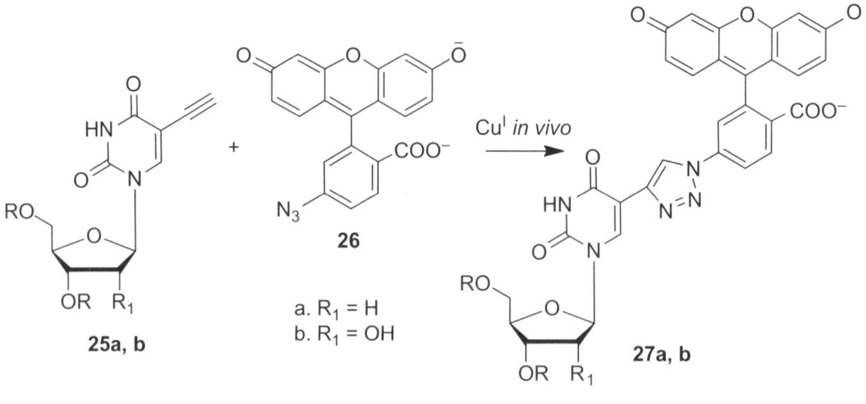

Figure 5.5   A method for click DNA and RNA labelling in cells and whole animals using 5-ethynyl dU and a fluorescent azide.[24,46]

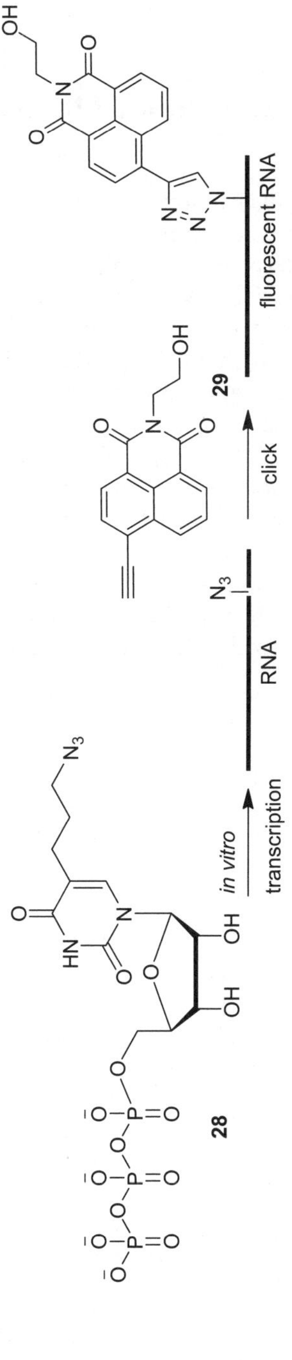

Figure 5.6 Click RNA labelling using T7 RNA polymerase, 5-azidopropyl UTP and a fluorescent alkyne.[47]

fluorescent labelling of RNA transcripts by click chemistry using a fluorescent alkyne (**29**). The authors also suggested the possibility of posttranscriptional labelling of cellular RNAs using this approach, although this has not yet been carried out (Figure 5.6).[47]

5.5 Other Click Reactions for Oligonucleotide Labelling

The 1,3-dipolar cycloaddition reaction between styrene- or norbornene-modified DNA and nitrile oxide derivatives of labels has been used to label oligonucleotides.[48,49] A styrene-derivatised dU phosphoramidite monomer (**30**) was incorporated during oligonucleotide synthesis and was reacted with various nitrile oxides, either on the synthesis column or post-synthetically, to form 3,5-disubstituted oxazolines such as **32** (Figure 5.7). The styrene was also introduced as a deoxyuridine triphosphate derivative (**31**) during PCR using KOD-XL polymerase, making it possible to label hundreds of sites in a 900-mer PCR product. It was also possible to combine the nitrile oxide-based modification of styrene oligonucleotides with the Cu(I)-catalysed alkyne–azide click reaction. A 300 bp PCR fragment was synthesized by using a mixture of the styrene triphosphate and the alkyne triphosphate, and a total of 154 thymidines were exchanged by alkyne and styrene thymidine derivatives in a statistical fashion. To introduce two different labels the Cu-catalysed click reaction was first performed using sugar azide (**7**). The sugar-modified PCR product was

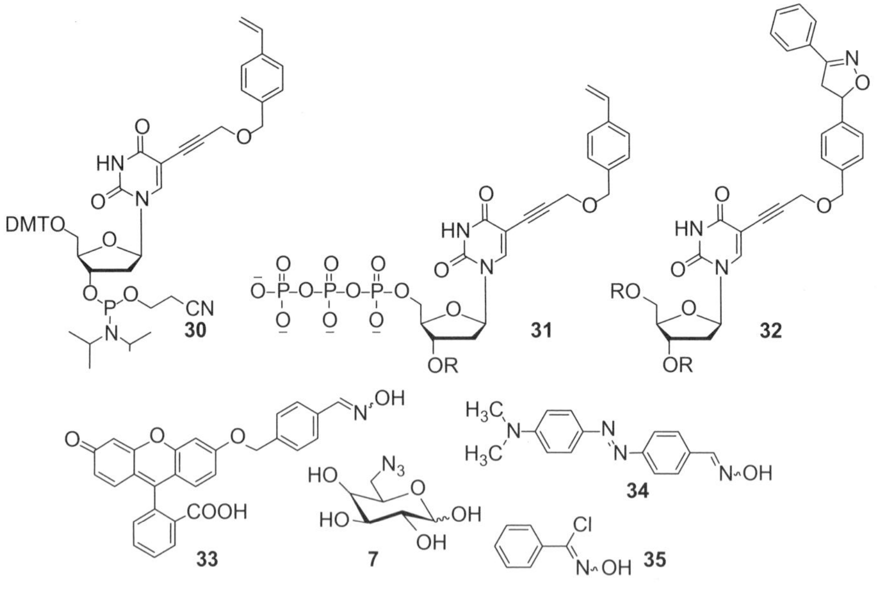

Figure 5.7 The click reaction between Styrene-derivatised DNA and nitrile oxides for DNA labelling.[49]

purified by ethanol precipitation and the nitrile oxide click reaction was then carried out using the nitrile oxide precursor **35** (Figure 5.7). The two click reactions were shown to be truly orthogonal, and remarkably no unreacted alkyne or styrene was found in the labelled amplicon.[49] Oximes such as **33** and **34** were also used in this study.

5.6 Click Chemistry for Oligonucleotide Strand Ligation

The enzymatic joining together of DNA strands (ligation) to produce a phosphodiester linkage is an essential biological process that occurs during DNA repair and replication. It is also used extensively as a tool in molecular biology. Purely chemical processes to carry out the same transformation have been developed previously,[50–52] but they have not been routinely adopted. Recently it has been shown that DNA strands can be joined together efficiently by the CuAAC reaction to give an artificial triazole backbone at the point of ligation (Figure 5.8A).[7,44] Interestingly careful design has produced a DNA backbone

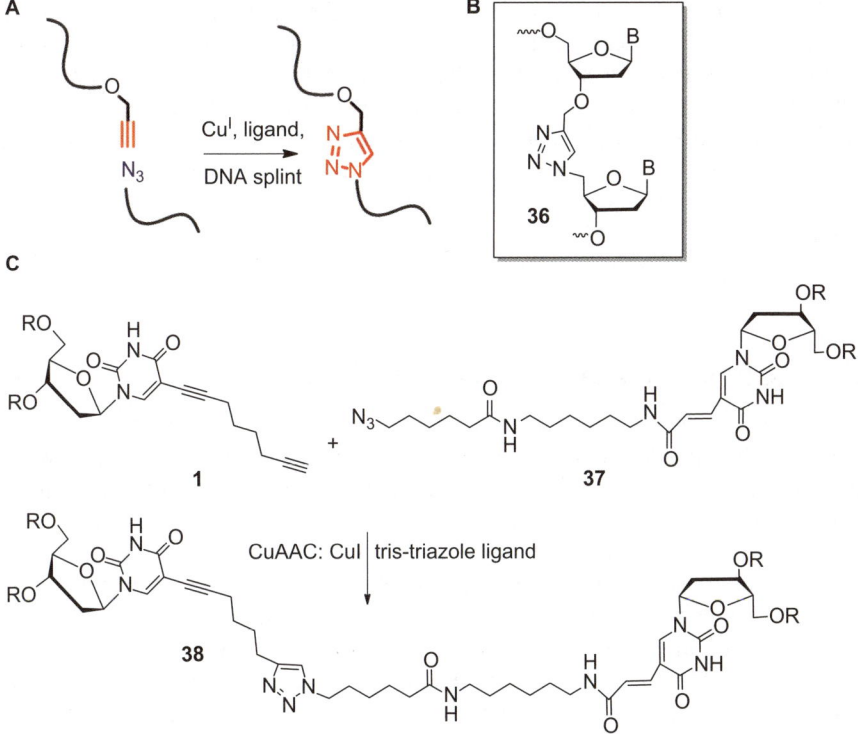

Figure 5.8 A: Cis-click DNA–RNA ligation. B: Biocompatible triazole backbone. C: Trans-click ligation. R = DNA or RNA strand.[22,25]

mimic that can be read accurately by DNA and RNA polymerases[25,53] and is functional *in vivo* (**36**, Figure 5.8B).[25] The surprising biocompatibility of the artificial triazole linkage, which bears no obvious resemblance to a phospho-diester group, has been explained by use of nuclear magnetic resonance (NMR),[54] and suggests potential applications in the field of gene synthesis. The advantage of chemical ligation over enzymatic ligation is that it can be carried out on a large scale, and it is compatible with a wide range of chemically modified nucleic acid analogues. RNA strands can also be ligated by the CuAAC reaction, either by intra-strand or inter-strand crosslinking (Figure 5.8C).[22] Ribozymes assembled in this manner are biologically active even if the modification is placed near the active site. Click chemistry is a promising approach to the synthesis of large RNA constructs or RNA analogues that are otherwise inaccessible by conventional solid-phase synthesis.

Das has developed methods of adding azide groups to the 5'- and 3'-ends of RNA. 5'-Azido RNA was made by transcription using T7 RNA polymerase, a DNA template and a 4:1 excess of 5'-azidoguanosine (**39**) over the normal dNTPs. Addition of a 3'-azide to single-stranded RNA was accomplished using 3'-azido-2',3'-dideoxyadenosine (**40**) and the enzyme poly(A) polymerase. The azide-functionalised RNA strands were used in click ligation reactions to yield two different artificial RNA backbones (**41** and **42** in Figure 5.9).[23]

The CuAAC reaction has been used to ligate artificial mimics of DNA. A highly efficient chemical ligation method has been developed for quantitative conjugation of PNA with DNA or PNA with PNA.[55] PNA oligomers with an alkyne at the C-terminus and an azide at the N-terminus were used in DNA-templated ligation reactions, enabling the synthesis of extended PNA sequences which are difficult to prepare by standard solid-phase methods. Templated PNA click ligation is sequence specific and capable of single nucleotide discrimination.

5.7 Click Chemistry in Nanotechnology

DNA is a valuable material in the field of nanotechnology owing to its ease of synthesis, highly selective assembly properties and addressability (*e.g.* by triplex formation and sequence-specific protein binding). It has been used as a scaffold to build nanostructures of great diversity and complexity for use in a wide variety of applications.[56] However, DNA nanoconstructs are thermo-dynamically unstable unless they are kept in a controlled environment of ambient temperature and aqueous buffer. This makes the purification of complex DNA nanostructures difficult or impossible, and limits their utility. To solve this problem a click-fixation technology based on the CuAAC reaction has been developed to produce stable DNA nanoconstructs which can be subsequently purified under denaturing conditions (Figure 5.10).[13] This approach could be expanded to build much larger addressable DNA nano-scaffolds in a hierarchical manner.

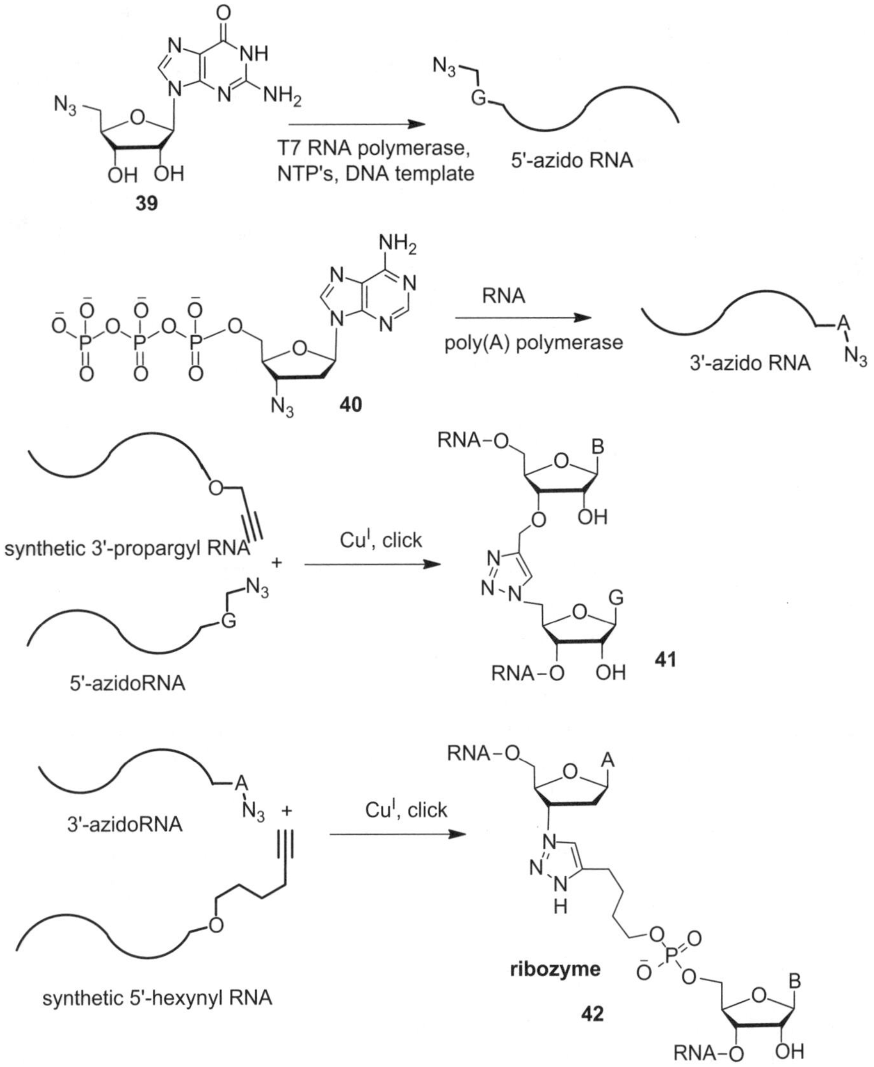

Figure 5.9 A: Synthesis of azide-functionalised RNA and click RNA ligation to produce artificial RNA backbone linkages at the ligation sites.[23]

Higher order DNA structures have also been the focus of click chemistry. Triplex DNA binders have been used in nanotechnology applications as external stimuli for controlling chemical reactivity in DNA-directed reactions. Efficient double click reactions can be carried out using a strong triplex binder such as a naphthylquinoline to create three-way branched non-symmetrical DNA nanostructures.[57] In addition, conformational isomers of G-quadruplexes have been trapped by performing click reactions between their termini,[58] leading to the discovery of a DNA–RNA hybrid G-quadruplex derived

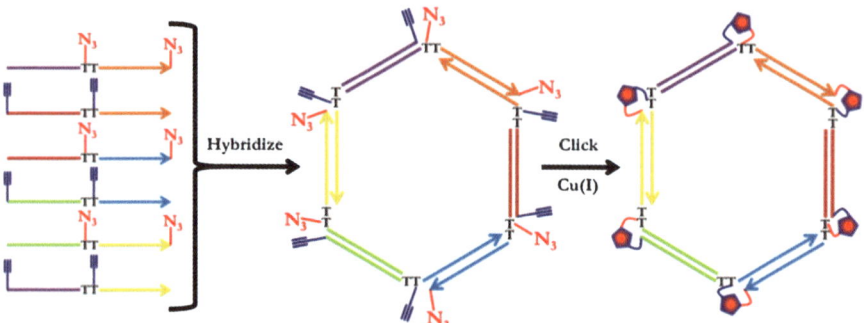

Figure 5.10   Fixation of a hexagonal DNA nanoconstruct using the CuAAC reaction. The nanostructure is cross-linked in six simultaneous click reactions, then purified by denaturing gel-electrophoresis. Each side of the hexagon is one turn of B-DNA.[13]

from the sequence of the human telomere. The click reaction allowed identification of the G-quadruplex reaction products in a complex solution. Traditional methods for the analysis of conformational variations in G-quadruplexes (such as NMR spectroscopy and X-ray crystallography) are not ideal because of the existence of equilibria between different quadruplex conformations. An efficient template-free technique for the inter-strand cross-linking of DNA has been developed using a 'bis-click' CuAAC reaction between alkyne-functionalised oligonucleotides and bifunctional azides.[59,60] The methodology is applicable to single-stranded DNA, duplexes and multistranded structures. Four-stranded DNA structures consisting of two cross-linked duplexes were obtained after hybridisation of the bis-click constructs. This is one of the many important contributions made to the field by the Seela group.[14,31,33,59–64]

5.8   Synthesis of Cyclic DNA

The CuAAC reaction has been used to synthesise very small cyclic DNA constructs from hairpin oligonucleotides functionalised with a 5′-terminal alkyne, a 3′-azide and containing a hexaethylene glycol loop.[9] As expected, they have very high thermodynamic stability as determined by UV melting. An NMR study on a cyclic GC–GC dinucleotide duplex (**43**, Figure 5.11) revealed the presence of stable hydrogen bonding between the base pairs. The structure with a single base pair (**44**) did not form inter-base hydrogen bonds, presumably as a result of the absence of stabilisation by base stacking. In a study on DNA drug binding, the mode of action of a novel threading intercalator was elucidated using a click-ligated cyclic duplex with both ends sealed to prevent entry of the DNA-binding molecule from the termini of the duplex.[65] Cyclic duplexes are of interest as potential therapeutic agents owing to their resistance to degradation in biological media, their high duplex stability and their small size relative to conventional duplexes. Large molecules do not enter cells

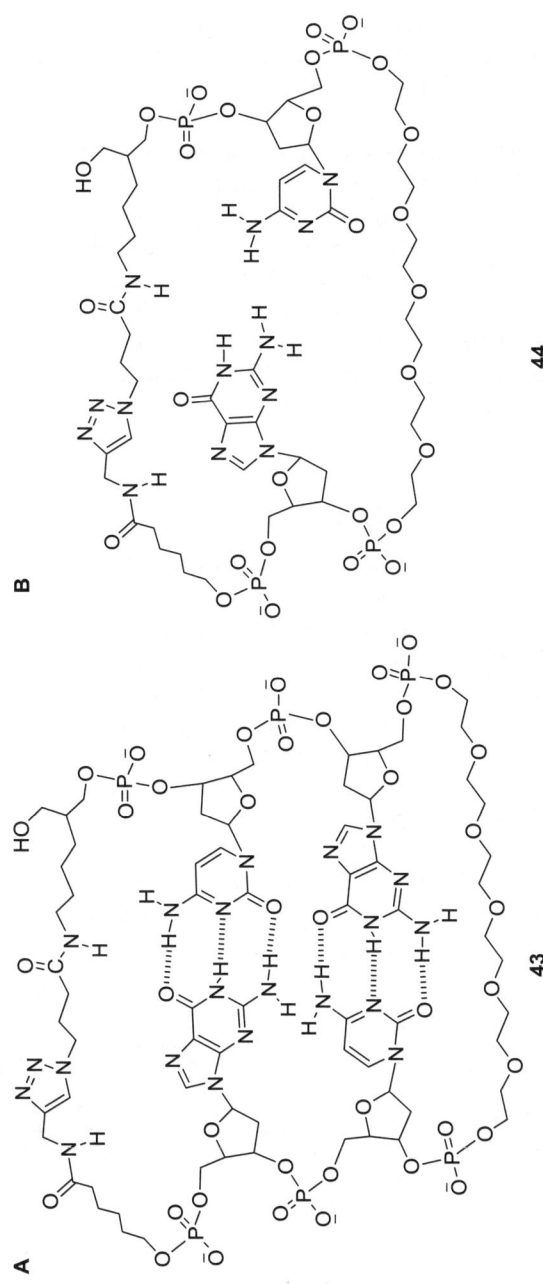

Figure 5.11 A: Stable cyclic GC/GC dinucleotide mini-duplex synthesised by click cyclisation of a hexaethylene glycol-linked hairpin loop. B: Cyclic construct with unstable GC base pair.[9]

efficiently, so methods to miniaturise therapeutic oligonucleotides are of great interest. In this context cyclic mini DNA duplexes have been investigated as decoys for transcription factors.[66] Alternative designs of cyclic dumbbell oligonucleotides have been synthesised by click chemistry and shown to have high duplex melting temperatures and strong resistance to degradation by snake venom phosphodiesterase. These oligonucleotides were designed to bind to the nuclear factor (NF)-κB p50 homo-dimer as *in vivo* decoys.[67]

At the other extreme of size, click chemistry has been applied to the assembly of a double-stranded DNA catenane consisting of six turns of B-DNA from two complementary synthetic oligonucleotides.[7] A stepwise synthetic strategy was adopted (Figure 5.12) in which each oligonucleotide was labelled with a 5′-alkyne and a 3′-azide. One oligonucleotide was cyclised in a non-templated CuAAC reaction and was used as a template to cyclise the second oligonucleotide in a templated click reaction. Tandem T.T mismatches were placed at intervals of 10 base pairs in the cyclic duplex to produce points of flexibility, without which the duplex would be too rigid to bend into a circle.

5.9   The SPAAC Reaction on DNA

The use of the CuAAC reaction in oligonucleotide labelling requires care because nucleic acids are unstable in the presence of Cu(I) and oxygen. The problem is alleviated by the use of a Cu(I)-binding ligand, normally a tris-triazole which protects the DNA from degradation and also accelerates the click reaction.[45] The addition of acetonitrile to the labelling buffer and exclusion of oxygen have also been shown to reduce DNA degradation.[23] However, despite the spectacular success of the CuAAC reaction in the DNA and RNA field, Cu(I) has one particular disadvantage. It is cytotoxic,[68] and is therefore not compatible with *in vivo* applications of nucleic acids, or for use in experiments that involve living cells. Unfortunately the copper-free DNA-templated AAC Huisgen reaction with terminal alkynes is exceedingly slow[69] unless highly activated alkynes are used, but these are unstable in aqueous media so

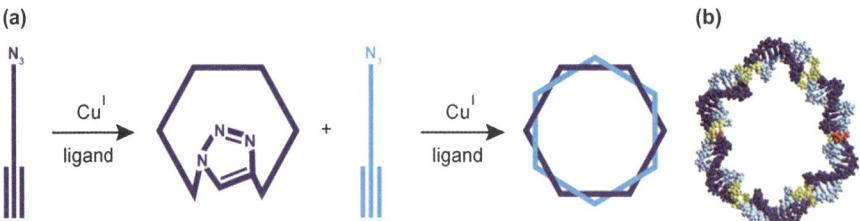

Figure 5.12   **a**: Formation of double-stranded DNA catenane (double-stranded 70-mer) from a cyclic template oligonucleotide and linear complementary strand. **b**: A double stranded DNA catenane in which T.T hinges are shown in yellow, click-ligated regions in red and the two DNA strands in light and dark blue.[7]

they cannot be used in biology. For carbohydrates and other biomolecules the issue of biocompatibility has been elegantly solved by the development of the ring strain-promoted azide–alkyne [3 + 2] cycloaddition reaction (SPAAC).[68,70–72] This is quite different from the CuAAC reaction because it involves the reaction between an azide and a strained alkyne, normally a cyclooctyne derivative, and is promoted by distortion of the sp-orbitals of the alkyne which cannot adopt their preferred 180° angle. Reaction to form a triazole (sp^2 hybridisation, preferred angle 120°) allows release of energy accompanied by the formation of a stable product. The strained eight-membered ring provides a compromise between good reactivity with azides and high stability in aqueous media. Smaller rings are too unstable and larger rings are not sufficiently reactive. Analogues of cyclooctyne have been synthesised which are modified in order to increase reactivity.[70,71] The SPAAC reaction has recently been adopted for nucleic acids, both in oligonucleotide labelling[73,74] and in copper-free DNA strand ligation (Figure 5.13).[75] In the context of DNA it is a very fast and clean reaction, but there is one minor disadvantage when compared with the CuAAC reaction; the synthesis of strained cyclooctynes is more complex than that of terminal alkynes. Unlike the CuAAC reaction which produces exclusively 1,4-triazoles, the SPAAC reaction can give

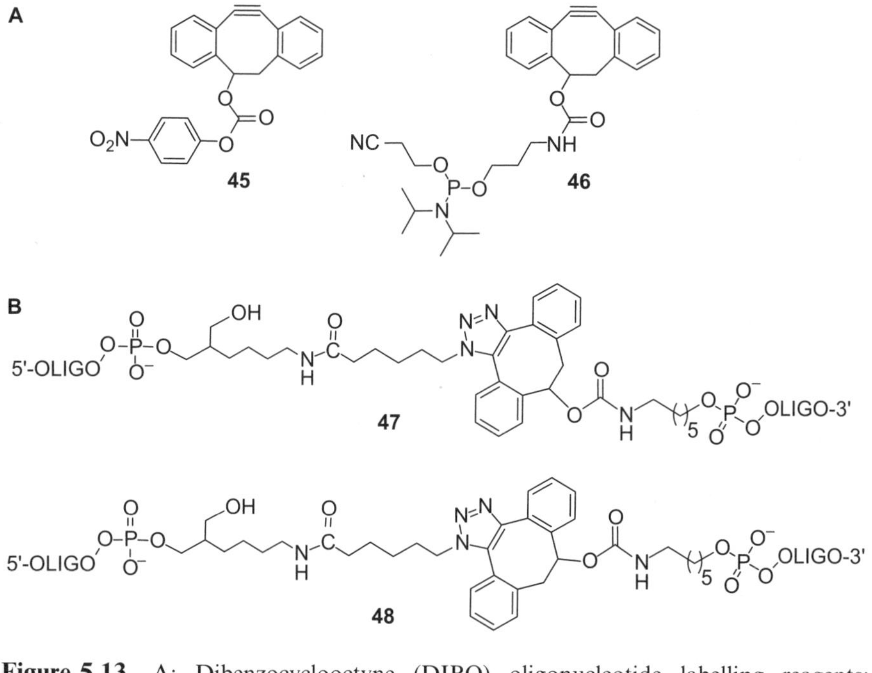

Figure 5.13 A: Dibenzocyclooctyne (DIBO) oligonucleotide labelling reagents: DIBO p-nitrophenyl carbonate (**45**) and phosphoramidite monomer for 5′-oligonucleotide addition (**46**). B: Click linkage between DNA strands formed from oligonucleotides with 3′- azide and 5′-DIBO. Both regioisomers of the dibenzocyclooctyl triazole are shown.[75]

rise to two regioisomeric products, shown for dibenzocyclooctyne in Figure 5.13 (**47**, **48**).

Dibenzocyclooctyne (DIBO) is remarkably stable to the standard conditions of oligonucleotide deprotection (5 h at 55 °C in concentrated aqueous ammonia), so it is suitable for incorporation into DNA as a phosphoramidite monomer.[75] It is noteworthy that this strained alkyne is very stable in the presence of nucleophiles, yet reacts quantitatively with an azide within 5 min at room temperature if the components are held together by DNA templating. The Diels–Alder reaction has recently been used for DNA strand ligation and this too is a very fast reaction.[76]

5.10   Conclusion

Click chemistry has been used to great advantage in the nucleic acid field for *in vitro* and *in vivo* labelling and strand ligation. The applications are growing rapidly as orthogonal click reactions are discovered and developed. It is highly likely that new click reactions will be discovered, allowing even more ambitious and creative chemistry to be carried out on nucleic acids and applied in biology and nanotechnology. Of particular interest is the prospect of carrying out such chemical reactions in live cells.

Acknowledgements

AHE-S and TB have received funding from the European Union's Seventh Framework Programme (FP7/2007-2013) under grant agreement N° 201418 (READNA) and from the UK BBSRC, for the sLoLa project "extending the boundaries of nucleic acid chemistry."

References

1. H. C. Kolb, M. G. Finn and K. B. Sharpless, *Angew. Chem. Int. Edit.*, 2001, **40**, 2004–2021.
2. C. W. Tornoe, C. Christensen and M. Meldal, *J. Org. Chem.*, 2002, **67**, 3057–3064.
3. V. V. Rostovtsev, L. G. Green, V. V. Fokin and K. B. Sharpless, *Angew. Chem. Int. Edit.*, 2002, **41**, 2596–2599.
4. R. Huisgen, *Angew. Chem. Int. Edit.*, 1963, **2**, 633–645.
5. F. F. Amblard, J.-H. Cho and R. F. Schinazi, *Chem. Rev.*, 2009, **109**, 4207–4220.
6. A. H. El-Sagheer and T. Brown, *Chem. Soc. Rev.*, 2010, **39**, 1388–1405.
7. R. Kumar, A. H. El-Sagheer, J. Tumpane, P. Lincoln, L. M. Wilhelmsson and T. Brown, *J. Am. Chem. Soc.*, 2007, **129**, 6859–6864.
8. P. Kocalka, A. H. El-Sagheer and T. Brown, *Chembiochem.*, 2008, **9**, 1280–1285.

9. A. H. El-Sagheer, R. Kumar, S. Findlow, J. M. Werner, A. N. Lane and T. Brown, *Chembiochem.*, 2008, **9**, 50–52.

10. J. Gierlich, G. A. Burley, P. M. E. Gramlich, D. M. Hammond and T. Carell, *Org. Lett.*, 2006, **8**, 3639–3642.

11. D. I. Rozkiewicz, J. Gierlich, G. A. Burley, K. Gutsmiedl, T. Carell, B. J. Ravoo and D. N. Reinhoudt, *Chembiochem.*, 2007, **8**, 1997–2002.

12. M. Fischler, U. Simon, H. Nir, Y. Eichen, G. A. Burley, J. Gierlich, P. M. E. Gramlich and T. Carell, *Small.*, 2007, **3**, 1049–1055.

13. E. P. Lundberg, A. H. El-Sagheer, P. Kocalka, L. M. Wilhelmsson, T. Brown and B. Norden, *Chem. Commun.*, 2010, **46**, 3714–3716.

14. F. Seela and X. Ming, *Helv. Chim. Acta.*, 2008, **91**, 1181–1200.

15. P. Kocalka, N. K. Andersen, F. Jensen and P. Nielsen, *Chembiochem.*, 2007, **8**, 2106–2116.

16. T. Fujino, N. Yamazaki and H. Isobe, *Tetrahedron Lett.*, 2009, **50**, 4101–4103.

17. H. Isobe, T. Fujino, N. Yamazaki, M. Guillot-Nieckowski and E. Nakamura, *Org. Lett.*, 2008, **10**, 3729–3732.

18. A. Nuzzi, A. Massi and A. Dondoni, *Qsar & Combinatorial Sci.*, 2007, **26**, 1191–1199.

19. R. Lucas, P. H. Elchinger, P. A. Faugeras and R. Zerrouki, *Nucleoside, Nucleotide & Nucleic Acid.*, 2010, **29**, 168–177.

20. R. Lucas, R. Zerrouki, R. Granet, P. Krausz and Y. Champavier, *Tetrahedron.*, 2008, **64**, 5467–5471.

21. J. Vergnaud, P. Faugeras, V. Chaleix, Y. Champavier and R. Zerrouki, *Tetrahedron Lett.*, 2011, **52**, 6185–6189.

22. A. H. El-Sagheer and T. Brown, *Proc. Natl. Acad. Sci. USA*, 2010, **107**, 15329–15334.

23. E. Paredes and S. R. Das, *Chembiochem.*, 2011, **12**, 125–131.

24. A. Salic and T. J. Mitchison, *Proc. Natl. Acad. Sci. USA*, 2008, **105**, 2415–2420.

25. A. H. El-Sagheer, A. P. Sanzone, R. Gao, A. Tavassoli and T. Brown, *Proc. Natl. Acad. Sci. USA*, 2011, **108**, 11338–11343.

26. C. B. Reese, *Org. Biomol. Chem.*, 2005, **3**, 3851–3868.

27. M. H. Caruthers, *Accounts Chem. Res.*, 1991, **24**, 278–284.

28. R. T. Ranasinghe and T. Brown, *Chem. Commun.*, 2005, 5487–5502.

29. R. T. Ranasinghe and T. Brown, *Chem. Commun.*, 2011, **47**, 3717–3735.

30. J. Shendure, R. D. Mitra, C. Varma and G. M. Church, *Nature Reviews Genetics.*, 2004, **5**, 335–344.

31. F. Seela and V. R. Sirivolu, *Org. Biomol. Chem.*, 2008, **6**, 1674–1687.

32. F. Seela and V. R. Sirivolu, *Chem. Biodivers.*, 2006, **3**, 509–514.

33. F. Seela and V. R. Sirivolu, *Helv. Chim. Acta.*, 2007, **90**, 535–552.

34. S. Berndl, N. Herzig, P. Kele, D. Lachmann, X. H. Li, O. S. Wolfbeis and H. A. Wagenknecht, *Bioconj. Chem.*, 2009, **20**, 558–564.

35. P. M. E. Gramlich, C. T. Wirges, A. Manetto and T. Carell, *Angew. Chem. Int. Edit.*, 2008, **47**, 8350–8358.

36. J. Gierlich, K. Gutsmiedl, P. M. E. Gramlich, A. Schmidt, G. A. Burley and T. Carell, *Chem.-Eur. J.*, 2007, **13**, 9486–9494.

37. F. Seela, V. R. Sirivolu and P. Chittepu, *Bioconj. Chem.*, 2008, **19**, 211–224.

38. P. M. E. Gramlich, S. Warncke, J. Gierlich and T. Carell, *Angew. Chem. Int. Edit.*, 2008, **47**, 3442–3444.
39. P. M. E. Gramlich, C. T. Wirges, J. Gierlich and T. Carell, *Org. Lett.*, 2008, **10**, 249–251.
40. M. Gerowska, L. Hall, J. Richardson, M. Shelbourne and T. Brown, *Tetrahedron.*, 2012, **68**, 857–864.
41. J. Steger, D. Graber, H. Moroder, A. S. Geiermann, M. Aigner and R. Micura, *Angew. Chem. Int. Edit.*, 2010, **49**, 7470–7472.
42. A. Kiviniemi, P. Virta and H. Lonnberg, *Bioconj. Chem.*, 2008, **19**, 1726–1734.
43. M. Aigner, M. Hartl, K. Fauster, J. Steger, K. Bister and R. Micura, *ChemBioChem.*, 2011, **12**, 47–51.
44. A. H. El-Sagheer and T. Brown, *J. Am. Chem. Soc.*, 2009, **131**, 3958–3964.
45. T. R. Chan, R. Hilgraf, K. B. Sharpless and V. V. Fokin, *Org. Lett.*, 2004, **6**, 2853–2855.
46. C. Y. Jao and A. Salic, *Proc. Natl. Acad. Sci. USA*, 2008, **105**, 15779–15784.
47. H. Rao, A. A. Sawant, A. A. Tanpure and S. G. Srivatsan, *Chem. Commun.*, 2012, **48**, 498–500.
48. K. Gutsmiedl, C. T. Wirges, V. Ehmke and T. Carell, *Org. Lett.*, 2009, **11**, 2405–2408.
49. K. Gutsmiedl, D. Fazio and T. Carell, *Chem.-Eur. J.*, 2010, **16**, 6877–6883.
50. K. J. Luebke and P. B. Dervan, *J. Am. Chem. Soc.*, 1989, **111**, 8733–8735.
51. N. G. Dolinnaya, M. Blumenfeld, I. N. Merenkova, T. S. Oretskaya, N. F. Krynetskaya, M. G. Ivanovskaya, M. Vasseur and Z. A. Shabarova, *Nucleic Acids Res.*, 1993, **21**, 5403–5407.
52. N. I. Sokolova, D. T. Ashirbekova, N. G. Dolinnaya and Z. A. Shabarova, *Bioorg. Khimiya*, 1987, **13** , 1286–1288.
53. A. H. El-Sagheer and T. Brown, *Chem. Commun.*, 2011, **47**, 12057–12058.
54. A. Dallmann, A. H. El-Sagheer, L. Dehmel, C. Mügge, C. Griesinger, N. P. Ernsting and T. Brown, *Chem.-Eur. J.*, DOI:10.1002/chem.201102979.
55. X. H. Peng, H. Li and M. Seidman, *Eur. J. Org. Chem.*, 2010, 4194–4197.
56. N. C. Seeman, *Annu. Rev. Biochem.*, 2010, **79**, 65–87.
57. M. F. Jacobsen, J. B. Ravnsbaek and K. V. Gothelf, *Org. Biomol. Chem.*, 2010, **8**, 50–52.
58. Y. Xu, Y. Suzuki and M. Komiyama, *Angew. Chem. Int. Edit.*, 2009, **48**, 3281–3284.
59. H. Xiong and F. Seela, *J. Org. Chem.*, 2011, **76**, 5584–5597.
60. S. S. Pujari, H. Xiong and F. Seela, *J. Org. Chem.*, 2010, **75**, 8693–8696.
61. V. R. Sirivolu, P. Chittepu and F. Seela, *Chembiochem.*, 2008, **9**, 2305–2316.
62. F. Seela, H. Xiong, P. Leonard and S. Budow, *Org. Biomol. Chem.*, 2009, **7**, 1374–1387.
63. F. Seela and V. R. Sirivolu, *Nucleosides Nucleotides & Nucleic Acids.*, 2007, **26**, 597–601.
64. P. Chittepu, V. R. Sirivolu and F. Seela, *Bioorg. Med. Chem.*, 2008, **16**, 8427–8439.

65. P. Nordell, F. Westerlund, A. Reymer, A. H. El-Sagheer, T. Brown, B. Norden and P. Lincoln, *J. Am. Chem. Soc.*, 2008, **130**, 14651–14658.
66. A. H. El-Sagheer and T. Brown, *Int. J. Peptide Res.Therapeut.*, 2008, **14**, 367–372.
67. M. Nakane, S. Ichikawa and A. Matsuda, *J. Org. Chem.*, 2008, **73**, 1842–1851.
68. J. C. Jewett and C. R. Bertozzi, *Chem. Soc. Rev.*, 2010, **39**, 1272–1279.
69. A. H. El-Sagheer and T. Brown, *Pure Appl. Chem.*, 2010, **82**, 1599–1607.
70. J. C. Jewett, E. M. Sletten and C. R. Bertozzi, *J. Am. Chem. Soc.*, 2010, **132**, 3688–3690.
71. P. V. Chang, J. A. Prescher, E. M. Sletten, J. M. Baskin, I. A. Miller, N. J. Agard, A. Lo and C. R. Bertozzi, *Proc. Natl. Acad. Sci. USA*, 2010, **107**, 1821–1826.
72. E. M. Sletten and C. R. Bertozzi, *Org. Lett.*, 2008, **10**, 3097–3099.
73. P. van Delft, N. J. Meeuwenoord, S. Hoogendoorn, J. Dinkelaar, H. S. Overkleeft, G. A. van der Marel and D. V. Filippov, *Org. Lett.*, 2010, **12**, 5486–5489.
74. K. N. Jayaprakash, C. G. Peng, D. Butler, J. P. Varghese, M. A. Maier, K. G. Rajeev and M. Manoharan, *Org. Lett.*, 2010, **12**, 5410–5413.
75. M. Shelbourne, X. Chen, T. Brown and A. H. El-Sagheer, *Chem. Commun.*, 2011, **47**, 6257–6259.
76. A. H. El-Sagheer, V. V. Cheong and T. Brown, *Org. Biomol. Chem.*, 2011, **9**, 232–235.

CHAPTER 6

Therapeutic Applications of Nucleic Acid Aptamer Conjugates

DAVID H. J. BUNKA AND PETER G. STOCKLEY*

Astbury Centre for Structural Molecular Biology, University of Leeds, Leeds, LS2 9JT, UK
*Email: p.g.stockley@leeds.ac.uk

6.1 Introduction

The 'birth' of nucleic acid aptamers, and the process through which they are isolated, is largely credited to two independent groups. Tuerk and Gold showed that it was possible to isolate an RNA stem–loop with increased affinity for bacteriophage T4 DNA polymerase, after partial randomisation of the natural binding sequence, and iterative cycles of target protein binding and preferential amplification.[1] Their publication first coined the term 'Systematic Evolution of Ligands by Exponential Enrichment' (SELEX). In the same year, Ellington and Szostak demonstrated that it is possible to isolate specific RNA sequences with an affinity for defined organic dyes from a highly degenerate starting library.[2] They introduced the term 'aptamer' for the end product of such selections.

In the 20-plus years since these pioneering reports, there has been an explosion in the use and adaptation of *in vitro* selection of aptamers to fulfil a wide range of applications. This chapter will focus on developments in the isolation and modification of aptamers with the aim of creating therapeutically

RSC Biomolecular Sciences No. 26
DNA Conjugates and Sensors
Edited by Keith R Fox and Tom Brown
© The Royal Society of Chemistry 2012
Published by the Royal Society of Chemistry, www.rsc.org

useful molecules. Indeed one aptamer, Pegaptanib[3] (also known as Macugen™), has already received regulatory approval for clinical use. This anti-vascular endothelial growth factor (VEGF) aptamer received FDA approval in December 2004 for the treatment of the wet form of age-related macular degeneration. Since then, many other aptamers have progressed through to late stage clinical trial (reviewed in[4–6]). This progress may seem slow, but when one considers that it typically takes 10–15 years for a drug candidate to pass through development, clinical trial and regulatory approval before finally becoming commercially available, it is actually surprising that aptamers, which are relatively novel drug entities, have reached this stage so quickly. When compared to the development of monoclonal antibody therapeutics; the first of which was approved for clinical use in 1986 (11 years after the first description of these antibodies) and the second 8 years later, it becomes clear that we are on the cusp of an aptamer-based therapeutic revolution. This is exemplified by the number of aptamers described in scientific journals isolated with the ultimate goal of therapeutic application. These range from aptamers against neurodegenerative targets such as the β-secretase BACE-1,[7] the Alzheimer's disease associated Aβ protein[8,9] and prion protein PrP,[10–12] to cardiovascular disease-related targets such as phospholamban,[13] P-selectin,[14,15] platelet-derived growth factor,[16] integrin αvβ3,[17] advanced glycation end products,[18] CXCL10[19] and vasopressin,[20] numerous viral targets,[21] including those associated with influenza virus,[22,23] human immunodeficiency virus (HIV),[24–31] hepatitis C virus (HCV)[32–35] and herpes simplex virus (HSV),[36] as well as a whole range of cancer associated targets[37,38] such as acute myeloid leukaemia,[39,40] several forms of lymphoma,[41] and cancer of the lung,[42] liver,[43] breast[44] and prostate.[45] In addition, several review articles are published each year covering aspects of aptamer therapeutic development.[46,47]

Nucleic acid aptamers can be developed for therapeutic application in much the same way as monoclonal antibodies. However, aptamers have several distinct advantages, the most obvious being that no organisms are required for their isolation. This opens up the potential for isolation of aptamers under a much more diverse range of conditions and against targets which would not be amenable to antibody generation. For example, it is impossible to isolate antibodies from animals under anything other than physiological temperatures, pH, salinity, *etc.* Highly toxic compounds are also nearly impossible to raise antibodies against for obvious reasons. The introduction of counter-selection steps to the process of aptamer isolation also facilitates the isolation of more specific aptamers, which are able to discriminate between closely related targets. Examples of this include the anti-theophylline aptamer which discriminates against caffeine by 10 000-fold,[48] even though the two molecules differ only by a single methyl group, and the adenosine triphosphate aptamer which discriminates against adenosine monophosphate;[49] the two molecules differing only by their 5′-phosphorylation states. Aptamer isolation *in vitro* allows control over the precise nature of the target interaction. By carefully choosing a suitable combination of selection and counter-selection targets, aptamers can be directed to a particular molecular surface, thus interfering with a specific protein–protein interaction. Such specificity is rarely seen with antibodies and should reduce the likelihood of non-target interactions; meaning that aptamers should have considerably fewer of the side effects common to other therapeutic agents.

Estimates suggest that >60% of current drug targets are cell-surface proteins,[50] with receptors playing the most prominent role, because they are readily accessible (compared with intracellular targets) and trigger cellular responses to environmental cues through a cascade of interactions. In addition, the expression of many cell surface proteins is up-regulated within cancerous cells, making them obvious potential targets for therapeutic aptamers. Although many 'traditional' SELEX protocols involve the use of soluble, purified target(s), alternative methods have been developed to isolate aptamers in more complex and 'natural' environments. This may either be as part of an extracted tissue sample,[51] or indeed, as in one of the earliest examples, on the surface of an organism, such as the African trypanosome.[52] More recent examples involve the expression of aptamer targets as cell-surface proteins[53,54] or in human plasma.[55] Cell-SELEX[56] allows the selection of aptamers against specific cellular targets without significant prior knowledge of the nature of those targets,[57–59] and may also generate aptamers for multiple cellular targets in parallel. Such experiments can reveal novel, accessible cell-surface markers, speeding up the process of biomarker discovery. It is also especially useful because, although many proteins are up-regulated, the differences in their expression profiles make it highly unlikely that all cells within a tumour have identical cell surface proteins. A panel of aptamers which recognises a range of targets on the cell surface may therefore prove more beneficial than a high affinity aptamer against a single cellular target. This principle has been demonstrated directly using magnetic nanoparticles (MNPs) coupled to one, two, three or four different aptamer sequences (against the same cell line).[60] The results indicated that increasing the number of different sequences on the MNPs led to improved detection of the target cells, when compared with an equal concentration of a single-aptamer sequence. An extension of the Cell-SELEX strategy allows the direct selection of aptamers which have been taken up by the target cells. This involves removal of cell-surface bound aptamers using a protease such as trypsin, before recovery of the internalised aptamers.[61] Aptamers isolated in this way should be more suitable for use as targeted therapeutic agents.

6.2   Alternative Chemistries Improve Oligonucleotide Stability, Bioavailability and Distribution

Most current therapeutic targets are either free in serum or displayed on the surface of cells and accessible from the bloodstream. For this reason, most therapeutic aptamers to date have been administered by intravenous, subcutaneous, or intravitreal routes. Other routes, such as inhalation, should also be possible but have not yet been demonstrated.

In order to persist in serum, oligonucleotide aptamers must overcome a series of challenges including a nuclease rich medium, clearance through renal filtration as well as uptake and accumulation in tissues such as the liver, kidney, spleen, lymph nodes, and bone marrow.[62] Unfortunately, data on the biodistribution and elimination of therapeutic aptamers are limited to those published for Macugen™, and a few studies involving test aptamers.[63-65] However, the findings are consistent with those for other oligonucleotides, and show that elimination is almost entirely by renal clearance of intact aptamers or their degradation products.

To combat these biological barriers, a wide variety of surprisingly simple chemical modifications have been employed, the most common of which will be mentioned here (reviewed in more detail elsewhere[5,66]).

6.2.1 Nucleotide Modification

Oligonucleotide synthesis has become a routine process in many laboratories. This is exemplified by the number of companies who provide an oligonucleotide synthesis service or supply reagents for doing so. Based on chemistries established in the mid 1980s,[67] it is now a trivial matter to synthesise and purify milligram quantities of oligonucleotides, up to (and in some cases greater than) ~ 100 nucleotides in length. Synthesis is readily scaled up to grams and even kilograms for commercial manufacture. Similarly, it has become routine in many laboratories to prepare DNA or RNA aptamers with a wide variety of modified nucleotides. For a review of the basic steps of solid-phase oligonucleotide synthesis, the recent improvements to the process and coverage of simple modifications see ref.[68] The majority of these modifications aim to produce nucleic acids which are more resistant to nuclease degradation. The most commonly used include sugar modifications,[69,70] such as 2'-fluoro, 2'-amino, 2'-O-methyl, 2'-O-methoxyethyl or 2'-O-dimethyl allyl ribose, or nucleotide base modifications, including the addition of propenyl, 5-(N aminoalkyl), methyl, trifluoromethyl, or phenyl groups. Backbone modifications such as substitution of one of the non-bridging phosphoryl oxygens, *e.g.* with sulfur to create a phosphorothioate,[71] or to a methyl phosphonate, have also been shown to enhance binding affinities, through improved aptamer stability and nuclease resistance.[72-74] A selection of these modified nucleotides can be seen in Figure 6.1. Inclusion of these modified nucleotides during the SELEX process has been made possible by the identification of several polymerase variants capable of utilising them as substrates.[75,76]

The addition of locked nucleic acid (LNA) triphosphates during PCR and *in vitro* transcription has also shown promise for improving aptamer stability and affinity.[77-79] These LNAs have a methylene bridge between the 2'-oxygen and 4'-carbon, which locks the ribose in the 'endo' conformation. As well as improving resistance to nucleases, incorporation of LNAs at specific sites within aptamers has been shown to improve affinity and function.[80] This is thought to be due to the associated reduction in conformational dynamics, leading to a more stable active conformation. Several DNA and RNA

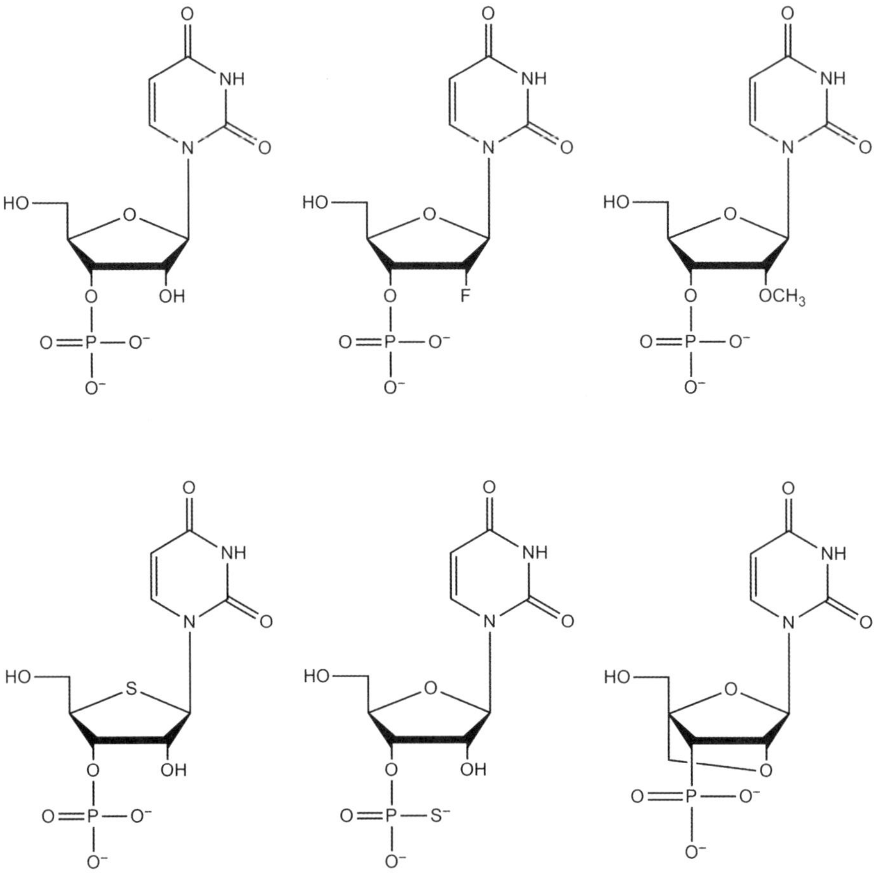

Figure 6.1   Chemical structure of 2′-OH nucleotides and examples of the most commonly used stabilising nucleotide modifications, including 2′-fluoro, 2′-O-methyl, 4-thio, phosphorothioate and locked nucleic acid bases (from top left to bottom right, respectively).

polymerases have recently been shown to incorporate LNAs successfully when they are spaced out within the aptamer.[81,82] However, further work is required to identify polymerases capable of directly amplifying a fully modified aptamer.

In a similar approach, recent reports have demonstrated heredity and evolution in six nucleic acid polymer variants, not found in Nature. These so called xeno-nucleic acids (XNAs) are made up of nucleobases attached to a backbone consisting of an alternative sugar moiety. These XNAs were also successfully used to isolate high affinity aptamers, using engineered XNA polymerases.[182,183]

Other nucleotide modifications have been created with the aim of generating aptamers with other functions. Incorporation of uracil analogues such as 5-iodouracil (5-IU) or 5-bromouracil (5-BrU) allows covalent cross-linking of aptamers to their targets following activation with ultraviolet (UV) laser light, both during and after SELEX experiments.[83,84] Such irreversible binding events allow low picomolar concentrations of target to be bound specifically in the presence of serum with very little non-specific cross-linking.[85]

It is important to note that while these modifications may stabilise aptamers against a wide variety of degrading enzymes, as yet uncharacterised nucleases may be present when aptamers are used in a clinical setting. It is therefore important to consider additives to the aptamer formulation. For example, a known anti-HIV aptamer, which blocks receptor uptake, was synthesised for use as part of a trial microbicide. Despite carrying a range of 'stabilising' nucleotide modifications, the aptamer was degraded within minutes of administration, by novel nucleases.[86] The addition of inhibitory Zn^{2+} ions improved aptamer stability, without significantly altering its target affinity and functional effects. Presumably similar additions will work with other anti-viral aptamers such as those reported for HSV-2.[36]

6.2.2 End Modification

6.2.2.1 Inverted Caps

As previously mentioned, nuclease activity is a key obstacle to the therapeutic application of aptamers. As well as nucleotide modifications incorporated throughout the length of the aptamer, novel functional groups may be added as the final step of the solid-phase synthesis. For example, the addition of an inverted nucleotide at the 3'-terminus creates an aptamer which has two 5' but no 3'-termini, as shown in Figure 6.2. Given that 3'-exonuclease activity in serum is much higher than 5'-exonuclease activity, this relatively simple modification has been shown to increase serum half-life of an unmodified DNA aptamer from a few minutes to approximately 1 h.[87] Others have avoided this problem by eliminating the aptamer termini altogether by circularising their aptamers via intra- or intermolecular ligation.[88] Aptamer constructs generated in this way exhibited serum half-lives exceeding 10 h.

6.2.2.2 Thiols and Amines

While each of the modifications mentioned above imparts a useful feature to the resulting aptamer, they do not allow for the addition of significantly larger (and more useful) modifications. The introduction of nucleotides carrying chemically reactive groups, such as primary amines, thiols or aldehyde precursors, gives the resulting aptamer numerous potential points for attachment of a wide range of functional conjugates. Several extensive reviews on current bioconjugation techniques describe these reactions in more detail,[89] so they will only be covered briefly here. An aptamer may be readily created which carries a thiol (SH) group at the 5'-terminus, 3'-terminus or at a chosen point with the aptamer. These groups then readily react with other thiol-containing compounds (forming a stable S–S linkage), or more commonly a sulfur–carbon bond with more reactive maleimide groups, as shown in Scheme 6.1.

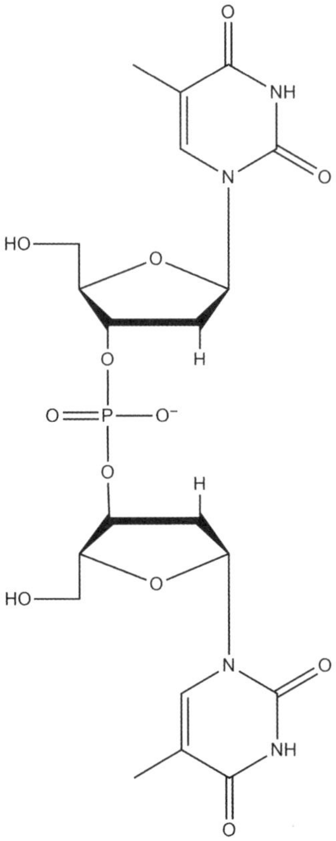

Figure 6.2   Chemical structure of a 3′-3′ inverted dT linkage used to create oligonucleotides which lack a 3′-terminus.

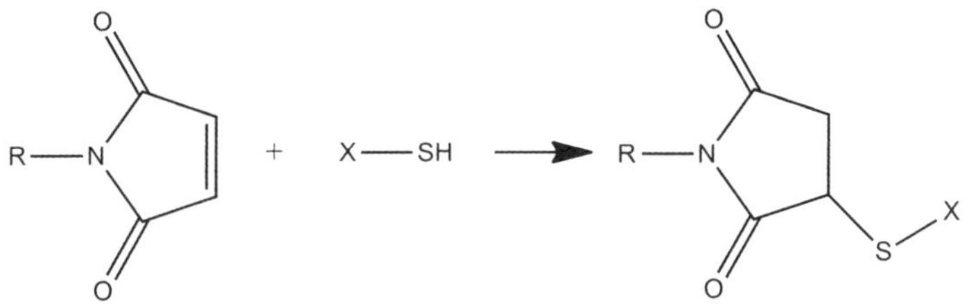

Scheme 6.1   Chemical coupling of a thiol group on an oligonucleotide (X) to a maleimide group on the desired conjugate (R).

Similarly, aptamers carrying primary amines are capable of reacting with
N-hydroxysuccinimide (NHS) groups on the conjugation target (shown in
Scheme 6.2). Nucleotides carrying these reactive functionalities can be enzymatically incorporated during the SELEX process or added at specific locations
during solid-phase synthesis. This avoids the losses in binding activity often
seen when such conjugates are added to multiple sites on proteins.

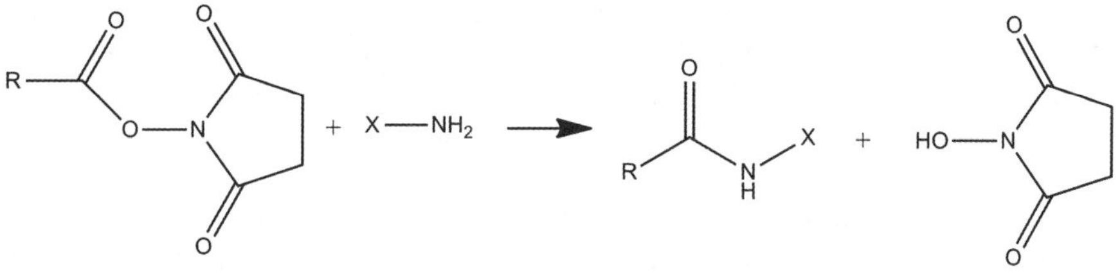

Scheme 6.2 Chemical coupling of an amine group on an oligonucleotide (X) to an *N*-hydroxysuccinimide group on the desired conjugate (R).

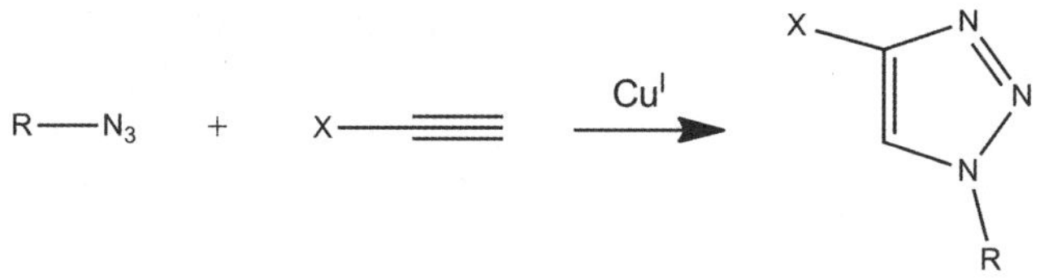

Scheme 6.3 'Click chemistry' reaction: coupling of an alkyne group on an oligonucleotide (X) to an azide group on the desired conjugate (R).

6.2.2.3 'Click' Chemistry

'Click chemistry' refers to a group of simple, rapid, versatile, site-specific and high-yield reactions. The most commonly used 'Click' reaction is the Huisgen 1,3-dipolar cycloaddition of azides and alkynes, to form 1,2,3-triazoles (shown in Scheme 6.3). The discovery of Cu(I)-catalysis of this process[90] allows a wide range of applications in bioconjugation.[91,92] Several reviews have been published, covering the basics of click chemistry,[93,94] its use in therapeutic applications[95,96] and utility with nucleic acids.[97] We mention it briefly here because of its increasing importance.

Given that both azides and alkynes are completely inert to most chemical functionalities, the CuAAC [Cu(I)-catalysed azide and alkyne cycloaddition] reaction is highly specific, with very few undesired side reactions. This is beneficial because it greatly simplifies purification of the desired product. In addition, 'Click' reactions are usually fusion or condensation processes, meaning that they leave no by-products other than water. They are also highly thermodynamically favoured, making the reaction irreversible. CuAAC has additional benefits in that it occurs readily in a wide range of solvents including water, and at a range of temperatures (0–160 °C), pH (5–12) and biological conditions (*e.g.* a reducing environment[98]). These properties make 'Click' chemistry ideal for nucleic acid aptamer conjugation to a range of functional biomolecules.[99–101] Indeed, this 'Click' reaction has been used for many applications such as ligation of single strands of DNA, cross-linking complementary strands, cyclising single and double strands, labelling oligonucleotides with reporter groups, attaching DNA to surfaces, producing analogues of DNA with modified nucleobases and backbones, synthesising large, chemically modified RNA constructs and creating biochemically active PCR templates. Importantly, it has been shown that synthesised nucleosides, each containing a single terminal alkyne on their aromatic base, are readily incorporated into oligonucleotides and that the properties of these oligonucleotides and their duplexes are not significantly different from the unmodified equivalents.[102]

Unfortunately, owing to the toxic nature of copper, the therapeutic application of 'Click' chemistry may be limited, without significant steps to remove this catalyst. However, recent developments have negated the need for a catalyst. Substitution of simple alkynes with cyclic groups containing a 'strained' alkyne allows the Click reaction to occur without the need for Cu(I). This was demonstrated with the use of cyclooctyne derivatives in a Cu(I)-free DNA ligation.[103]

In terms of therapeutic development, the resulting 1,2,3-triazoles make ideal bioconjugate linkers because they are extremely water soluble; they are similar to amide bonds, but they are not as readily hydrolysed and they are also extremely rigid, reducing the likelihood of steric effects of either part of the resulting conjugate.[93]

6.2.2.4   Polyethylene Glycol (PEG)

As previously mentioned, not only must a therapeutic aptamer survive in a nuclease-rich environment, it must also persist in the body long enough to act upon the desired target. Length-minimised, functional aptamers typically have a molecular mass between 5 and 20 kDa, meaning that they readily pass through the renal glomerulus, which has a molecular mass cut-off for filtration between 30 and 50 kDa. Conjugation of polymers in this size range to a functional aptamer would therefore be expected to reduce renal clearance rates dramatically. One of the most common conjugates used to achieve this is polyethylene glycol (PEG),[104–108] because it has low toxicity, is non-immunogenic, has high solubility in water and is commercially available in a range of molecular weights and with a large variety of terminal functional groups suitable for aptamer conjugation.[70,87,109–111] A number of FDA approved drugs in clinical practice use PEG for improved pharmaceutical properties,[112] and its conjugation to nucleic acids has been shown to improve their circulating half-life from 5–10 minutes, to 1 day in mice,[113–115] as well as improving biodistribution.[63] However, as one might expect, conjugation of a moiety of this size may interfere with aptamer function. A careful balance must therefore be struck between improved pharmacokinetics and the inherent reduction in aptamer activity. The precise position and size of the conjugate must therefore be assessed for each aptamer. For a review of the general procedures for studying aptamer pharmacokinetics, biodistribution, *etc.* and the effects of common conjugates, see ref.[116]

The availability of PEG with different molecular weights affords the opportunity to modulate the rate of renal clearance, depending on the desired aptamer function. For example, aptamers which are required to be active for short periods may be left unPEGylated to ensure that they are rapidly cleared after administration. This may be useful for aptamers used as imaging reagents. Similarly, the anti-thrombin aptamer ARC183 is a short un-PEGylated DNA aptamer with a plasma half-life of approximately 10 min. This allows a rapid return of normal coagulation when administration is stopped.[117] Aptamers intended for longer term treatments can be conjugated to 40 kDa PEG, resulting in half-lives of approximately 75 h in *Cynomolgus* monkeys. An aptamer with an intermediate half-life, such as the anti-vWF (von Willebrand factor) aptamer ARC1779, carries a 20 kDa PEG conjugate; ARC1779 has a half-life of 2–4 h.[118]

In addition to the more conventional PEG architectures such as linear and branched PEGs, studies have been carried out using PolyPEG structures based on a polymethacrylate backbone with PEG 'teeth' attached by ester bonds, to give a 'comb-like' architecture. The comb-shaped polymer has been shown to have benefits over linear PEGs, including reduced viscosity.[107] The attachment of PEG to the backbone *via* ester bonds also means that the conjugated groups could degrade to small, readily excreted units, avoiding the potential toxicological problems due to vacuole formation seen with linear PEG.[119,120] Comparison of these various PEG architectures, attached via a maleimide to a 3′-thiol-modified DNA aptamer, showed that branched PEG (2 × 20 kDa) and linear PEG (20 kDa) interfere with the binding of the aptamer to its target, decreasing the affinity when compared with the unmodified aptamer.[121] In contrast, the comb-like PolyPEG conjugates displayed similar binding affinities to that of the unmodified aptamer. This suggests that it is the physical size and conformation of the 20 kDa linear PEG and 40 kDa branched PEG that cause the reduction in affinity, because it is not observed with either branched 20 kDa (2 × 10 kDa) PEG or the larger PolyPEG polymers.

6.2.2.5 Cholesterol and Other Derivatives

A few alternative conjugation ligands which have demonstrated some reduction in renal clearance include attachment of carrier proteins such as streptavidin,[64] association with liposomes through conjugation of lipid tags[122] or attachment of cholesterol derivatives.[109] While the affect on renal clearance seems to be less than for that seen for PEG conjugates,[123] they do pave the way for other therapeutically useful strategies.

6.2.2.6 Fluorophores

An obvious choice of functional conjugate to attach to aptamers is a fluorescent probe. Aptamers conjugated to quantum-dots, fluorophores, quenchers, FRET-pairs, *etc.* open up a plethora of opportunities for development of biosensors and other aptamer-based detection systems. These will not be discussed here but are extensively covered elsewhere in this volume.

6.3 Oligonucleotide Conjugates as Novel Targeted Therapies

A lack of target specificity is a common problem for many therapeutic agents. In the case of harsh chemotherapeutic agents used to treat cancer, this often leads to life-threatening side effects for the patient. This problem could be significantly reduced if the therapeutic agent were more effectively delivered to the desired site of action, such as a tumour. Aptamers have been shown to target specifically many cancer associated tissues, without showing significant binding to related materials. This would result in an increased local drug concentration around the disease sites, and may therefore increase the rate of uptake of the drug via mechanisms such as receptor-mediated endocytosis. This makes aptamers ideal candidates for development as part of a targeted therapeutic agent (or 'smart drug') by direct action or by acting as escort molecules to deliver intracellular therapeutic agents.[124] In addition, if the aptamer binds to an internalised cell-surface protein, any conjugated cargo can be directly delivered into the diseased cell. A few examples of aptamers used in this way will be described here and are outlined in Figure 6.3.

6.3.1 Aptamer Smart Drugs

As one can imagine, the specificity of aptamer–target interactions means that aptamers can be effectively used to 'hit' disease associated targets, without

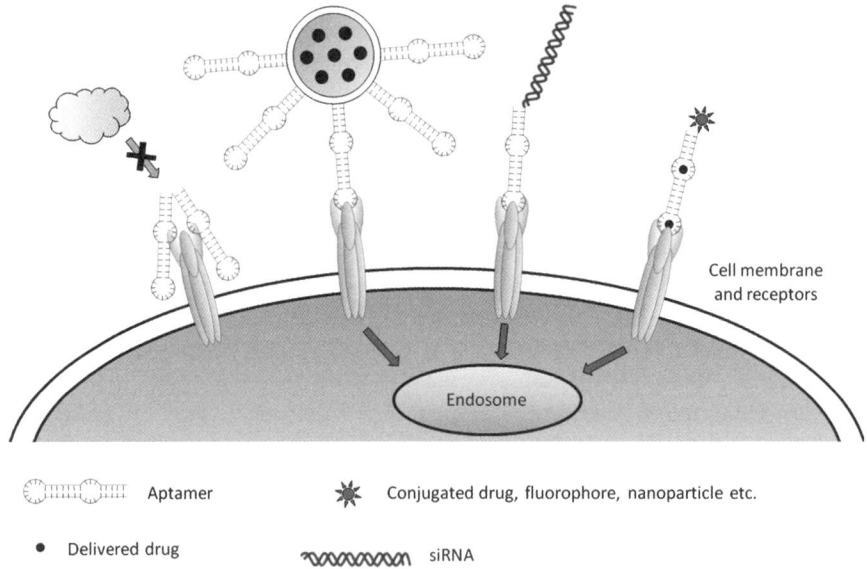

Figure 6.3   Diagram showing various applications of aptamer conjugates as targeted therapeutics. Aptamers may directly hinder disease-associated target interactions, thus acting as 'smart drugs' (far left), they may be conjugated to the surface of larger, cargo-loaded delivery vehicles, thus acting as a targeting molecule (middle left), they may form part of a chimeric oligonucleotide, delivering a second functional nucleic acid to the target site (middle right), or they may be conjugated to several different molecules to serve as a multifunction drug delivery and/or imaging reagent (far right).

affecting similar molecules. In this sense, aptamers can be used directly as 'smart-drugs'. Many aptamers have shown potential in this regard. A few recent examples will be outlined here.

An aptamer has been identified which binds the serum glycoprotein alpha-fetoprotein (AFP). This protein is found in fetal serum but is almost undetectable in healthy adult serum. However, AFP is re-expressed and excreted by hepatocellular carcinoma cells and is therefore used as a biomarker associated with hepatocarcinoma. The AFP-specific aptamer was shown to suppress AFP-induced proliferation of hepatocarcinoma cells as well as AFP-mediated oncogene expression.[125] This effect is suggested to be through sequestration of AFP, and hence inhibition of the initiation step of AFP-associated signal transduction.

Similarly, an anti-epidermal growth factor receptor (EFGR) aptamer was shown to bind to any cell expressing EFGR on its surface. Aptamer binding was shown to inhibit both EGFR activation and any subsequent EGFR-mediated signalling pathways. This resulted in the inhibition of the anti-apoptotic STAT3 and consequently induced cell death. This effect was also seen in cell lines that are resistant to approved doses of other EGFR inhibitors such as gefitinib and cetuximab.[126]

6.3.2   Oligonucleotide Hybrids and Chimeras

Unlike the examples above, many aptamers do not have a direct therapeutic effect. However, the creation of aptamer hybrids (or chimeras) allows the functional capability of the aptamer to be combined with that of another molecule.[127] There are many possible types of chimera; aptamer–aptamer, aptamer–siRNA and even aptamer–enzyme chimeras, suitable for a range of applications (reviewed in refs[128,129]), but here we will focus on therapeutic applications.

One of the first reports of aptamer–siRNA multimers used the motor 'pRNA' of bacteriophage phi29.[130] The pRNA readily forms dimers, trimers and higher order assemblies through 'hand-in-hand' interactions between single-stranded loops.[131] These were used to assemble trimers of chimeric pRNAs carrying either FITC, anti-CD4 siRNA or a known anti-CD4 aptamer.[28] A correlation between CD4 expression and internalisation/silencing was observed in mouse models, suggesting that the CD4 aptamer mediated cell-specific delivery of siRNA.[130]

In contrast to these complex multimers, aptamer–siRNA chimeras with simpler designs have also been developed. For example, a prostate specific membrane antigen (PSMA) aptamer–siRNA chimera was shown to target the expression of survival genes *plk1* and *bcl-2* in PSMA-expressing prostate cancer cells, resulting in gene silencing, reduced cell proliferation, and cell death. No effects were seen with PSMA-negative cells.[132] Modifications known to enhance the silencing activity of siRNA have also been incorporated into this chimera, resulting in increased activity.[133]

An anti-gp120 aptamer conjugated to siRNA targeting the *tat/rev* common exon was shown to have potent anti-HIV activity.[134] This chimera successfully inhibits models of HIV infection and gp120-mediated cell fusion. To overcome the development of viral resistance to the siRNA, the chimera design was modified to allow the siRNA of choice to be attached to the aptamer via simple Watson–Crick base pairing to an adjoining 'sticky-bridge'. This adapted chimera successfully bound to gp120-expressing cells, down-regulated gene expression, and suppressed HIV replication in cell culture.[135] Other anti-gp120 and anti-CD4 aptamers have been shown to deliver anti-HIV siRNAs successfully, resulting in the simultaneous inhibition of HIV infection and replication.[31,136]

As mentioned, the chimera need not be another nucleic acid. Conjugation of an active enzyme to a targeting aptamer allows the possibility of enzyme replacement therapy (ERT). As the name suggests, this is a therapeutic application in which an absent or deficient enzyme is supplemented with a functional counterpart. A known transferrin receptor (TfR) DNA aptamer was conjugated to the lysosomal enzyme α-L-iduronidase.[137] The aptamer–enzyme conjugates bound to TfR on the surface of mouse fibroblasts and were successfully internalised by receptor-mediated endocytosis. Inside the lysosome, the functional enzyme successfully counteracted the glycosaminoglycan accumulation seen in α-L-iduronidase deficient cells. The results demonstrate the potential of aptamers to deliver enzymes specifically into target cells; this is a first step towards an aptamer-mediated enzyme replacement therapy.

6.3.3 Aptamers Delivering Toxic Cargo

In addition to aptamer chimeras, several reports have demonstrated the utility of aptamers conjugated to known drug molecules for delivery of therapeutic agents to a specific cell line or tissue.[138] A few examples will be given here to demonstrate the range of this application. One of the most commonly used aptamers for development of these delivery systems targets the PSMA presented on the surface of human prostate cancer cells.[45] The detection of specific endocytosis of PSMA into cells, *via* clathrin-coated pits,[139] suggests that PSMA-specific aptamers might form components of potential drug delivery vehicles and that similar routes may be used to target other cell lines. An example of this approach involved linking a PSMA-specific aptamer to a gelonin derivative.[140] Gelonin is a toxin that acts within the ribosome and disrupts protein biosynthesis. The gelonin derivative used lacks the natural cellular translocational domain, preventing direct uptake by cells. Upon conjunction with the PSMA aptamer, the gelonin variant was internalised by PSMA-presenting tumour cells, resulting in cell death. This demonstrated the ability of a PSMA aptamer to deliver a therapeutically relevant toxin as payload into target cells.

Another commonly used toxin is the chemotherapeutic agent doxorubicin (Dox). Several studies have already demonstrated that the Dox C-13 hydrazone derivative has comparable cytotoxic effects to Dox, but can be conjugated to a carrier and released at pH 4.5–5.5.[141] In the context of drug delivery, the endosomal compartment is usually acidic. An aptamer–Dox hydrazone conjugate may therefore be used to carry the toxic cargo and release it once taken up by the endosome. This was demonstrated with another commonly used DNA aptamer (referred to as Sgc8c) which targets the transmembrane receptor protein tyrosine kinase 7 (PTK7) and can distinguish between the target leukaemia cells and normal cells.[142] Sgc8c was successfully conjugated to the Dox derivative and shown to be highly cytotoxic to target cells, while the aptamer or Dox alone had no effect.[143]

Others have also used the aptamer–Dox pair but have shown that conjugation was not required because Dox can intercalate into the aptamer structure.[144] As an extension of this, the PSMA aptamer 'A10' was immobilised on the surface of fluorescent quantum dots (QDs) that were then incubated with Dox, to give a QD–A10–Dox conjugate.[145] Interestingly, intercalation of Dox quenched QD fluorescence, but this was restored upon Dox release. When incubated with PSMA-positive cells, uptake of these multifunctional nanoparticle conjugates could be detected through restoration of QD fluorescence. Additionally, the released Dox inhibited cell growth. This is an elegant demonstration of simultaneous detection and treatment of cancerous cells with aptamer conjugates. Quantum dots provide an additional advantage because multiple copies of an aptamer (or several different aptamers) may be coupled to a single QD. This produces a multivalent targeting agent with greatly enhanced avidity.

This approach has also been applied to the cell surface glycoprotein Mucin 1 (Muc1). This is an attractive target for aptamer-based anti-cancer targeting and imaging reagents, because its expression in tumours is increased by at least 10-fold compared with that in normal tissue. Anti-Muc1 aptamers 'AptA' and 'AptB'[146] were successfully conjugated to quantum dots (QDs) and Dox.[147] Greater accumulation of aptamer–QD was seen in a mouse model of human ovarian cancer compared with non-targeted QDs.

Similar approaches to create multifunctional nanoparticles using the aptamer A10–Dox conjugated to superparamagnetic iron oxide nanoparticles (SPION) or gold nanoparticles (GNPs) have also proven successful in simultaneous imaging and treatment of prostate cancer cells.[148,149] A recent report of aptamers that specifically bind the cancer cell line MCF-10AT1 showed that dye-labelled aptamer 'KMF2-1a' was specifically internalised by target breast cancer cells by endocytosis.[44] A complex of streptavidin and biotin-labelled KMF2-1a was also specifically delivered into the target breast cancer cells, suggesting that this approach of simultaneous diagnosis and therapy may be applicable to many other diseases.

As a further extension of this multi-function use of aptamers, the anti-PSMA aptamer was conjugated to the surface of a polyethyleneimine and polyethylene glycol (PEI–PEG) delivery vehicle, created to carry an anti-Bcl-xL short hairpin RNA (shRNA). The aptamer was also incubated with Dox, to give an additional toxic cargo. This 'polyplex' was successfully targeted to prostate cancer cells and triggered apoptosis both *in vitro* and *in vivo*.[148]

An innovative adaptation of aptamers in targeted therapy involves the generation of phototoxic aptamers for use in photodynamic therapy (PDT) against specific cancer cells.[150] Short *O*-glycan peptides, specifically expressed only on the surface of many cancer cell types, were targeted with DNA aptamers. The photodynamic agent chlorin e6 was conjugated to a 5′-amine group introduced on the resulting aptamers. A 500-fold increase in light-induced cytotoxicity was seen relative to the drug alone, which resulted in tissue-specific killing of cancer cells.

A related approach utilised the anti-nucleolin DNA aptamer 'AS1411', which has previously been shown to be taken up by nucleolin-presenting cancer cells.[151] The aptamer was conjugated with the porphyrin derivative 5,10,15,20-tetrakis(1-methylpyridinium-4-yl)porphyrin (TMPyP4). MCF7 breast cancer cells and normal endothelial M10 cells were treated with the aptamer–TMP conjugate and showed significantly higher TMPyP4 accumulation in MCF7 breast cancer cells than in normal epithelial cells.[152] Subsequent exposure to light resulted in significantly more damage in the MCF7 cells than the control cells.

6.3.4   Aptamer-targeted Delivery Vehicles

As shown above, a wide range of molecular species have been transported using aptamers as nanoscale delivery vehicles. These include siRNAs, drugs/toxins, enzymes, photodynamic molecules, and radionuclides. However, these can only be delivered a few at a time. In order to improve the delivery of greater doses of therapeutically active compounds, aptamers have been used to target much larger delivery vehicles carrying far greater cargoes.

An example of aptamer targeted delivery of cargo-loaded nanoparticles involved encapsulation of the cisplatin prodrug Pt(IV) within PLGA–PEG nanoparticles. These particles were coated with the PSMA aptamer A10.[153] After cell-specific targeting and uptake of the aptamer–nanoparticle conjugates, the internalised Pt(IV) was released and reduced to its platinum(II) form, cisplatin; this resulted in an 80-fold increase in cell death compared with free cisplatin or conjugates lacking the aptamer portion. Studies have also shown that these Pt(IV)–nanoparticle–aptamer conjugates have improved pharmacokinetics, biodistribution, tolerability, and efficiency compared with cisplatin,[154] suggesting a potential role for such targeted conjugates in the treatment of human prostate cancer. Another example involved conjugation of a nucleolin-specific aptamer to a cisplatin-loaded liposome, and led to delivery of the drug to tumours.[155]

An alternative approach involved modifying the end of the Ramos cell specific aptamer, 'TDO5',[41,156] with a hydrophobic tail attached via a PEG linker.[157] This tail results in the spontaneous self-assembly of micelle-like nanoparticles. Interestingly, the aptamer micelles showed higher affinity for their target than the aptamer alone. This increased affinity was ascribed to the multivalent effect from multiple aptamer binding events, because competition experiments demonstrated that the aptamer micelles have a significantly lower k_{off} than the aptamer alone, consistent with multivalent binding. Furthermore, assembly of the aptamers with micelles allowed them to enter the target cells by fusion of the micelle with the membranes of target cells. This may allow cell targeting using cell surface markers that are not readily internalised. Additionally, these micelles could be loaded with imaging reagents or drugs to give novel simultaneous diagnostic and therapeutic reagents.

We have previously shown that virus-like particles (VLPs) can be specifically assembled around toxins or therapeutic oligonucleotides, when conjugated with RNAs encompassing a viral assembly initiation signal. These VLPs can hold many copies of the 'drug' molecule and can be targeted to cells specifically by covalent decoration of peptides and proteins.[158,159] This idea has now been extended using cell-specific DNA aptamers as targeting ligands.[160]

6.4 Oligonucleotide Conjugates as Imaging Reagents

Following administration, aptamer distribution is rapid, as is uptake by cellular targets. This is followed by a rapid clearance from the body, leading to high signal to 'noise' ratios making them ideal for use as *in vivo* imaging reagents. One of the earliest reports of an aptamer applied in this way used the Tenascin-C aptamer, TTA1.[111] The aptamer was modified with 2′-F Pyrimidines, 2′-O Me purines and a 3′-3′ inverted nucleotide cap to improve serum stability, then labelled with ^{99m}Tc, conjugated via an incorporated 5′-amine. The ^{99m}Tc-labelled aptamer displayed both rapid tumour uptake (maximised at ∼10 minutes) and rapid blood clearance (half-life < 2 minutes). Rapid tumour uptake and blood clearance yielded a tumour-to-blood ratio of 50 : 1 within 3 hours and 180 : 1 within 16 hours, enabling clear tumour imaging.

The nucleolin aptamer AS1411 has also been used as the basis for a multi-functional imaging reagent. Aptamer conjugation to cobalt–ferrite nano-particles, surrounded by fluorescent rhodamine, created an imaging agent referred to as MF-AS1411.[161] This product was also labelled with ^{67}Ga to form MFR-AS1411, which has the potential for use in magnetic resonance imaging (MRI), optical, and nuclear imaging. Fluorescent imaging confirmed aptamer-conjugated nanoparticle uptake in C6 tumours in injected animals. Some non-specific uptake in the liver and intestines was also observed. This clearly has potential for application with other aptamers once optimised to reduce non-specific aptamer uptake.

Other studies described *in vivo* imaging of Muc-1-positive tumours using monomeric and tetrameric aptamer conjugates.[162,163] To prepare a tetravalent complex, the macrocyclic chelator DOTA or carboxy porphyrin was used. The aptamer was conjugated with the chelator via amide bonds formed between the carboxylic acids of the chelator and the 5′-amino group of the modified oligonucleotide. The conjugates showed high coupling efficiency, with more than 90% aptamer labelling.

Several preclinical studies of aptamer-targeted imaging agents have been most successful in cancer, but other disease processes have also been studied.[164–166]

6.5   Nanoparticles

Studies have shown that bioconjugation of oligonucleotides to gold nano-particles via covalent bonds could occur easily via click chemistry.[167] However, immobilising large proteins (such as antibodies) on the surface of the nano-particles can result in aggregation and impair the performance of the nano-materials. In addition, the complexity and expense of antibody production and more difficult routes of bioconjugation affect the ultimate utility of antibodies paired with nanomaterials.

Radiofrequency magnetic field (RFMF) is widely used for MRI and pho-tothermal therapy.[168,169] Previous reports have demonstrated remote control of DNA structures by heating conjugated gold nanoparticles by irradiation using RFMF. This results in the denaturation of the conjugated DNA, but leaves the surrounding molecules relatively unaffected.[170] Aptamer structure and function could also be controlled in this way and would have several advantages over UV light irradiation. Although there are reports describing the remote control of aptamers by UV light,[171,172] deep tissue penetration is difficult and limits their application *in vitro*, without specialised equipment.[173] Given that RFMF can penetrate deep tissues, it is more broadly applicable in a clinical setting. In addition, the heat denaturing effect of the irradiated gold nanoparticles on the conjugated aptamer does not affect its chemical structure (as UV irradiation can). This means that the aptamer can refold into a functional state. As an example of this application, the anti-thrombin aptamer was synthesised including a 5′-thiol group to allow conjugation to mono-maleimide-modified gold nanoparticles.[174] This work demonstrated that the inhibitory effects of the aptamer could be modulated by RFMF, allowing precise control over thrombin activity.

6.6   Toxicology

As with other therapeutics, oligonucleotide toxicity can manifest itself through a variety of on-target or off-target mechanisms. Off-target effects of oligonu-cleotides have been extensively studied during the development of antisense oligonucleotide therapeutics,[175–177] and include anticoagulation, complement activation and innate immune stimulation. Given that aptamer therapeutics is a relatively new field, there is very limited information regarding the toxicological properties of aptamers, except for the published clinical trial data for Macugen™.

Oligonucleotide-induced anticoagulation has been described, and results in a measurable prolongation of coagulation times.[178] However, this effect is gen-erally minor and has not been associated with observed bleeding. Stimulation of the innate immune system has also been observed for some antisense oligonucleotides, resulting in a new class of intentionally immunostimulatory oligonucleotides.[179] However, these interactions can lead to undesired immune stimulation. Innate immune activation occurs as a consequence of the activation of Toll-like receptor 3 (TLR3), TLR7, TLR8 or TLR9. TLR3 responds to double-stranded RNA, TLR7 and TLR8 respond to single-stranded RNA, and TLR9 responds to unmethylated CG motifs in DNA (CpG motifs). As aptamers are highly structured, they show very little activation of these receptors. In addition, the 2'-modified nucleotides used in aptamers do not lead to Toll-like receptor responses.[180] Complement activation has also been observed with therapeutic oligonucleotides through their interaction with various proteins of the complement pathway. Studies conducted in primates have shown that these effects are generally mild and only occur at high oligonucleotide concentrations.[176]

It is important to note that the effects described here have been observed for therapeutic oligonucleotides and may not necessarily apply to aptamers. Aptamers have significantly different structures and chemical compositions, and are frequently associated with other molecules such as PEG. These differences can dramatically change the properties relative to unmodified oligonucleotides. In addition, the 2' modifications commonly used in aptamers have a significant effect on their toxicology; for example, 2'-*O*-methyl aptamers may be less toxic because these substituted nucleotides occur naturally in ribosomal RNA.[181] Given that aptamers are highly specific to their chosen target(s), the likelihood of off-target interactions should be minimised, reducing the chance of effects such as those described above.

6.7 Perspectives

In this chapter we have described a few examples of the increasing number of approaches used to produce novel therapeutic molecules from simple target-binding oligonucleotides. These include modifications which improve oligonucleotide pharmacokinetic and pharmacodynamic properties, as well as allowing their conjugation with therapeutically active compounds, to give specifically targeted therapies. Developments in this field are now making significant head-way in both pre-clinical and clinical trials, aided by the recent expiry of the initial broad spectrum patents covering the SELEX process. With these successes and the increasing number of chemical functionalities which may be readily added to isolated aptamer sequences, it is highly likely that oligonucleotide-based therapies will begin to fill many of the currently unmet clinical needs, ensuring continued growth in the field of nucleic acid based therapies.

References

1. C. Tuerk and L. Gold, *Science*, 1990, **249**, 505–510.
2. A. D. Ellington and J. W. Szostak, *Nature*, 1990, **346**, 818–822.

3. E. W. Ng, D. T. Shima, P. Calias, E. T. Cunningham, Jr., D. R. Guyer and A. P. Adamis, *Nat. Rev. Drug Discov.*, 2006, **5**, 123–132.

4. D. H. Bunka, O. Platonova and P. G. Stockley, *Curr. Opin. Pharmacol.*, 2010, **10**, 557–562.

5. G. Kaur and I. Roy, *Expert Opin. Investig. Drugs.*, 2008, **17**, 43–60.

6. A. D. Keefe and S. T. Cload, *Curr. Opin. Chem. Biol.*, 2008, **12**, 448–456.

7. A. Rentmeister, A. Bill, T. Wahle, J. Walter and M. Famulok, *RNA*, 2006, **12**, 1650–1660.

8. F. Rahimi and G. Bitan, *J. Vis. Exp.*, 2010, http://www.jove.com/.

9. F. Ylera, R. Lurz, V. A. Erdmann and J. P. Furste, *Biochem. Biophys. Res. Commun.*, 2002, **290**, 1583–1588.

10. D. F. Bibby, A. C. Gill, L. Kirby, C. F. Farquhar, M. E. Bruce and J. A. Garson, *J. Virol. Methods*, 2008, **151**, 107–115.

11. N. M. Sayer, M. Cubin, A. Rhie, M. Bullock, A. Tahiri-Alaoui and W. James, *J. Biol. Chem.*, 2004, **279**, 13102–13109.

12. S. J. Xiao, P. P. Hu, X. D. Wu, Y. L. Zou, L. Q. Chen, L. Peng, J. Ling, S. J. Zhen, L. Zhan, Y. F. Li and C. Z. Huang, *Anal. Chem.*, 2010, **82**, 9736–9742.

13. Y. Tanaka, T. Honda, K. Matsuura, Y. Kimura and M. Inui, *J. Pharmacol. Exp. Ther.*, 2009, **329**, 57–63.

14. D. R. Gutsaeva, J. B. Parkerson, S. D. Yerigenahally, J. C. Kurz, R. G. Schaub, T. Ikuta and C. A. Head, *Blood*, 2011, **117**, 727–735.

15. R. D. Jenison, S. D. Jennings, D. W. Walker, R. F. Bargatze and D. Parma, *Antisense Nucleic Acid Drug Dev.*, 1998, **8**, 265–279.

16. O. Leppanen, N. Janjic, M. A. Carlsson, K. Pietras, M. Levin, C. Vargeese, L. S. Green, D. Bergqvist, A. Ostman and C. H. Heldin, *Arterioscler. Thromb. Vasc. Biol.*, 2000, **20**, E89–E95.

17. J. Mi, X. Zhang, P. H. Giangrande, J. O. McNamara, 2nd, S. M. Nimjee, S. Sarraf-Yazdi, B. A. Sullenger and B. M. Clary, *Biochem. Biophys. Res. Commun.*, 2005, **338**, 956–963.

18. Y. Higashimoto, S. Yamagishi, K. Nakamura, T. Matsui, M. Takeuchi, M. Noguchi and H. Inoue, *Microvasc. Res.*, 2007, **74**, 65–69.

19. M. L. Marro, D. A. Daniels, A. McNamee, D. P. Andrew, T. D. Chapman, M. S. Jiang, Z. Wu, J. L. Smith, K. K. Patel and K. L. Gearing, *Biochemistry*, 2005, **44**, 8449–8460.

20. K. P. Williams, X. H. Liu, T. N. Schumacher, H. Y. Lin, D. A. Ausiello, P. S. Kim and D. P. Bartel, *Proc. Natl. Acad. Sci. USA*, 1997, **94**, 11285–11290.

21. M. Ellenbecker, L. Sears, P. Li, J. M. Lanchy and J. Stephen Lodmell, *Antiviral Res.* 2012, **93**, 330–339.

22. S. K. Choi, C. Lee, K. S. Lee, S. Y. Choe, I. P. Mo, R. H. Seong, S. Hong and S. H. Jeon, *Mol. Cells*, 2011, **32**, 527–533.

23. S. C. Gopinath, T. S. Misono, K. Kawasaki, T. Mizuno, M. Imai, T. Odagiri and P. K. Kumar, *J. Gen. Virol.*, 2006, **87**, 479–487.

24. A. K. Dey, M. Khati, M. Tang, R. Wyatt, S. M. Lea and W. James, *J. Virol.*, 2005, **79**, 13806–13810.

25. M. A. Ditzler, D. Bose, N. Shkriabai, B. Marchand, S. G. Sarafianos, M. Kvaratskhelia and D. H. Burke, *Nucleic Acids Res.*, 2011, **39**, 8237–8247.
26. D. M. Held, J. D. Kissel, J. T. Patterson, D. G. Nickens and D. H. Burke, *Front. Biosci.*, 2006, **11**, 89–112.
27. S. Kelley, S. Boroda, K. Musier-Forsyth and B. I. Kankia, *Biophys. Chem.*, 2011, **155**, 82–88.
28. E. Kraus, W. James and A. N. Barclay, *J. Immunol.*, 1998, **160**, 5209–5212.
29. N. Li, Y. Wang, A. Pothukuchy, A. Syrett, N. Husain, S. Gopalakrisha, P. Kosaraju and A. D. Ellington, *Nucleic Acids Res.*, 2008, **36**, 6739–6751.
30. J. Zhou, H. Li, J. Zhang, S. Piotr and J. Rossi, *J. Vis. Exp.*, 2011, http://www.jove.com/.
31. J. Zhou, Y. Shu, P. Guo, D. D. Smith and J. J. Rossi, *Methods*, 2011, **54**, 284–294.
32. K. Fukuda, Y. Toyokawa, K. Kikuchi, K. Konno, R. Ishihara, C. Fukazawa, S. Nishikawa and T. Hasegawa, *Nucleic Acids Symp. Ser. (Oxf.)*, 2008, 205–206.
33. K. Kikuchi, T. Umehara, F. Nishikawa, K. Fukuda, T. Hasegawa and S. Nishikawa, *Biochem. Biophys. Res. Commun.*, 2009, **386**, 118–123.
34. K. Konno, S. Fujita, M. Iizuka, S. Nishikawa, T. Hasegawa and K. Fukuda, *Nucleic Acids Symp. Ser. (Oxf.)*, 2008, 493–494.
35. F. Nishikawa, K. Funaji, K. Fukuda and S. Nishikawa, *Oligonucleotides*, 2004, **14**, 114–129.
36. M. D. Moore, D. H. Bunka, M. Forzan, P. G. Spear, P. G. Stockley, I. McGowan and W. James, *J. Gen. Virol.*, 2011, **92**, 1493–1499.
37. A. S. Barbas and R. R. White, *Curr. Opin. Investig. Drugs*, 2009, **10**, 572–578.
38. L. Cerchia, J. Hamm, D. Libri, B. Tavitian and V. de Franciscis, *FEBS Lett.*, 2002, **528**, 12–16.
39. J. L. Barton, D. H. Bunka, S. E. Knowling, P. Lefevre, A. J. Warren, C. Bonifer and P. G. Stockley, *Nucleic Acids Res.*, 2009, **37**, 6818–6830.
40. K. Sefah, Z. W. Tang, D. H. Shangguan, H. Chen, D. Lopez-Colon, Y. Li, P. Parekh, J. Martin, L. Meng, J. A. Phillips, Y. M. Kim and W. H. Tan, *Leukemia*, 2009, **23**, 235–244.
41. Z. Tang, D. Shangguan, K. Wang, H. Shi, K. Sefah, P. Mallikratchy, H. W. Chen, Y. Li and W. Tan, *Anal. Chem.*, 2007, **79**, 4900–4907.
42. H. W. Chen, C. D. Medley, K. Sefah, D. Shangguan, Z. Tang, L. Meng, J. E. Smith and W. Tan, *ChemMedChem*, 2008, **3**, 991–1001.
43. D. Shangguan, L. Meng, Z. C. Cao, Z. Xiao, X. Fang, Y. Li, D. Cardona, R. P. Witek, C. Liu and W. Tan, *Anal. Chem.*, 2008, **80**, 721–728.
44. K. Zhang, K. Sefah, L. Tang, Z. Zhao, G. Zhu, M. Ye, W. Sun, S. Goodison and W. Tan, *ChemMedChem*, 2012, **7**, 79–84.
45. S. E. Lupold, B. J. Hicke, Y. Lin and D. S. Coffey, *Cancer Res.*, 2002, **62**, 4029–4033.
46. J. C. Burnett and J. J. Rossi, *Chem. Biol.*, 2012, **19**, 60–71.
47. X. Ni, M. Castanares, A. Mukherjee and S. E. Lupold, *Curr. Med. Chem.*, 2011, **18**, 4206–4214.

48. G. R. Zimmermann, R. D. Jenison, C. L. Wick, J. P. Simorre and A. Pardi, *Nat. Struct. Biol.*, 1997, **4**, 644–649.
49. P. L. Sazani, R. Larralde and J. W. Szostak, *J. Am. Chem. Soc.*, 2004, **126**, 8370–8371.
50. J. P. Overington, B. Al-Lazikani and A. L. Hopkins, *Nat. Rev. Drug Discov.*, 2006, **5**, 993–996.
51. S. Li, H. Xu, H. Ding, Y. Huang, X. Cao, G. Yang, J. Li, Z. Xie, Y. Meng, X. Li, Q. Zhao, B. Shen and N. Shao, *J. Pathol.*, 2009, **218**, 327–336.
52. M. Homann and H. U. Goringer, *Nucleic Acids Res.*, 1999, **27**, 2006–2014.
53. S. P. Ohuchi, T. Ohtsu and Y. Nakamura, *Biochimie*, 2006, **88**, 897–904.
54. S. M. Shamah, J. M. Healy and S. T. Cload, *Acc. Chem. Res.*, 2008, **41**, 130–138.
55. S. Fitter and R. James, *J. Biol. Chem.*, 2005, **280**, 34193–34201.
56. K. Sefah, D. Shangguan, X. Xiong, M. B. O'Donoghue and W. Tan, *Nat. Protoc.*, 2010, **5**, 1169–1185.
57. L. Cerchia and V. de Franciscis, *Trends Biotechnol.*, 2010, **28**, 517–525.
58. K. T. Guo, G. Ziemer, A. Paul and H. P. Wendel, *Int. J. Mol. Sci.*, 2008, **9**, 668–678.
59. D. Shangguan, Z. Cao, L. Meng, P. Mallikaratchy, K. Sefah, H. Wang, Y. Li and W. Tan, *J. Proteome Res.*, 2008, **7**, 2133–2139.
60. C. D. Medley, S. Bamrungsap, W. Tan and J. E. Smith, *Anal. Chem.*, 2011, **83**, 727–734.
61. Z. Xiao, E. Levy-Nissenbaum, F. Alexis, A. Luptak, B. A. Teply, J. M. Chan, J. Shi, E. Digga, J. Cheng, R. Langer and O. C. Farokhzad, *ACS Nano.*, 2012, **6**, 696–704.
62. R. S. Geary, R. Z. Yu, T. Watanabe, S. P. Henry, G. E. Hardee, A. Chappell, J. Matson, H. Sasmor, L. Cummins and A. A. Levin, *Drug Metab. Dispos.*, 2003, **31**, 1419–1428.
63. R. M. Boomer, S. D. Lewis, J. M. Healy, M. Kurz, C. Wilson and T. G. McCauley, *Oligonucleotides*, 2005, **15**, 183–195.
64. H. Dougan, D. M. Lyster, C. V. Vo, A. Stafford, J. I. Weitz and J. B. Hobbs, *Nucl. Med. Biol.*, 2000, **27**, 289–297.
65. C. K. Dyke, S. R. Steinhubl, N. S. Kleiman, R. O. Cannon, L. G. Aberle, M. Lin, S. K. Myles, C. Melloni, R. A. Harrington, J. H. Alexander, R. C. Becker and C. P. Rusconi, *Circulation*, 2006, **114**, 2490–2497.
66. G. Mayer, *Angew Chem. Int. Ed. Engl.*, 2009, **48**, 2672–2689.
67. M. H. Caruthers, *Science*, 1985, **230**, 281–285.
68. Y. S. Sanghvi, *Curr. Protoc. Nucleic Acid Chem.*, 2011, **Chapter 4**, Unit 4 1 1–22.
69. L. S. Green, D. Jellinek, C. Bell, L. A. Beebe, B. D. Feistner, S. C. Gill, F. M. Jucker and N. Janjic, *Chem. Biol.*, 1995, **2**, 683–695.
70. J. Ruckman, L. S. Green, J. Beeson, S. Waugh, W. L. Gillette, D. D. Henninger, L. Claesson-Welsh and N. Janjic, *J. Biol. Chem.*, 1998, **273**, 20556–20567.
71. F. Eckstein and G. Gish, *Trends Biochem. Sci.*, 1989, **14**, 97–100.

72. M. L. Andreola, C. Calmels, J. Michel, J. J. Toulme and S. Litvak, *Eur. J. Biochem.*, 2000, **267**, 5032–5040.

73. S. Jhaveri, B. Olwin and A. D. Ellington, *Bioorg. Med. Chem. Lett.*, 1998, **8**, 2285–2290.

74. D. J. King, J. G. Safar, G. Legname and S. B. Prusiner, *J. Mol. Biol.*, 2007, **369**, 1001–1014.

75. J. Chelliserrykattil and A. D. Ellington, *Nat. Biotechnol.*, 2004, **22**, 1155–1160.

76. R. Sousa and R. Padilla, *EMBO J.*, 1995, **14**, 4609–4621.

77. K. S. Schmidt, S. Borkowski, J. Kurreck, A. W. Stephens, R. Bald, M. Hecht, M. Friebe, L. Dinkelborg and V. A. Erdmann, *Nucleic Acids Res.*, 2004, **32**, 5757–5765.

78. R. N. Veedu, B. Vester and J. Wengel, *J. Am. Chem. Soc.*, 2008, **130**, 8124–8125.

79. R. N. Veedu and J. Wengel, *Mol. Biosyst.*, 2009, **5**, 787–792.

80. A. R. Kore, M. Hodeib and Z. Hu, *Nucleosides Nucleotides Nucleic Acids*, 2008, **27**, 1–17.

81. R. N. Veedu, B. Vester and J. Wengel, *Chembiochem*, 2007, **8**, 490–492.

82. R. N. Veedu, H. V. Burri, P. Kumar, P. K. Sharma, P. J. Hrdlicka, B. Vester and J. Wengel, *Bioorg. Med. Chem. Lett.*, 2010, **20**, 6565–6568.

83. E. N. Brody, M. C. Willis, J. D. Smith, S. Jayasena, D. Zichi and L. Gold, *Mol. Diagn.*, 1999, **4**, 381–388.

84. K. B. Jensen, B. L. Atkinson, M. C. Willis, T. H. Koch and L. Gold, *Proc. Natl. Acad. Sci. USA*, 1995, **92**, 12220–12224.

85. M. C. Golden, B. D. Collins, M. C. Willis and T. H. Koch, *J. Biotechnol.*, 2000, **81**, 167–178.

86. M. D. Moore, J. Cookson, V. K. Coventry, B. Sproat, L. Rabe, R. D. Cranston, I. McGowan and W. James, *J. Biol. Chem.*, 2011, **286**, 2526–2535.

87. J. Floege, T. Ostendorf, U. Janssen, M. Burg, H. H. Radeke, C. Vargeese, S. C. Gill, L. S. Green and N. Janjic, *Am. J. Pathol.*, 1999, **154**, 169–179.

88. D. A. Di Giusto and G. C. King, *J. Biol. Chem.*, 2004, **279**, 46483–46489.

89. J. Kalia and R. T. Raines, *Curr. Org. Chem.*, 2010, **14**, 138–147.

90. V. V. Rostovtsev, L. G. Green, V. V. Fokin and K. B. Sharpless, *Angew Chem. Int. Ed. Engl.*, 2002, **41**, 2596–2599.

91. C. W. Tornoe, C. Christensen and M. Meldal, *J. Org. Chem.*, 2002, **67**, 3057–3064.

92. Q. Wang, T. R. Chan, R. Hilgraf, V. V. Fokin, K. B. Sharpless and M. G. Finn, *J. Am. Chem. Soc.*, 2003, **125**, 3192–3193.

93. V. D. Bock, H. Hiemstra and J. H. van Maarseveen, *Eur. J. Org. Chem.*, 2006, 51–68.

94. H. C. Kolb, M. G. Finn and K. B. Sharpless, *Angew Chem. Int. Ed. Engl.*, 2001, **40**, 2004–2021.

95. C. D. Hein, X. M. Liu and D. Wang, *Pharm. Res.*, 2008, **25**, 2216–2230.

96. H. C. Kolb and K. B. Sharpless, *Drug Discov. Today*, 2003, **8**, 1128–1137.

97. A. H. El-Sagheer and T. Brown, *Chem. Soc. Rev.*, 2010, **39**, 1388–1405.

98. W. H. Zhan, H. N. Barnhill, K. Sivakumar, T. Tian and Q. Wang, *Tetrahedron Lett.*, 2005, **46**, 1691–1695.

99. K. Gogoi, M. V. Mane, S. S. Kunte and V. A. Kumar, *Nucleic Acids Res.*, 2007, **35**, e139.

100. P. M. Gramlich, S. Warncke, J. Gierlich and T. Carell, *Angew Chem. Int. Ed. Engl.*, 2008, **47**, 3442–3444.

101. S. H. Weisbrod and A. Marx, *Chem. Commun. (Camb.)*, 2008, 5675–5685.

102. F. Seela, V. R. Sirivolu and P. Chittepu, *Bioconjug. Chem.*, 2008, **19**, 211–224.

103. M. Shelbourne, X. Chen, T. Brown and A. H. El-Sagheer, *Chem. Commun.*, 2011, **47**, 6257–6259.

104. P. Caliceti and F. M. Veronese, *Adv. Drug Deliv. Rev.*, 2003, **55**, 1261–1277.

105. J. M. Harris and R. B. Chess, *Nat. Rev. Drug Discov.*, 2003, **2**, 214–221.

106. M. J. Roberts, M. D. Bentley and J. M. Harris, *Adv. Drug Deliv. Rev.*, 2002, **54**, 459–476.

107. S. M. Ryan, G. Mantovani, X. Wang, D. M. Haddleton and D. J. Brayden, *Expert Opin. Drug Deliv.*, 2008, **5**, 371–383.

108. F. M. Veronese and G. Pasut, *Drug Discov. Today*, 2005, **10**, 1451–1458.

109. J. M. Healy, S. D. Lewis, M. Kurz, R. M. Boomer, K. M. Thompson, C. Wilson and T. G. McCauley, *Pharm. Res.*, 2004, **21**, 2234–2246.

110. K. L. Heredia, T. H. Nguyen, C. W. Chang, V. Bulmus, T. P. Davis and H. D. Maynard, *Chem. Commun. (Camb.)*, 2008, 3245–3247.

111. B. J. Hicke, A. W. Stephens, T. Gould, Y. F. Chang, C. K. Lynott, J. Heil, S. Borkowski, C. S. Hilger, G. Cook, S. Warren and P. G. Schmidt, *J. Nucl. Med.*, 2006, **47**, 668–678.

112. C. S. Fishburn, *J. Pharm. Sci.*, 2008, **97**, 4167–4183.

113. P. E. Burmeister, S. D. Lewis, R. F. Silva, J. R. Preiss, L. R. Horwitz, P. S. Pendergrast, T. G. McCauley, J. C. Kurz, D. M. Epstein, C. Wilson and A. D. Keefe, *Chem. Biol.*, 2005, **12**, 25–33.

114. T. Kawaguchi, H. Asakawa, Y. Tashiro, K. Juni and T. Sueishi, *Biol. Pharm. Bull.*, 1995, **18**, 474–476.

115. S. R. Watson, Y. F. Chang, D. O'Connell, L. Weigand, S. Ringquist and D. H. Parma, *Antisense Nucleic Acid Drug Dev.*, 2000, **10**, 63–75.

116. P. R. Bouchard, R. M. Hutabarat and K. M. Thompson, *Ann. Rev. Pharmacol. Toxicol.*, 2010, **50**, 237–257.

117. A. DeAnda, Jr., S. E. Coutre, M. R. Moon, C. M. Vial, L. C. Griffin, V. S. Law, M. Komeda, L. L. Leung and D. C. Miller, *Ann. Thorac. Surg.*, 1994, **58**, 344–350.

118. J. C. Gilbert, T. DeFeo-Fraulini, R. M. Hutabarat, C. J. Horvath, P. G. Merlino, H. N. Marsh, J. M. Healy, S. Boufakhreddine, T. V. Holohan and R. G. Schaub, *Circulation*, 2007, **116**, 2678–2686.

119. A. Bendele, J. Seely, C. Richey, G. Sennello and G. Shopp, *Toxicol. Sci.*, 1998, **42**, 152–157.

120. C. D. Conover, R. Linberg, C. W. Gilbert, K. L. Shum and R. G. Shorr, *Artif. Organs.*, 1997, **21**, 1066–1075.
121. C. Da Pieve, P. Williams, D. M. Haddleton, R. M. Palmer and S. Missailidis, *Bioconjug. Chem.*, 2010, **21**, 169–174.
122. M. C. Willis, B. D. Collins, T. Zhang, L. S. Green, D. P. Sebesta, C. Bell, E. Kellogg, S. C. Gill, A. Magallanez, S. Knauer, R. A. Bendele, P. S. Gill and N. Janjic, *Bioconjug. Chem.*, 1998, **9**, 573–582.
123. J. Soutschek, A. Akinc, B. Bramlage, K. Charisse, R. Constien, M. Donoghue, S. Elbashir, A. Geick, P. Hadwiger, J. Harborth, M. John, V. Kesavan, G. Lavine, R. K. Pandey, T. Racie, K. G. Rajeev, I. Rohl, I. Toudjarska, G. Wang, S. Wuschko, D. Bumcrot, V. Koteliansky, S. Limmer, M. Manoharan and H. P. Vornlocher, *Nature*, 2004, **432**, 173–178.
124. B. J. Hicke and A. W. Stephens, *J. Clin. Invest.*, 2000, **106**, 923–928.
125. Y. J. Lee and S. W. Lee, *Biochem. Biophys. Res. Commun.*, 2012, **417**, 521–527.
126. C. L. Esposito, D. Passaro, I. Longobardo, G. Condorelli, P. Marotta, A. Affuso, V. de Franciscis and L. Cerchia, *PLoS One*, 2011, **6**, e24071.
127. D. H. Burke and J. H. Willis, *RNA*, 1998, **4**, 1165–1175.
128. J. R. Kanwar, K. Roy and R. K. Kanwar, *Crit. Rev. Biochem. Mol. Biol.*, 2011, **46**, 459–477.
129. D. Peer and J. Lieberman, *Gene Ther.*, 2011, **18**, 1127–1133.
130. S. Guo, N. Tschammer, S. Mohammed and P. Guo, *Hum. Gene Ther.*, 2005, **16**, 1097–1109.
131. D. Shu, L. P. Huang, S. Hoeprich and P. Guo, *J. Nanosci. Nanotechnol.*, 2003, **3**, 295–302.
132. J. O. McNamara, 2nd, E. R. Andrechek, Y. Wang, K. D. Viles, R. E. Rempel, E. Gilboa, B. A. Sullenger and P. H. Giangrande, *Nat. Biotechnol.*, 2006, **24**, 1005–1015.
133. J. P. Dassie, X. Y. Liu, G. S. Thomas, R. M. Whitaker, K. W. Thiel, K. R. Stockdale, D. K. Meyerholz, A. P. McCaffrey, J. O. McNamara, 2nd and P. H. Giangrande, *Nat. Biotechnol.*, 2009, **27**, 839–849.
134. J. Zhou, H. Li, S. Li, J. Zaia and J. J. Rossi, *Mol. Ther.* 2008, **16**, 1481–1489.
135. J. Zhou, P. Swiderski, H. Li, J. Zhang, C. P. Neff, R. Akkina and J. J. Rossi, *Nucleic Acids Res.*, 2009, **37**, 3094–3109.
136. C. P. Neff, J. Zhou, L. Remling, J. Kuruvilla, J. Zhang, H. Li, D. D. Smith, P. Swiderski, J. J. Rossi and R. Akkina, *Sci. Transl. Med.*, 2011, **3**, 66ra66.
137. C. H. Chen, K. R. Dellamaggiore, C. P. Ouellette, C. D. Sedano, M. Lizadjohry, G. A. Chernis, M. Gonzales, F. E. Baltasar, A. L. Fan, R. Myerowitz and E. F. Neufeld, *Proc. Natl. Acad. Sci. USA*, 2008, **105**, 15908–15913.
138. J. R. Kanwar, R. R. Mohan, R. K. Kanwar, K. Roy and R. Bawa, *Nanomedicine (Lond.)*, 2010, **5**, 1435–1445.
139. H. Liu, P. Moy, S. Kim, Y. Xia, A. Rajasekaran, V. Navarro, B. Knudsen and N. H. Bander, *Cancer Res.*, 1997, **57**, 3629–3634.

140. T. C. Chu, J. W. Marks, 3rd, L. A. Lavery, S. Faulkner, M. G. Rosenblum, A. D. Ellington and M. Levy, *Cancer Res.*, 2006, **66**, 5989–5992.

141. D. Willner, P. A. Trail, S. J. Hofstead, H. D. King, S. J. Lasch, G. R. Braslawsky, R. S. Greenfield, T. Kaneko and R. A. Firestone, *Bioconjug. Chem.*, 1993, **4**, 521–527.

142. D. Shangguan, Y. Li, Z. Tang, Z. C. Cao, H. W. Chen, P. Mallikaratchy, K. Sefah, C. J. Yang and W. Tan, *Proc. Natl. Acad. Sci. USA*, 2006, **103**, 11838–11843.

143. Y. F. Huang, D. Shangguan, H. Liu, J. A. Phillips, X. Zhang, Y. Chen and W. Tan, *Chembiochem*, 2009, **10**, 862–868.

144. V. Bagalkot, O. C. Farokhzad, R. Langer and S. Jon, *Angew Chem. Int. Ed. Engl.*, 2006, **45**, 8149–8152.

145. V. Bagalkot, L. Zhang, E. Levy-Nissenbaum, S. Jon, P. W. Kantoff, R. Langer and O. C. Farokhzad, *Nano Lett.*, 2007, **7**, 3065–3070.

146. C. S. Ferreira, C. S. Matthews and S. Missailidis, *Tumour Biol.*, 2006, **27**, 289–301.

147. R. Savla, O. Taratula, O. Garbuzenko and T. Minko, *J. Control. Release*, 2011, **153**, 16–22.

148. D. Kim, Y. Y. Jeong and S. Jon, *ACS Nano*, 2010, **4**, 3689–3696.

149. A. Z. Wang, V. Bagalkot, C. C. Vasilliou, F. Gu, F. Alexis, L. Zhang, M. Shaikh, K. Yuet, M. J. Cima, R. Langer, P. W. Kantoff, N. H. Bander, S. Jon and O. C. Farokhzad, *ChemMedChem*, 2008, **3**, 1311–1315.

150. C. S. Ferreira, M. C. Cheung, S. Missailidis, S. Bisland and J. Gariepy, *Nucleic Acids Res.*, 2009, **37**, 866–876.

151. S. Soundararajan, W. Chen, E. K. Spicer, N. Courtenay-Luck and D. J. Fernandes, *Cancer Res.*, 2008, **68**, 2358–2365.

152. Y. A. Shieh, S. J. Yang, M. F. Wei and M. J. Shieh, *ACS Nano*, 2010, **4**, 1433–1442.

153. S. Dhar, F. X. Gu, R. Langer, O. C. Farokhzad and S. J. Lippard, *Proc. Natl. Acad. Sci. USA*, 2008, **105**, 17356–17361.

154. S. Dhar, N. Kolishetti, S. J. Lippard and O. C. Farokhzad, *Proc. Natl. Acad. Sci. USA*, 2011, **108**, 1850–1855.

155. Z. Cao, R. Tong, A. Mishra, W. Xu, G. C. Wong, J. Cheng and Y. Lu, *Angew Chem. Int. Ed. Engl.*, 2009, **48**, 6494–6498.

156. P. Mallikaratchy, Z. Tang, S. Kwame, L. Meng, D. Shangguan and W. Tan, *Mol. Cell. Proteomics*, 2007, **6**, 2230–2238.

157. Y. Wu, K. Sefah, H. Liu, R. Wang and W. Tan, *Proc. Natl. Acad. Sci. USA*, 2010, **107**, 5–10.

158. M. Wu, W. L. Brown and P. G. Stockley, *Bioconjug. Chem.*, 1995, **6**, 587–595.

159. M. Wu, T. Sherwin, W. L. Brown and P. G. Stockley, *Nanomedicine*, 2005, **1**, 67–76.

160. N. Stephanopoulos, G. J. Tong, S. C. Hsiao and M. B. Francis, *ACS Nano.*, 2010, **4**, 6014–6020.

161. D. W. Hwang, H. Y. Ko, J. H. Lee, H. Kang, S. H. Ryu, I. C. Song, D. S. Lee and S. Kim, *J. Nucl. Med.*, 2010, **51**, 98–105.

162. K. E. Borbas, C. S. M. Ferreira, A. Perkins, J. I. Bruce and S. Missailidis, *Bioconjug. Chem.*, 2007, **18**, 1205–1212.
163. C. Da Pieve, A. C. Perkins and S. Missailidis, *Nucl. Med. Biol.*, 2009, **36**, 703–710.
164. J. Charlton, J. Sennello and D. Smith, *Chem. Biol.*, 1997, **4**, 809–816.
165. Z. Q. Cui, Q. Ren, H. P. Wei, Z. Chen, J. Y. Deng, Z. P. Zhang and X. E. Zhang, *Nanoscale*, 2011, **3**, 2454–2457.
166. M. V. Yigit, D. Mazumdar and Y. Lu, *Bioconjug. Chem.*, 2008, **19**, 412–417.
167. J. L. Brennan, N. S. Hatzakis, T. R. Tshikhudo, N. Dirvianskyte, V. Razumas, S. Patkar, J. Vind, A. Svendsen, R. J. Nolte, A. E. Rowan and M. Brust, *Bioconjug. Chem.*, 2006, **17**, 1373–1375.
168. J. Cardinal, J. R. Klune, E. Chory, G. Jeyabalan, J. S. Kanzius, M. Nalesnik and D. A. Geller, *Surgery*, 2008, **144**, 125–132.
169. A. Ito, Y. Kuga, H. Honda, H. Kikkawa, A. Horiuchi, Y. Watanabe and T. Kobayashi, *Cancer Lett.*, 2004, **212**, 167–175.
170. K. Hamad-Schifferli, J. J. Schwartz, A. T. Santos, S. Zhang and J. M. Jacobson, *Nature*, 2002, **415**, 152–155.
171. M. C. Buff, F. Schafer, B. Wulffen, J. Muller, B. Potzsch, A. Heckel and G. Mayer, *Nucleic Acids Res.*, 2010, **38**, 2111–2118.
172. V. Mikat and A. Heckel, *RNA*, 2007, **13**, 2341–2347.
173. K. Konig, *J. Microsc.* 2000, **200**, 83–104.
174. K. Taira, K. Abe, T. Ishibasi, K. Sato and K. Ikebukuro, *J. Nucleic Acids*, 2011, 103872.
175. C. A. Farman and D. J. Kornbrust, *Toxicol. Pathol.*, 2003, **31 Suppl**, 119–122.
176. S. P. Henry, G. Beattie, G. Yeh, A. Chappel, P. Giclas, A. Mortari, M. A. Jagels, D. J. Kornbrust and A. A. Levin, *Int. Immunopharmacol.*, 2002, **2**, 1657–1666.
177. J. K. Marquis and J. M. Grindel, *Curr. Opin. Mol. Ther.*, 2000, **2**, 258–263.
178. J. P. Sheehan and H. C. Lan, *Blood*, 1998, **92**, 1617–1625.
179. W. Barchet, V. Wimmenauer, M. Schlee and G. Hartmann, *Curr. Opin. Immunol.*, 2008, **20**, 389–395.
180. D. Yu, D. Wang, F. G. Zhu, L. Bhagat, M. Dai, E. R. Kandimalla and S. Agrawal, *J. Med. Chem.*, 2009, **52**, 5108–5114.
181. C. M. Smith and J. A. Steitz, *Cell*, 1997, **89**, 669–672.
182. V. B. Pinheiro, A. I. Taylor, C. Cozens, M. Abramov, M. Renders, S. Zhang, J. C. Chaput, J. Wengel, S.-Y. Peak-Chew, S. H. McLaughlin, P. Herdewijn and P. Holliger, Synthetic Genetic Polymers Capable of Heredity and Evolution, *Science*, 2012, **336**, 341–344.
183. G. F. Joyce, Toward an Alternative Biology, *Science*, 2012, **336**, 307–308.

CHAPTER 7

pH Sensitive DNA Devices

SONALI SAHA AND YAMUNA KRISHNAN*

National Centre for Biological Sciences, TIFR, GKVK, Bellary Road, Bangalore 560065, India
*Email: Yamuna@ncbs.res.in

7.1 Introduction

Solution pH regulates the protonation state of molecules and hence changes their physicochemical properties such as chemical reactivity,[1] solubility,[2–4] conductance,[5–8] and spectral properties to name a few.[9,10] Such pH-mediated physicochemical changes at the small molecule level also translate to larger biomolecules that comprise many protonatable side groups or residues where these residue-level changes ultimately alter the structure and function of the relevant biomacromolecule. The pH is known to affect biomolecules in terms of their phase transitions,[11] structural transitions,[12] biomolecular interactions[13,14] and hence their functions such as catalytic activity.[15–17] General acid–base catalysis carried out by both protein and RNA in living cells, in which enzyme functional groups donate and accept protons, is the most common example.[18–22] Effective general acid–base catalysis requires atoms with pK_as near the pH of the reaction, which in the case of biological systems is close to pH 7.4. The protonated or ionised forms of nucleobases are important in RNA structure and also in RNA function for nucleobase-mediated catalysis.[23,24]

The pH-switchable physicochemical characteristics of biomolecules are utilised by nature to toggle cellular processes using pH modulation such as receptor–cargo dissociation,[25] enzymatic degradation,[25] entry of toxins and viruses,[26] post-translational modifications and sorting of proteins,[27–30] cellular

RSC Biomolecular Sciences No. 26
DNA Conjugates and Sensors
Edited by Keith R Fox and Tom Brown
© The Royal Society of Chemistry 2012
Published by the Royal Society of Chemistry, www.rsc.org

proliferation,[31] cell migration,[32,33] apoptosis,[34] signalling, *etc.*[35,36] The pH-regulated cellular events also influence biological phenomena at the organismal level such as embryogenesis, yolk uptake, spermatogenesis, neurodegeneration, excretion and cell–cell fusion.[37]

The nucleic acid scaffold lends itself to exploitation of the same principles as those present in naturally occurring pH-induced structural transitions, to enable the engineering of artificial pH-switchable molecular devices. Similar to RNA, the DNA scaffold can also undergo pH-triggered structural transitions. However, the physiological significance of these is yet to be established convincingly. DNA is favoured over RNA as a nucleic acid scaffold for building artificial devices owing to its inherently superior thermal and hydrolytic stability. Here we will cover these pH-switchable DNA structural motifs and describe their functional evolution into artificially engineered pH-stimulated devices.

7.2 Triplexes

7.2.1 Structural Features of Triplexes

Triplexes were first reported in 1957 by Felsenfeld *et al.*, who found that poly (U) and poly (A) formed a stable 2:1 complex in the presence of $MgCl_2$.[38] Pyrimidine oligonucleotides fit into the major groove of duplex DNA with a strand orientation parallel to the purine-rich strand of the Watson–Crick base paired duplex to form a local pyrimidine–purine*pyrimidine triple-helical structure via Hoogsteen base pairing.[39–45] Triplexes usually consist of two pyrimidine and one purine strand (YR*Y) or *vice versa* (YR*R). YR*Y triplexes consist of isomorphous base pairs TA*T and CG*C$^+$ (Figure 7.1a and b). Importantly, CG*C$^+$ triplexes, unlike TA*T triplexes, require the protonation of the N3 of cytosine in the third strand which makes the former favourable under acidic conditions. Triplexes are also stabilised by CG*G, TA*A and CG*A$^+$ base pairing (Figure 7.1c, d and e). CG*A$^+$ base pairs form at acidic pH.[46]

The canonical intermolecular triplex consists of either (a) three independent oligonucleotide strands or (b) a long DNA duplex carrying a homopurine–homopyrimidine insert and the corresponding triplex-pairing oligonucleotide sequence.[40,47–50] Different topologies of triplexes have been experimentally verified.[45] The intramolecular triple-stranded DNA structure formed by a purine [d(R_n)] and pyrimidine [d(Y_n)] rich duplex DNA sequence has been described as H-DNA by Frank-Kamenetskii and co-workers.[51] In this unusual structure, half of the d(Y_n) strand forms a Watson–Crick duplex with the (dR_n) strand, while the other half of (dY_n) folds back and binds in the major groove of the duplex via Hoogsteen base pairs. The structural features of DNA triplexes have been confirmed using different methods such as chemical and enzymatic probing, affinity cleavage and gel electrophoresis.[45]

Fibre X-ray diffraction of poly (dT)–poly (dA)–poly (dT) revealed that the axial rise per base in triplexes is 3.26 Å with a helical twist of 30°.[52] A pitch of 12

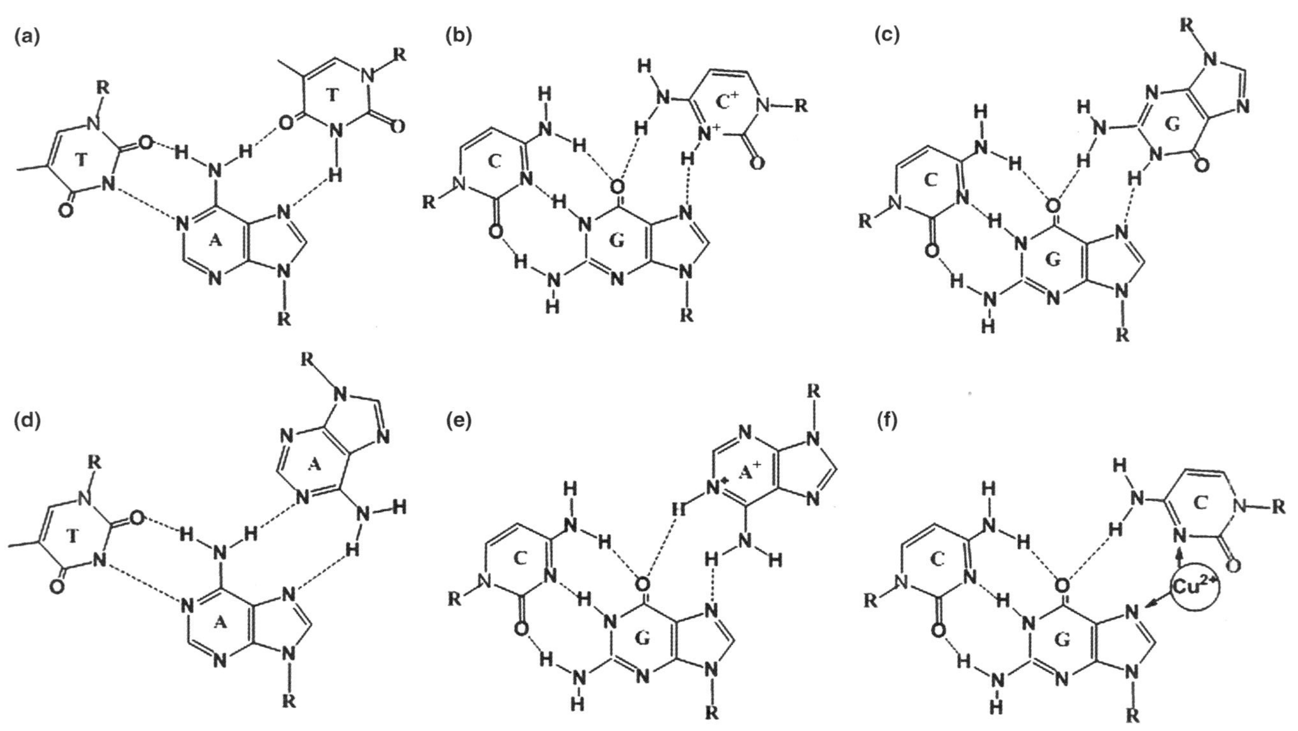

Figure 7.1 Base triplets involved in triplex formation: (a) TA*T, (b) CG*C⁺, (c) CG*G, (d) TA*A, (e) CG*A⁺ and (f) C^Cu G*C.

base pairs per turn, dislocation of the axis by almost 3 Å and small base-tilts result in a large and deep major groove that is capable of accommodating the third strand. Subsequently, two-dimensional nuclear magnetic resonance (2D NMR) and infrared spectroscopy confirmed that nucleotides in the triplex structure adopted a C2′ *endo* sugar pucker as seen in B-DNA.[53–55] The NMR studies on inter- and intramolecular triplexes unambiguously showed the requirement for cytosine protonation in the formation of CG*C$^+$ triplets, with the involvement of Hoogsteen base pairing and the parallel orientation of the third strand.[42,44,56,57]

An important factor in the stability of pH sensitive triplexes is protonation at the N3 position of cytosine in the third strand. This enables the formation of CG*C$^+$ base triplets and provides partial backbone charge neutralisation. Triplexes containing CG*C$^+$ base pairs are stable in acidic to neutral solutions, but dissociate as the pH increases.[40,48,50,58–60] Substitution of the cytosines in the third strand with 5-methylcytosines (m^{5}C) increases the apparent stability and consequently the upper limits of pH requirements for triplex formation.[59–62] In addition, Cu^{2+} can promote the formation and stabilisation of a pyrimidine triplex under nearly physiological conditions, bypassing the protonation requisite of the third strand cytosines.[63,64] The classical Hoogsteen hydrogen bond between the N3 atom of the protonated cytosine in the third strand and the N7 atom of the guanine in the duplex is replaced by a copper ion as shown in Figure 7.1f. In the C^{Cu}G*C base triplet, Cu^{2+} ions are chelated by N7 of guanine and N3 of the third strand cytosine.

The pH dependence of strand dissociation constants is a result of the rapid acid–base equilibrium of pyrimidine single strands. The association rate for triplex formation decreases with increasing pH in accordance with the dissociation constants. The triplex formation is a second-order reaction at low pH, whereas it can be interpreted as a third-order reaction at neutral pH, suggesting that different triplex formation pathways are observed depending on the pH.[65] The chemical reactivity of triplexes indicates that long purine and pyrimidine sequences (>36 base pairs) can adopt intramolecular triplexes and thereby reduce the dependence on low pH for structure formation.[66]

7.2.2 Molecular Devices Based on Triplexes

Structural transitions involving triplexes have been used to construct DNA nanomachines in which controlled molecular scale motion can be achieved by conformational changes within the DNA assembly that are induced by alteration in environmental pH.[67] These nanomachines are based on the reversible formation and dissociation of a DNA triplex containing CG*C$^+$ triplets across a range of solution pH, between pH 5.0 and 8.0. The conformational change of the machine has been monitored by Fluorescence Resonance Energy Transfer (FRET). The design of the nanomachine consists of three strands. At pH 8.0, the nanomachine forms an open state complex with three 15-base-pair duplexes and a single-stranded region, shown in green

(Figure 7.2a), which adopts a random coil configuration. Upon lowering the pH, this single-stranded region associates with the duplex to form a triplex, which leads to increase in FRET. This operation is highly reversible, with just 47% CG*C$^+$ triplet base composition being adequate to stabilise the triplex in acidic solution.

Conformational changes of triplexes have also been observed using DNA gold nanoparticle (DNA–AuNP) conjugates.[68,69] A binary mixture of AuNPs was separately functionalised with duplex forming, as well as pyrimidine-rich, strands using well-established thiol chemistry. At neutral pH, the binary mixture of these functionalised AuNPs remained as isolated particles in solution and showed a characteristic plasmon band.

At pH 5.0, a fraction of the cytosines in the pyrimidine-rich sequence, shown in green (Figure 7.2b), become protonated and associate with the DNA duplex to form a stable DNA triplex which leads to DNA–AuNP aggregation. Upon aggregation, surface plasmons on the AuNPs couple with each other and the

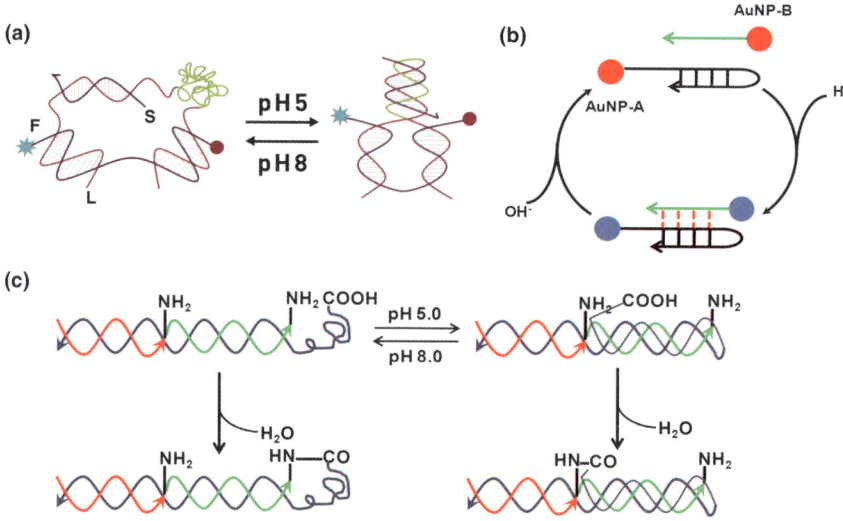

Figure 7.2   (a) The construction and operation of a DNA nanomachine. The machine consists of three DNA strands: a strand with a fluorescent label (F, shown as purple line), a long strand (L, shown as pink line), and a short strand (S, shown as purple line). The blue filled stars and red filled circles represent rhodamine green and quencher-1 (BHQ-1), respectively. A DNA triplex involving the S and L strands forms and dissociates reversibly.[67] (b) Triplex–gold nanoparticle (AuNP) conjugates and their assembly. AuNP-A functionalised with DNA strands that can form an intramolecular duplex (black) and an AuNP-B functionalised with a triplex forming DNA (green). At acidic pH, the triplex-driven assembly of AuNPs yielded three-dimensional networks of AuNPs.[68] (c) Schematic showing redirection of chemical reactions by a conformational switch that alters the location of the carboxyl group. This is achieved by formation and dissociation of a DNA triplex induced by a pH change.[70]

resonance wavelength undergoes a red-shift. This is associated with visible colour change and establishes the basis of a colorimetric pH sensor based on triplex formation.

A strategy has been developed to redirect chemical reactions on the basis of a DNA duplex–triplex transition.[70] Three functional groups are conjugated to three DNA single strands, respectively, which associate with each other to form a complex (Figure 7.2c). When the pH is altered to give acidic conditions, the formation of a triplex positions the two reactive functional groups so as to promote a chemical reaction between them. Thus a change in environment brings about a structural transition in a DNA assembly, which results in a chemical conjugation event.

7.3 i-Motifs

7.3.1 Structural Features of i-Motifs

In 1993, Leroy *et al.* reported a four-stranded nucleic acid motif formed from C-rich sequences containing intercalated $C \cdot C^+$ base pairs, namely the i-motif.[71] The i-motif is a four-stranded complex formed from two parallel stranded duplexes held together by $C \cdot C^+$ base pairs (Figure 7.3a) that associate non-covalently with the base pairs of the component parallel duplex, being fully intercalated with each other. This led to the motif being called the

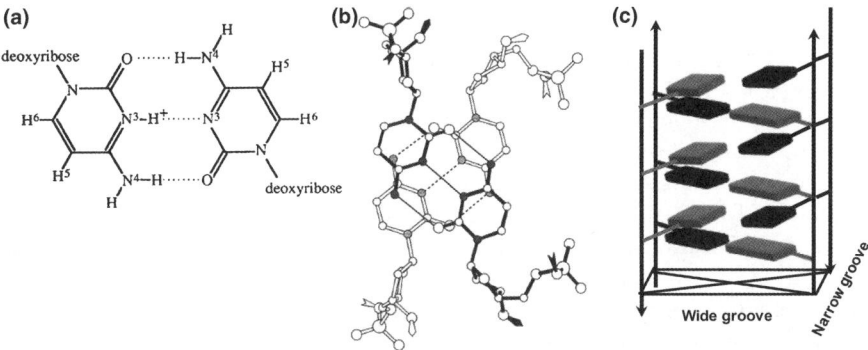

Figure 7.3 (a) Scheme of hydrogen bonding observed in hemiprotonated cytosines ($C \cdot C^+$ base pairs). The imino proton switches rapidly between the two bases. (b) Stacking of adjacent $C \cdot C^+$ base pairs in i-tetraplexes or i-motifs. Note the complementary arrangement of the exocyclic keto and amino groups and absence of stacking of the aromatic nucleobase heterocycles. The $C \cdot C^+$ base pair of one parallel duplex is drawn with solid bonds and associated hydrogen bonds between the cytosines are shown with solid lines. The adjacent $C \cdot C^+$ base pair from the intercalated duplex is depicted with open bonds, and hydrogen bonds are shown with broken lines. Nitrogens are highlighted by stippling. (c) Schematic showing the two grooves formed by intercalation of both parallel duplexes.

i-motif, or intercalated motif. At acidic or mildly acidic pH, the cytosines undergo hemi-protonation to form $C \cdot C^+$ base pairs, originally reported in crystals of acetyl cytosines and later in polydeoxy- and polyribonucleotides.[72–75] Hence, the stability of the i-motif depends on solution pH.[76,77] In the i-motif, both the parallel duplexes are oriented anti-parallel to each other such that each base pair is face-to-face with its neighbours (Figure 7.3b). Crystallographic data on d-CCCC and d-CCCT oligodeoxynucleotides confirmed that the individual parallel stranded duplexes are right-handed and underwound, and that both the constituent duplexes are 'zipped together' in an antiparallel fashion, giving rise to two narrow and two wide grooves (Figure 7.3c).[78,79] All four strands in the i-motif appear identical on NMR time scales. Owing to intercalation, unusually close intermolecular contacts between the sugar–phosphate backbones in the narrow grooves result in unfavourable electrostatic repulsion, which needs to be shielded by cations or bridging water molecules for i-motif stability.[80,81] It has been shown that these close contacts between hydrogen atoms from adjacent sugar moieties in the two strands in narrow grooves are the consequence of stabilising C–H...O type hydrogen bonds between neighbouring deoxyriboses.[82]

Investigation of the solution structure of i-motif DNA at various pHs revealed that the i-motif is structurally dynamic over a wide pH range and adopts multiple conformations ranging from the folded i-motif to a random coil conformation.[83]

In the case of PNA, in which the sugar–phosphate component of DNA is replaced with a neutral N-(2-aminoethyl)glycine spacer, the i-motif has comparable thermal stability to the DNA analogue, but exists over a very narrow pH range.[84] This high sensitivity to solution pH is probably due to the lack of electrostatic energy gain corresponding to partial neutralization of backbone charge by the protonated cytosines.[80]

In order to understand narrow groove interactions in i-motif formation, the concept of the hybrid i-motif was developed, in which a 1 : 1 binary mixture of nucleic acid analogues of i-motif forming sequences (DNA–PNA, RNA–PNA and DNA–RNA), with varying sugar–phosphate backbones, has been used to form hybrid i-motifs. Interestingly, such a binary mixture yields a unique population of hybrid i-motifs instead of a statistical mixture of C-rich nucleic acid analogues. Therefore, comparison of this observed structure with the other theoretical possibilities that could have arisen from the binary mixture revealed the stabilising factors that promote i-motif formation.[85–87] In the PNA_2–DNA_2 i-motif, the PNA–DNA heteroduplexes are formed and oriented such that both DNA strands occupy one narrow groove and the PNA strands occupy the other groove, indicating the importance of sugar–sugar interactions between the DNA backbones, despite the unfavourable collateral electrostatic repulsion.[85] This hybrid i-motif shows increased thermal stability compared with the DNA_4 and PNA_4 i-motifs, probably due to reduction in the electrostatic repulsion associated with multi-stranded structures. The hybrid i-motif is an electrically neutral complex, compared with the net negatively charged DNA_4 i-motif and net positively charged PNA_4 i-motif. The hybrid i-motif also exists over a pH regime intermediate between the parent DNA_4 and PNA_4 i-motifs, probably owing to the lower number of DNA backbones, which confers a limited ability to tolerate pH changes.

In the RNA_2–PNA_2 hybrid i-motif, RNA–PNA duplex intercalation reveals that the ribose–ribose interaction in the narrow groove is still more favourable, despite both the electrostatic repulsion between the ribose phosphates and the steric clash between the 2′ hydroxyls.[86] Interestingly, NMR data indicate that the partially intercalated topology for the RNA_2–PNA_2 hybrid i-motif is analogous to the M form of the RNA_4 i-motif, avoiding one of the six possible 2′-hydroxyl/2′-hydroxyl contacts.[88]

A binary mixture of C-rich DNA and RNA kinetically gives rise to DNA_2–RNA_2 with intermediate stability, which eventually disproportionates to form the thermodynamically favoured DNA_4 i-motif.[87] Similar to the RNA_2–PNA_2 hybrid, the DNA_2–RNA_2 hybrid i-motif displays two different sugar backbones that can self-sort into narrow grooves, indicating the importance of the specificity of sugar–sugar interaction over steric clash between the 2′ hydroxyls. In summary, these studies have shown that the intercalation topology of the i-motifs depends on sugar–sugar interactions in the narrow groove and that these interactions have a subtle specificity that drives i-motif topology.

Given that the pK_a value of N3 in cytidine monophosphate is 4.4–4.5,[89] the maximum number of $C \cdot C^+$ base pairs would be expected to form at a pH equal to the pK_a of cytosine N3. The thermal stability of the $d(TC_5)$ i-motif as a function of pH revealed that it exists over a pH regime 4.2–6.8 with maximum stability at pH 4.5, the pK_a of cytosine. Long lengths of cytosine repeats increase the thermal stability of i-motifs even at higher pH.[77,90] The pH-dependent structural transition from an unfolded single stranded C-rich segment to the folded i-motif form has been studied for different lengths of human c-*myc* locus using Circular Dichroism (CD) spectra, and native gel electrophoresis shows that thermal stability increases with length of the i-motif at even higher pH.[91] The Krishnan group has shown that the C-rich oligonucleotides can self-associate into long wire-like structures called I-wires made up of slipped i-motifs.[92] Leroy *et al.* have investigated the mechanism of I-wire formation and shown that these supramolecular structures form depending on the monomer concentration used.[92,93] These higher order structures show high thermal stability and resistance to pH denaturation attributed to an increased number of $C \cdot C^+$ base pairs. Other than the potential applications of these nanowires in nanotechnology, their formation suggested an association of incompletely intercalated i-motifs, as proposed by the Krishnan group, which were subsequently identified as the intermediates in the i-motif formation pathway by Leroy.[94]

A study on human telomeric DNA sequences showed that under near-physiological conditions of pH, temperature and salt concentration, telomeric DNA is predominantly in a duplex form.[95] However, pH plays a major structural role and low pH stabilises the folded i-motif structure of the C-rich strand, thus acting as a competitor for duplex formation.

7.3.2 Molecular Devices Based on i-Motifs

In 2002, Alberti and Mergny showed that, for G-rich strands, the transition between duplex and quadruplex states induced by the addition of a "DNA fuel" strand can function as a nanoswitch over a number of cycles.[96] This study inspired the concept of developing an analogous nanomachine using the conformational switch between an i-motif and a duplex of C-rich strands triggered by pH modulation. Subsequently, Liu *et al.* constructed a molecular switch based on the fast transition between i-motif and duplex that is driven by a pH toggle (Figure 7.4a).[97] This operation generates only water and salt as by-products. Hence, this nanomachine can undergo many working cycles between open and closed forms with high processivity. These studies show that the relative stability of i-motifs increases at mildly acidic pH when compared with mismatched Watson–Crick duplexes and they can switch structurally with the aid of a pH toggle. In this study, the original buffer conditions had to be restored to achieve cyclical operation.

However, continuous operation of these devices necessitates keeping track of the state of the devices and the external addition of acid or base at specific times. The Simmel group took a conceptually different approach to demonstrate the conformational transition of a single-stranded DNA molecule, driven autonomously by an oscillating chemical reaction occurring in a continuous reactor pre-designed into the system.[98] Demonstration of this autonomous motion in molecular devices is a key advancement in this field. Here, the

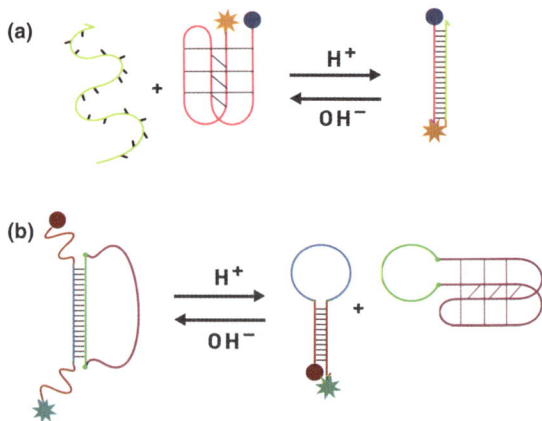

Figure 7.4   (a) An i-motif based molecular machine that uses a pH toggle. At acidic pH the C-rich strand (pink line) folds into an i-motif leaving behind the complementary strand (green line). At high pH, it unfolds to form a Watson–Crick duplex. The blue filled circles and dark yellow filled stars represent the quencher and the fluorophore respectively.[97] (b) Formation of an i-motif controls the motion of another DNA assembly such as a molecular beacon. Stars indicate the fluorophore and spheres represent quenchers used to monitor the molecular motion.[109]

conformational switch of a C-rich single strand was monitored using fluorescence quenching and showed a sharp transition between pH 5.5 and 6.5. In this setup, the oscillations slowly died out because of the continuous decrease in reactant concentrations. In principle, an infinite number of homogeneous pH oscillations require continuous filling, combined with the simultaneous removal of waste materials. To overcome this limitation, Simmel's group operated the chemical oscillator in a reactor with outlets to drive the DNA nanomachine, which was immobilised on a gold-coated glass surface.[99] They showed that the surface-bound DNA nanomachines functioned with improved efficiency and showed a highly processive autonomous conformational switching.

It can be toggled indefinitely by other stimuli such as UV irradiation that each can induce a change of solution pH.[100,101] Liu *et al.* showed conformational switching of i-motif DNA by a coupled light-induced OH^- emitter system in non-contact mode. In the presence of UV light, malachite green carbinol bases (MGCB) give out OH^- ions and lead to an increase in the solution pH. The rise in solution pH deforms i-motif structures into random coils with deprotonation of the $C \cdot C^+$ base pairs. In the dark condition, the malachite green (MG) cation recombines with the OH^- ions and returns to the MGCB, making the solution acidic. Consequently, the C-rich oligonucleotides switch back to the i-motif conformation. However, reversible cycling happens over much longer time scales. These authors have also used an alternating electric field to accelerate cooperative molecular motion for the surface-immobilised i-motif based device, leading to a remarkable shortening of the closing times.[102] One of the important findings from these studies was that i-motif based nanomachines are functional despite immobilisation on a 2D surface.

Advanced designs of i-motif based switches consist of, coaxially, a stacked DNA duplex which undergoes a bending motion transduced by i-motif formation (Figure 7.5).[103,104] Liu's group has constructed a DNA nano-triangle with a C-rich oligonucleotide in one arm.[104] Changes in environmental pH trigger the C-rich strand to fold into an i-motif and hence compact the triangle (Figure 7.5b). However, it generates single-strand waste in its closed state. The Krishnan group developed an i-motif based DNA nanomachine, namely the I-switch, in which coaxially stacked duplex segments contain C-rich overhangs with two cytosine stretches in each (Figure 7.5a).[103] These remain single stranded at higher pH (open state) and combine intramolecularly to form an i-motif at acidic pH (closed state). Here the transition between the closed and open state is completely intramolecular and is monitored by FRET; a more quantitative method than the fluorescence quenching used in most of the previously mentioned devices. All these features make the I-switch suitable to function as a pH sensor within a complex intracellular environment. The study showed that the I-switch can enter cells via a specific receptor-mediated endocytic pathway, without any further modification, reside in endosomes and quantitatively report the pH change during endosomal maturation. The I-switch has also been targeted to endosomal recycling pathways by conjugation to a given endocytic ligand. A subsequent study showed that the I-switch

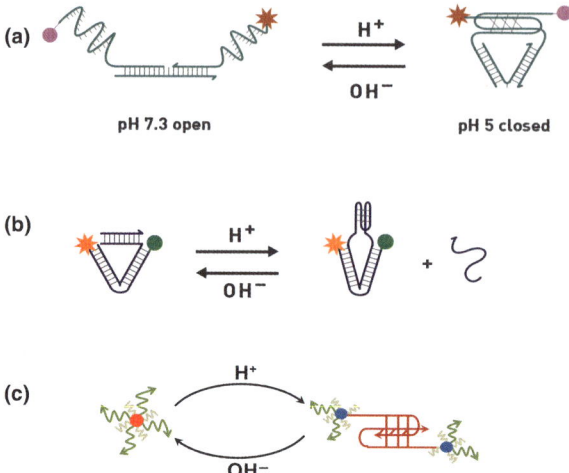

pH 7.3 open   **pH 5 closed**

Figure 7.5   (a) I-switch: now used for biological applications. It is a three-strand construct containing a duplex region with C-rich overhangs at both ends modified with a FRET pair (violet circle and brown star). At acidic pH, formation of an i-motif by the C-rich overhangs brings the two ends of the duplex closer, leading to FRET.[103] (b) DNA nano-triangle with C-rich strand in one arm (green lines). At acidic pH it dissociates from its complementary duplex and folds into an i-motif, scrunching the triangle.[104] (c) AuNP aggregation driven by interparticle i-motif formation at acidic pH.[106]

can be specifically targeted to a given receptor *in vivo* and is able to sense pH in real-time along a given endocytic pathway within the cells of a living organism.[105] The I-switch represents the first demonstration of the functionality of a rationally designed DNA nanomachine inside living systems. Thus pH-sensitive nucleic acid devices represent a class of DNA devices that have great potential within living systems.

Structural transitions of i-motifs have been shown to generate different outputs as a functional readout of environmental pH. The i-motif based self assembly has been used to drive pH-dependent aggregation of AuNPs that leads to changes in the optical properties of AuNPs and associated visible colour change (Figure 7.5c). This colorimetric assay has been exploited as an efficient pH sensor *in vitro* with an impressive precision of 0.04 pH units.[106–108] A study showed that the formation of an i-motif could disassemble certain forms of Watson–Crick interactions under isothermal conditions in response to pH variations that control the motion of another DNA nanomachine such as a molecular beacon (see Figure 7.4b).[109] The nanomechanical motion caused by the folding and unfolding of the i-motif has also been transduced to position a fluorophore proximal and distal from a gold-coated surface.[110]

C-Rich oligonucleotides immobilized on one side a of cantilever form a self-folded i-motif upon change in solution pH and induce repulsive in-plane surface forces or compressive surface stress which causes a cantilever to bend.[111] These experiments provide the first direct experimental demonstrations that DNA-based nanodevices can actually be used to generate force.

The structural transition between duplex and i-motif can also be described as an enthalpy-driven transformation from one given ordered state into another. To achieve enthalpy-driven wettability of the surface, DNA modified with a fluoride-containing hydrophobic group was immobilised onto a gold surface using gold–thiol chemistry. Upon modulation of the solution pH, the DNA strand could conceal and expose the fluoride-containing hydrophobic group by virtue of a pH-triggered conformational change of the i-motif structure into an extended structure, such as either a stretched single-stranded structure or a double-stranded structure, leading to intelligent switching of surface properties.[112]

Covalently modified Single-Walled Carbon Nanotube (SWCNTs) with i-motif forming oligonucleotide sequences showed controllable/switchable electrochemical activity and capacitor behaviour, due to changes in the charge transport upon stacking or unstacking of DNA strands, during formation of a pH-induced i-motif (Figure 7.6a). The SWCNT attachments significantly affect the thermodynamic properties of the DNA strand by improving the stability of the i-motif structure through hydrophobic interactions.[113] In an interesting study, i-motif forming sequences immobilised on the surface of a conical pore have been shown to control the closing and opening of the pore, which functions as an ion channel mimic. At neutral pH, the random coil conformation of i-motif forming sequences allows the passage of ions, whereas at acidic pH, the

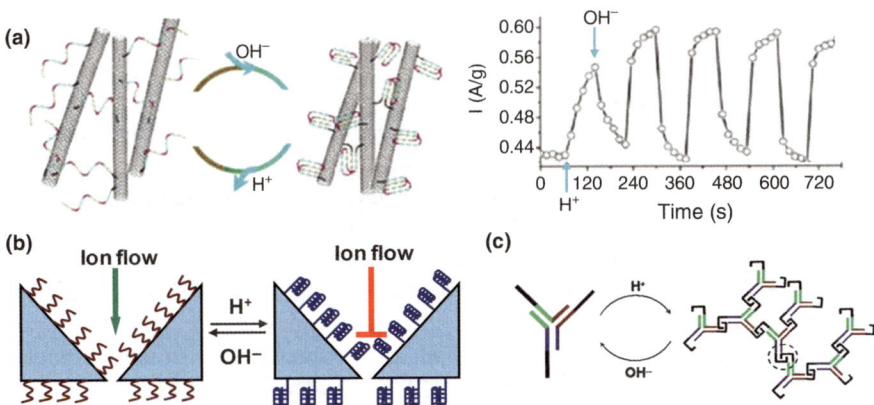

Figure 7.6 (a) pH-switchable electrochemical activity. SWCNT covalently modified with i-motif forming sequences. At low pH, i-motif formation aligns the tubes and changes their electrical properties.[113] (b) Ion channel mimic. C-rich DNA strands immobilised on the surface of a conical pore. Single-strand conformation allows passage of ions but i-motif formation at acidic pH blocks the pore.[114] (c) Three-way junction with C-rich overhangs at the each end form 3D networks via intermolecular i-motif formation.[115]

i-motifs formed by these sequences block the nanopore (Figure 7.6b). Here the gating performance was determined by measuring the ionic current across the channel in an environment with varying proton content.[114]

In 2009, Cheng *et al.* reported a pH-responsive pure DNA hydrogel free from ligating enzymes (see Figure 7.10c, below).[115,116] They designed a Y-shaped three-way junction complex with interlocking domains containing two cytosine-rich stretches at each terminus that can form intramolecular i-motif structures. At neutral pH, these Y-shaped building blocks remain as monomeric units owing to electrostatic repulsion. When the pH is decreased to 5, the interlocking domains act as 'sticky ends' owing to the formation of an intermolecular i-motif between adjacent Y-shaped structures, resulting in a hydrogel. The transition from solution state to gel state happens over a very short time scale. Such hydrogels could potentially be used as pH-toggled delivery agents.

7.4   Π-Helix and A-Motifs

7.4.1   Structural Characteristics of Poly rA and Poly dA Oligomers

Among the different homopolymeric nucleic acid sequences, single-stranded polyadenylic acid (poly A or poly rA) has attracted considerable attention because of its biological significance.[117,118] On the basis of sedimentation constants and viscosities, Fresco and Doty observed that poly rA undergoes a structural change when the pH of the medium is changed from neutral to acidic conditions.[119] Subsequently, early work on the structural elucidation of poly rA using methods such as X-ray diffraction,[120,121] UV spectroscopy,[122–124] optical rotatory dispersion (ORD),[125–129] electronic circular dichroism (ECD),[124,130–132,137] NMR[133–135] and more recently AFM concluded that poly rA exhibits two different structures depending on the pH.[136,143]

The structure of poly rA at neutral pH was suggested to be a right-handed single-stranded helix, stabilised by the stacked array of bases oriented nearly perpendicular to the helix axis with no evidence for internucleotide hydrogen bonding.[124,130,132] At acidic pH, however, poly rA exhibits a right-handed double helical structure with parallel strands and stacked protonated bases,[119,120,123] tilted with respect to the helical axis.[120,125] The double helical form of poly rA contains eight nucleotide pairs per helix turn.[131] On the basis of the X-ray diffraction studies, Rich *et al.* proposed that the double-stranded structure is stabilised by two factors: (i) the hydrogen bonds between two protonated adenine bases, and (ii) electrostatic attraction between the positively charged protons on the N1 position of the adenines and the negatively charged phosphate groups (Figure 7.7a).[120] Both parallel strands are held together by $AH^+–H^+A$ base pairing shown in Figure 7.7a. There are four hydrogen bonds per base pair, two of which are between H10 and N7; the other two are between H10 and phosphate oxygen.[120] The additional positive charge shown on N1 due to adenosine protonation is in close proximity to the

(a)

(b)

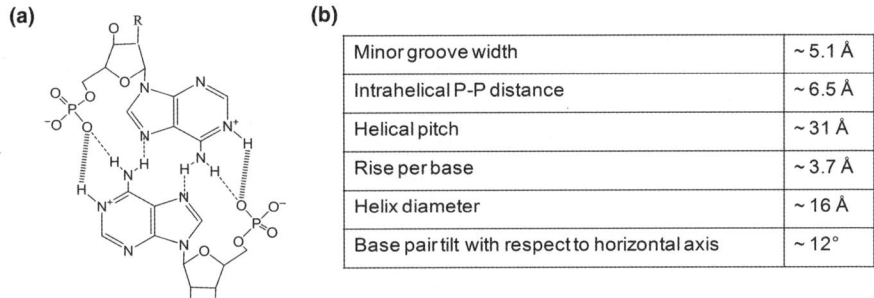

Minor groove width	~ 5.1 Å
Intrahelical P-P distance	~ 6.5 Å
Helical pitch	~ 31 Å
Rise per base	~ 3.7 Å
Helix diameter	~ 16 Å
Base pair tilt with respect to horizontal axis	~ 12°

Figure 7.7 (a) AH^+–H^+A base pairing scheme proposed in poly rA (R = OH) and poly dA (R = H).[120,149] (b) Summary of helical parameters of the A-motif obtained from Molecular Dynamics simulations.[149]

negatively charged phosphate group of the opposite backbone and stabilises the molecule electrostatically. This has been described as an 'inner salt' because the polymer no longer has an overall charge. Later, on the basis of solution phase experiments showing correlations between hydrogen-ion titration data and thermal denaturation curves, Holcomb and Timasheff reaffirmed that poly rA can form a double helical structure when the nucleobases are protonated and that these protons interact electrostatically with the negatively charged phosphates.[138,120]

This structural transition from the single-stranded to the double-stranded structure of poly rA was reported to occur at a pH near the pK_a of 5.87 for poly rA, as inferred from UV absorption, ORD, and ECD.[139] This suggests that complete protonation is not a structural requirement for the acidic structure of poly rA as originally determined by Rich *et al.*[120] Depending on the protonation state of the molecule, three different acidic conformations have been observed for poly rA: B form, A form, and a 'frozen' form which can also be distinguished by circular dichroism.[140,141] The least acidic form (at higher pH), designated as the B form or as an 'intermediate' form, is present at a pH just below pK_a, associated with the partial protonation of adenine moieties.[140] At pH 5.81, ORD of poly rA indicates the existence of the B form.[139] Upon further acidification, the B form gradually converts into the A form, which is also known as the 'tightly packed' form. This is stable at complete protonation of adenine nucleobases, indicated by ORD at pH 4.00.[139] Finally at low pH (< 3.8), the most acidic form, known as the 'frozen' form, represents a grid-like aggregate consisting of alternating, variably sized, single-stranded regions linked with short double-stranded regions.[140,141] A slow change in solution pH can prevent the development of the frozen form.

Maggini *et al.* studied the kinetics of the conformational change of poly rA, induced by jumps of solution pH.[142] The process of double helix formation begins with a second-order step leading to mismatched double-strands, which in turn evolve to the final, more stable and more extensively H-bonded double-helical form, following a series of first-order steps.

Interestingly, the 2′ OH does not play any role in stabilising the double-helical form of poly rA at acidic pH. Further, at acidic pH, some short adenine-rich DNA sequences can form parallel duplexes consists of A–A base pairs.[144–146] Using spectroscopic techniques in bulk, as well as at single molecule level, it has been shown that at neutral pH poly dA also exists as a structured single helix, similar to poly rA.[147,148] These facts inspired the discovery of the A-motif because they suggested that poly dA might also be capable of forming a double-stranded helix at acidic pH, similar to that of poly rA. Thus Chakraborty *et al.* showed that, similar to poly rA, poly dA can undergo a pH-induced conformational transition in a highly reversible manner (Figure 7.8a).[149] Poly dA switches its conformation between a structured single helix at neutral pH and a parallel-stranded double helix at low pH; the latter has been termed the A-motif.

Using pH-dependent gel electrophoresis, CD spectroscopy and fluorescence quenching studies, both structural forms, at acidic pH and neutral pH, have been characterised. The switching between the two states is highly processive. High resolution structural information on the A-motif was obtained using 2D NMR, which revealed that the duplex was held by reverse Hoogsteen type AH^+–H^+A base pairs, similar to a poly rA duplex (see Figure 7.7a). The stability of the A-motif decreases with increasing salt, indicating that the poly dA duplex is stabilised by electrostatic interaction between the $N1$–H^+ of

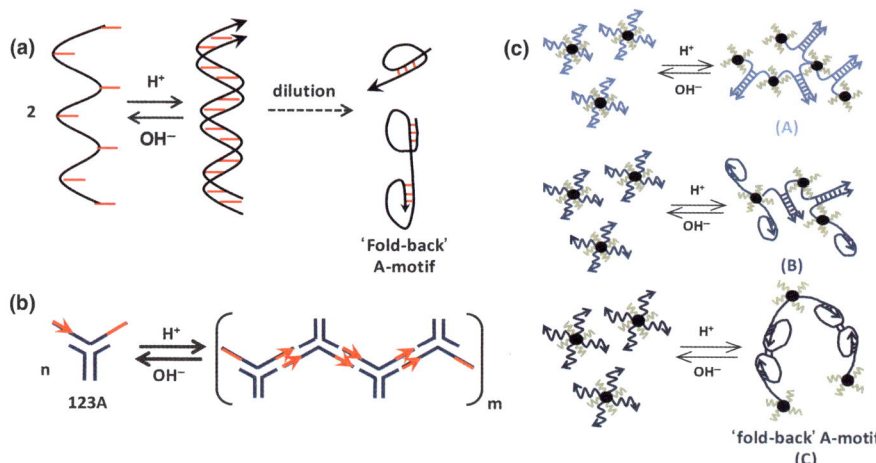

Figure 7.8 (a) A-motif formation from poly dA sequences.[149] 'Fold-back' A-motifs form at low strand concentrations under acidic pH.[151,153] (b) A-motifs as a pH-toggled glue for DNA architectures. 123A represents a pre-structured B-DNA motif (blue lines) bearing A-motif-forming overhangs (red lines), with pH reversible polymerization of 123A into 1D architecture via A-motif formation.[153] (c) Mode of association of AuNPs (black circles) modified with 5′ thiolated poly dA tracts of short, medium and long lengths shown in light blue, blue and dark blue lines respectively at acidic pH. Long poly dA tracts form 'fold-back' A-motifs and then associate into AuNP aggregates.[151]

adenosine and phosphate oxygen. Fluorescence quenching experiments indicated that slipped hybridisation is insignificant in short poly dA tracts. A well-defined sharp transition centred at pH 4.8, corresponding to the pK_a of adenines in poly dA observed in a pH titration study on dA_{15}, indicated a two-state transition. An atomistic model of the poly dA duplex obtained from a molecular dynamics simulation revealed that poly dA forms a right-handed, parallel-stranded duplex, previously referred to as a Π-helix, with eight base pairs per turn and 31.2 Å pitch (see Figure 7.7b). Molecular Dynamics (MD) simulations indicated that the $AH^+–H^+A$ base pairs show $\sim 12°$ tilt with respect to the horizontal axis, which is also a characteristic feature of the poly rA duplex (see Figure 7.7b). The MD simulated structure indicates that N1–H moves to within the H-bonding distance of O_2P for all base pairs despite the larger initial distance. Consistent with previous studies, MD simulations also showed that the poly dA single strand forms a nicely structured helix owing to stacking of the adenines, which is also reflected in its CD spectrum. The kinetics of this conformational transition is on millisecond time scales at sub-micromolar concentrations. In spite of its obvious lack of biological significance, poly dA has much potential for further application in nanotechnology as a building block or a sensor motif because of the superior thermal and hydrolytic stability of DNA over RNA.[150,151]

7.4.2 Molecular Devices Based on A-Motifs

The A-motif presents a new mode of directional strand hybridisation that can be switched on or off with a pH toggle and has potential as an alternative glue in Watson–Crick base pairing. It has been demonstrated that the judicious introduction of A-motif forming overhangs (AFOs) on pre-structured B-DNA building blocks could result in their pH-reversible assembly into a rigid DNA architecture (Figure 7.8b). These B-DNA building blocks exist as monomers at neutral pH because of electrostatic repulsion. However, by changing to acidic conditions these building blocks polymerise via the formation of A-motifs through their AFOs, without disrupting the structural integrity of the B-DNA motif.[150,152] This study revealed that the A-motif is compatible with and does not interfere with Watson–Crick base pairing in an assembly. Thus, a combination of B-DNA and A-motif forming sequences can be used to achieve active control over the directionality and processivity of the assembly of well-defined rigid DNA architectures, and to introduce new functionality into passive, rigid DNA structures. Thus, understanding the self-assembly of poly dA in as much detail as poly rA is also important and represents a currently understudied area. Using various spectroscopic methods, including fluorescence correlation spectroscopy (FCS) and single-molecule spectroscopy, Majima's group has observed an 'S' form of the A-motif with an intramolecular double-stranded region under dilute strand concentrations, similar to the 'frozen' form reported in the case of poly rA (Figure 7.8a).[153]

Simultaneously, pH-triggered self-association of poly dA has been observed using poly dA conjugated to AuNPs, as a function of length and mutation in the poly dA tract.[151] Formation of the A-motif in this case leads to a visible colour change and thus the system can be used as colorimetric pH sensor *in vitro*. This study also brings out the advantage of nucleic acid based pH sensors in which changing the sequence alters the stability of the resultant structured motif, depending on the base pairing efficiency, and thus a range of pH sensing may be accessed. The authors have also observed that the mode of AuNP association changes as a function of the poly dA tract (Figure 7.8c). Using intelligent incorporation of A→T mutations, they showed that longer poly dA tracts can 'fold-back' to form an intramolecular A-motif (Figure 7.8c).

This 'fold-back' A-motif could be analogous to the 'S' form A-motif described by the Majima group.[153] These fold-back A-motifs are observable only at low concentrations, and they were not possible to explore earlier with conventional biophysical techniques. These two above-mentioned studies used fluorescence techniques and colorimetric assays that are sensitive at 1000-fold lower concentrations than those explored earlier. Thus the application of nucleic acid motifs in the field of structural DNA nanotechnology can provide valuable structural insights about the motif.

7.5 Outlook

Protonation/deprotonation of nucleobases by changing pH can enable them to invoke unusual modes of hydrogen bonding or base pairing that result in the formation of non-B-DNA based structural motifs from the parent oligonucleotide sequence.[38,71,119,149] Such unusual structural motifs can be used in combination with B-DNA based structural modules that induce pH-toggled dynamics of the composite system. Such pH-sensitive DNA devices function with high processivity because each working cycle generates only salt and water as by-products, which are not toxic to the system. In other DNA devices, where the dynamics is introduced via strand displacement, double-stranded DNA waste is generated that toxifies the system and reduces processivity.[154,155] Further, the response times of pH-sensitive devices are on the timescale of 10–10^5 milliseconds,[97] and this is much faster than strand displacement-based devices that function on much longer timescales at best.[154,155] This is because of the fast diffusion of protons that leads to an intramolecular recognition event, as opposed to slower diffusing oligonucleotides that need to undergo an intermolecular recognition event. In cellular systems, with devices that operate solely based on nucleic acid recognition, the latter are slowed down to even longer timescales owing to a large number of sub-optimal sampling events that act as kinetic traps.[156] Thus it is not surprising that pH-sensitive DNA devices show quantitative recapitulation of their *in vitro* efficiency when they are introduced into cellular and organismal systems.[103,105]

In terms of device response times, the conformational changes undergone by nucleic acid-based assemblies are slow, unlike those seen in proteins.[157]

However, one can envisage accelerating these by incorporating protein or small molecule participators into nucleic acid-based devices. Despite the slow conformational changes of nucleic acid-based devices, an associated advantage is that, unlike proteins, nucleic acids *per se* induce low levels of immune response in biological hosts.[158–160] Thus, for biological applications, the introduction of non-nucleobase cofactors for synthetic devices has a trade-off between response times and device efficiency on the one hand and host immune reactions on the other. Nevertheless, pH-toggled DNA devices are promising candidates for investigating and exploiting different pH-dependent biological processes.

References

1. J. March, *Advanced Organic Chemistry: Reactions, Mechanisms, and Structure*, 4th edn, Wiley-Interscience, New York, USA, 1992.
2. P. Atkins, *Atkins' Physical Chemistry*, 7th edn, Oxford University Press, Oxford, UK, 2001.
3. C. A. S. Bergström, K. Luthman and P. Artursson, *Eur. J. Pharm. Sci.*, 2004, **22**, 387–398.
4. N. T. Hansen, I. Kouskoumvekaki, F .S. Jørgensen, S. Brunak and S. O. Jónsdóttir, *J. Chem. Inf. Model.*, 2006, **46**, 2601–2609.
5. D. Lee and T. Cui, *J. Vac. Sci. Technol. B*, 2009, **27**, 842–849.
6. X. Wei and A. J. Epstein, *Synthetic Metals*, 1995, **74**, 123–125.
7. J. Yue and A. J. Epstein, *J. Am. Chem. Soc.*, 1990, **112**, 2800–2801.
8. J. Yue, Z. H. Wang, K. R. Cromack, A. J. Epstein and A. G. MacDiarmid, *J. Am. Chem. Soc.*, 1991, **113**, 2665–2671.
9. M. M. Martin and L. Lindqvist, *J. Luminescence*, 1975, **10**, 381–390.
10. N. Klonis and W. H. Sawyer, *J. Fluorescence*, 1996, **6**, 147–157.
11. X. Cao, R. Bansil, K. R. Bhaskar, B. S. Turner, J. T. LaMont, N. Niu and N. H. Afdhal, *Biophys. J.*, 1999, **76**, 1250–1258.
12. D. A. Redfern and A. Gericke, *J. Lipid Res.*, 2005, **46**, 504–515.
13. C. D. Blundell, D. J. Mahoney, M. R. Cordell, A. Almond, J. D. Kahmann, A. Perczel, J. D. Taylor, I. D. Campbell and A. J. Day, *J. Biol. Chem.*, 2007, **282**, 12976–12988.
14. L. E. Goldsmith, M. Yu, L. H. Rome and H. G. Monbouquette, *Biochemistry*, 2007, **46**, 2865–2875.
15. R. F. Gesteland, *The RNA World*, 3rd edn, Cold Spring Harbor Laboratory Press, New York, USA, 2005.
16. A. Fersht, *Enzyme Structure and Mechanism*, 2nd edn, W H Freeman & Co (Sd), New York, USA, 1985.
17. D. M. J. Lilley and F. Eckstein, *Ribozymes and RNA Catalysis*, The Royal Society of Chemistry, Cambridge, UK, 2008.
18. M. J. Fedor and J. R. Williamson, *Nat. Rev. Mol. Cell. Biol.*, 2005, **6**, 399–412.
19. S. A. Strobel and J. C. Cochrane, *Curr. Opin. Chem. Biol.*, 2007, **11**, 636–643.

20. R. B. Silverman, *The Organic Chemistry of Enzyme-Catalyzed Reactions*, Academic Press, San Diego, 2000.
21. D. Herschlag, *J. Am. Chem. Soc.*, 1994, **116**, 11631–11635.
22. J. E. Thompson and R. T. Raines, *J. Am. Chem. Soc.*, 1994, **116**, 5467–5468.
23. P. C. Bevilacqua, T. S. Brown, S. Nakano and R. Yajima, *Biopolymers*, 2004, **73**, 90–109.
24. P. C Bevilacqua1and and R. Yajima, *Curr. Opin. Chem. Biol.*, 2006, **10**, 455–464.
25. A. S. Yang and B. Honig, *J. Mol. Biol.*, 1994, **237**, 602–614.
26. R. Montesano, J. Roth, A. Robert and L. Orci, *Nature*, 1982, **296**, 651–653.
27. P. Paroutis, N. Touret and S. Grinstein, *Physiology*, 2004, **19**, 207–215.
28. H. Palokangas, K. Metsikkö and K. Väänänen, *J. Biol. Chem.*, 1994, **269**, 17577–17585.
29. R. Guinea and L. Carrasco, *J. Virol.*, 1995, **69**, 2306–2312.
30. I.-M. Yu, W. Zhang, H. A. Holdaway, L. Li, V. A. Kostyuchenko, P. R. Chipman, R. J. Kuhn, M. G. Rossmann and J. Chen, *Science*, 2008, **319**, 1834–1837.
31. J. Pouysségur, A. Franchi, G. L'Allemain and S. Paris, *FEBS Lett.*, 1985, **190**, 115–119.
32. M. E. Meima, J. R. Mackley and D. L. Barber, *Curr. Opin. Nephrol. Hypertens.*, 2007, **16**, 365–372.
33. S. P. Denker and D. L. Barber, *J. Cell Biol.*, 2002, **159**, 1087–1096.
34. J. R. Schelling and Abu B. G. Jawdeh, *Am. J. Physiol., Renal Physiol.*, 2008, **295**, F625–F632.
35. M. Simons, W. J. Gault, D. Gotthardt, R. Rohatgi, T. J. Klein, Y. Shao, H. J. Lee, A. L. Wu, Y. Fang, L. M. Satlin, J. T. Dow, J. Chen, J. Zheng, M. Boutros and M. Mlodzik, *Nat. Cell Biol.*, 2009, **11**, 286–294.
36. J. R. Casey, S. Grinstein and J. Orlowski, *Nat. Rev. Mol. Cell Biol.*, 2010, **11**, 50–61.
37. S.-K. Lee, W. Li, S.-E. Ryu, T. Rhim and J. Ahnn, *Biochimica. Biophysica. Acta (BBA) - Bioenergetics*, 2010, **797**, 687–695.
38. G. Felsenfeld, D. R. Davies and A. Rich, *J. Am. Chem. Soc.*, 1957, **79**, 2023–2024.
39. J. C. Hanvey, J. Klysik and R. D. Wells, *J. Biol. Chem.*, 1988, **263**, 7386–7396.
40. H. E. Moser and P. B. Dervan, *Science*, 1987, **238**, 645–650.
41. D. Praseuth, L. Perrouault, T. Le Doan, M. Chassignol, N. Thuong and C. Hélène, *Proc. Natl. Acad. Sci. USA*, 1988, **85**, 1349–1353.
42. C. de los Santos, M. Rosen and D. Patel, *Biochemistry*, 1989, **28**, 7282–7289.
43. P. Rajagopal and J. Feigon, *Biochemistry*, 1989, **28**, 7859–7870.
44. P. Rajagopal and J. Feigon, *Nature*, 1989, **339**, 637–640.
45. M. D. Frank-Kamenetskii and S. M. Mirkin, *Annu. Rev. Biochem.*, 1995, **64**, 65–95.

46. V. A. Malkov, O. N. Voloshin, A. G. Veselkov, V. M. Rostapshov, I. Jansen, V. N. Soyfer and M. D. Frank-Kamenetskii, *Nucleic Acids Res.*, 1993, **21**, 105–111.

47. A. R. Morgan and R. D. Wells, *J. Mol. Biol.*, 1968, **37**, 63–80.

48. J. S. Lee, D. A. Johnson and A. R. Morgan, *Nucleic Acids Res.*, 1979, **6**, 3073–3091.

49. T. Le Doan, L. Perrouault, D. Praseuth, N. Habhoub, J. L. Decout, N. T. Thuong, J. L'homme and C. Hélène, *Nucleic Acids Res.*, 1987, **15**, 7749–7760.

50. V. I. Lyamichev, S. M. Mirkin, M. D. Frank-Kamenetskii and C. R. Cantor., *Nucleic Acids Res.*, 1988, **16**, 2165–2178.

51. S. M. Mirkin and M. D. Frank-Kamenetskii, *Annu. Rev. Biophys. Biomol. Struct.*, 1994, **23**, 541–576.

52. S. Arnott and E. Selsing, *J. Mol. Biol.*, 1974, **88**, 509–521.

53. F. B. Howard, H. T. Miles, K. Liu, J. Frazier, G. Raghunathan and V. Sasisekharan, *Biochemistry*, 1992, **31**, 10671–10677.

54. R. F. Macaya, P. Schultze and J. Feigon, *J. Am. Chem. Soc.*, 1992, **114**, 781–783.

55. G. Raghunathan, H. T. Miles and V. Sasisekharan, *Biochemistry*, 1993, **32**, 455–462.

56. M. M. Mooren, D. E. Pulleyblank, S. S. Wijmenga, M. J. Blommers and C. W. Hilbers, *Nucleic Acids Res.*, 1990, **18**, 6523–6529.

57. I. Radhakrishnan, X. Gao, C. de los Santos, D. Live and D.J. Patel, *Biochemistry*, 1991, **30**, 9022–9030.

58. L. J. Maher 3rd, P. B. Dervan and B. J. Wold, *Biochemistry*, 1990, **29**, 8820–8826.

59. T. J. Povsic and P. B. Dervan, *J. Am. Chem. Soc.*, 1989, **111**, 3059–3061.

60. G. E. Plum, Y. W. Park, S. F. Singleton, P. B. Dervan and K. J. Breslauer, *Proc. Natl. Acad. Sci. USA*, 1990, **87**, 9436–9440.

61. J. S. Lee, M. L. Woodsworth, L. J. Latimer and A. R. Morgan, *Nucleic Acids Res.*, 1984, **12**, 6603–6614.

62. S. F. Singleton and P. B. Dervan, *Biochemistry*, 1992, **31**, 10995–11003.

63. V. Horn, L. Lacroix, T. Gautier, M. Takasugi, J. L. Mergny and J. Lacoste, *Biochemistry*, 2004, **43**, 11196–11205.

64. C. Paris, F. Geinguenaud, C. Gouyette, J. Liquier and J. Lacoste, *Biophys J.*, 2007, **92**, 2498–2506.

65. H. Shindo, H. Torigoe and A. Sarai, *Biochemistry*, 1993, **32**, 8963–8969.

66. D. A. Collier and R. D. Wells, *J. Biol. Chem.*, 1990, **265**, 10652–10658.

67. Y. Chen, S.-H. Lee and C. Mao, *Angew. Chem. Int. Ed.*, 2004, **43**, 5335–5338.

68. Y. H. Jung, K.-B. Lee, Y.-G. Kim and I. S. Choi, *Angew. Chem. Int. Ed.*, 2006, **45**, 5960–5963.

69. Y. Chen and C. Mao, *Small*, 2008, **4**, 2191–2194.

70. Y. Chen and C. Mao, *J. Am. Chem. Soc.*, 2004, **126**, 13240–13241.

71. K. Gehring, J. L. Leroy and M. Guéron, *Nature*, 1993, **363**, 561–565.

72. R. E. Marsh, R. Bierstedt and E. L. Eichhorn, *Acta Crystallogr*, 1962, **15**, 310–316.

73. E. O. Akinrimisi, C. Sander and P. O. Ts'O, *Biochemistry*, 1963, **2**, 340–344.
74. R. B. Inman, *J. Mol. Biol.*, 1964, **9**, 624–637.
75. K. A. Hartman and A. Rich, *J. Am. Chem. Soc.*, 1965, **87**, 2033–2039.
76. J. L. Leroy, M. Guéron, J. L. Mergny and C. Hélène, *Nucleic Acids Res.*, 1994, **22**, 1600–1606.
77. J.-L. Mergny, L. Lacroix, X. Han, J.-L. Leroy and C. Hélène, *J. Am. Chem. Soc.*, 1995, **117**, 8887–8898.
78. L. Chen, L. Cai, X. Zhang and A. Rich, *Biochemistry*, 1994, **33**, 13540–13546.
79. C. H. Kang, I. Berger, C. Lockshin, R. Ratliff, R. Moyzis and A. Rich, *Proc. Natl. Acad. Sci. USA*, 1994, **91**, 11636–11640.
80. M. Guéron and J. L. Leroy, *Curr. Opin. Struct. Biol.*, 2000, **10**, 326–331.
81. I. Berger, C. Kang, A. Fredian, R. Ratliff, R. Moyzis and A. Rich, *Nat. Struct. Biol.*, 1995, **2**, 416–425.
82. I. Berger, M. Egli and A. Rich, *Proc. Natl. Acad. Sci. USA*, 1996, **93**, 12116–12121.
83. K. S. Jin, S. R. Shin, B. Ahn, Y. Rho, S. J. Kim and M. Ree, *J. Phys. Chem. B.*, 2009, **113**, 1852–1856.
84. Y. Krishnan-Ghosh, E. Stephens and S. Balasubramanian, *Chem. Commun.*, 2005, 5278–5280.
85. S. Modi, A. H. Wani and Y. Krishnan, *Nucleic Acids Res.*, 2006, **34**, 4354–4363.
86. S. Chakraborty, S. Modi and Y. Krishnan, *Chem. Commun.*, 2007, 70–72.
87. S. Chakraborty and Y. Krishnan, *Biochimie.*, 2008, **90**, 1088–1095.
88. K. Snoussi, S. Nonin-Lecomte and J. L. Leroy, *J. Mol. Biol.*, 2001, **309**, 139–153.
89. G. M. Blackburn and M. Gait, *Nucleic Acids in Chemistry and Biology*, 2nd edn, Oxford University Press, Oxford, 1996.
90. J. L. Mergny, *Biochemistry*, 1999, **38**, 1573–1581.
91. T. Simonsson, M. Pribylova and M. Vorlickova, *Biochem. Biophys. Res. Comm.*, 2000, **278**, 158–166.
92. H. B. Ghodke, R. Krishnan, K. Vignesh, G. V. P. Kumar, C. Narayana and Y. Krishnan, *Angew. Chem. Int. Ed.*, 2007, **46**, 2646–2649.
93. A. Laisné, D. Pompon and J.-L. Leroy, *Nucleic Acids Res.*, 2010, **38**, 3817–3826.
94. J.-L. Leroy, *Nucleic Acids Res.*, 2009, **37**, 4127–4134.
95. A. T. Phan and J.-L. Mergny, *Nucleic Acids Res.*, 2002, **30**, 4618–4625.
96. P. Alberti and J.-L. Mergny, *Proc. Natl. Acad. Sci. USA*, 2002, **100**, 1569–1573.
97. D. Liu and S. Balasubramanian, *Angew. Chem. Int. Ed.*, 2003, **42**, 5734–5736.
98. T. Liedl and F. C. Simmel, *Nano Lett.*, 2005, **5**, 1894–1898.
99. T. Liedl, M. Olapinski and F. C. Simmel, *Angew. Chem. Int. Ed.*, 2006, **45**, 5007–5010.

100. H. Liu, Y. Zhou, Y. Yang, W. Wang, L. Qu, C. Chen, D. Liu, D. Zhang and D. Zhu, *J. Phys. Chem. B.*, 2008, **112**, 6893–6896.

101. H. Liu, Y. Xu, F. Li, Y. Yang, W. Wang, Y. Song and D. Liu, *Angew. Chem. Int. Ed.*, 2007, **46**, 2515–2517.

102. Y. Mao, D. Liu, S. Wang, S. Luo, W. Wang, Y. Yang, Q. Ouyang and L. Jiang, *Nucleic Acids Res.*, 2007, **35**, e33.

103. S. Modi, D. Goswami, G. D. Gupta, S. Mayor and Y. Krishnan, *Nat. Nanotechnol.*, 2009, **4**, 325–330.

104. W. Wang, Y. Yang, E. Cheng, M. Zhao, H. Meng, D. Liu and D. Zhou, *Chem. Commun.*, 2009, 824–826.

105. S. Surana, J. M. Bhat, S. P. Koushika and Y. Krishnan, *Nat. Commun.*, 2011, **2**, 340.

106. W. Wang, H. Liu, D. Liu, Y. Xu, Y. Yang and D. Zhou, *Langmuir.*, 2007, **23**, 11956–11959.

107. J. Sharma, R. Chhabra, H. Yan and Y. Liu, *Chem. Commun.*, 2007, 477–479.

108. C. Chen, G. Song, J. Ren and X. Qu, *Chem. Commun.*, 2008, 6149–6151.

109. Y. Wang, X. Li, X. Liu and T. Li, *Chem. Commun.*, 2007, 4369–4371.

110. D. Liu, A. Bruckbauer, C. Abell, S. Balasubramanian, D.-J. Kang, D. Klenerman and D. Zhou, *J. Am. Chem. Soc.*, 2006, **128**, 2067–2071.

111. W. Shu, D. Liu, M. Watari, C. K. Riener, T. Strunz, M. E. Welland, S. Balasubramanian and R. A. McKendry, *J. Am. Chem. Soc.*, 2005, **127**, 17054–17060.

112. S. Wang, H. Liu, D. Liu, X. Ma, X. Fang and L. Jiang, *Angew. Chem. Int. Ed.*, 2007, **46**, 3915–3917.

113. S. R. Shin, C. K. Lee, S. H. Lee, S. I. Kim, G. M. Spinks, G. G. Wallace, I. So, J.-H. Jeong, T. M. Kang and S. Kim, *Chem. Commun.*, 2009, 1240–1242.

114. F. Xia, W. Guo, Y. Mao, X. Hou, J. Xue, H. Xia, L. Wang, Y. Song, H. Ji, Q. Ouyang, Y. Wang and L. Jiang, *J. Am. Chem. Soc.*, 2008, **130**, 8345–8350.

115. E. Cheng, Y. Xing, P. Chen, Y. Yang, Y. Sun, D. Zhou, L. Xu, Q. Fan and D. Liu, *Angew. Chem. Int. Ed.*, 2009, **48**, 7660–7663.

116. S. H. Um, J. B. Lee, N. Park, S. Y. Kwon, C. C. Umbach and D. Luo, *Nat. Mater.*, 2006, **5**, 797–801.

117. B. Alberts, *Molecular Biology of the Cell*, 5th edn, Garland Science, New York, 2008.

118. A. Sachs and E. Wahle, *J. Biol. Chem.*, 1993, **268**, 22955–22958.

119. J. R. Fresco and P. Doty, *J. Am. Chem. Soc.*, 1957, **79**, 3928–3929.

120. A. Rich, D. R. Davies, F. H. Crick and J. D. Watson, *J. Mol. Biol.*, 1961, **3**, 71–86.

121. V. Luzzati, A. Mathis, F. Masson and J. Witz, *J. Mol. Biol.*, 1964, **10**, 28–41.

122. J. R. Fresco and E. Klemperer, *Ann. N. Y. Acad. Sci.*, 1959, **81**, 730–741.

123. M. Leng and G. Felsenfeld, *J. Mol. Biol.*, 1966, **15**, 455–466.

124. E. Casassas, R. Tauler and I. Marques, *Macromolecules*, 1994, **27**, 1729–1737.

125. D. N. Holcomb and I. Tinoco Jr., *Biopolymers*, 1965, **3**, 121–133.

126. M. M. Warshaw, C. A. Bush and I. Tinoco Jr., *Biochem. Biophys. Res. Commun.*, 1965, **18**, 633–638.

127. P. K. Sarkar and J. T. Yang, *J. Biol. Chem.*, 1965, **240**, 2088–2093.

128. D. Poland, J. N. Vournakis and H. A. Scheraga, *Biopolymers*, 1966, **4**, 223–235.

129. A. M. Michelson, T. L. Ulbricht, T. R. Emerson and R. J. Swan, *Nature*, 1966, **209**, 873–874.

130. K. E. Van Holde, J. Brahms and A. M. Michelson, *J. Mol. Biol.*, 1965, **12**, 726–739.

131. J. Brahms, A. M. Michelson and K. E. Van Holde, *J. Mol. Biol.*, 1966, **15**, 467–488.

132. H. Hashizume and K. Imahori, *J. Biochem.*, 1967, **61**, 738–749.

133. F. E. Hruska and S. S. Danyluk, *J. Am. Chem. Soc.*, 1968, **90**, 3266–3267.

134. S. I. Chan and J. H. Nelson, *J. Am. Chem. Soc.*, 1969, **91**, 168–183.

135. D. B. Lerner and D. R. Kearns, *Biopolymers*, 1981, **20**, 803–816.

136. C. Ke, A. Loksztejn, Y. Jiang, M. Kim, M. Humeniuk, M. Rabbi and P. E. Marszalek, *Biophys. J.*, 2009, **96**, 2918–2925.

137. J. Brahms, *Nature*, 1964, **202**, 797–798.

138. D. N. Holcomb and S. N. Timasheff, *Biopolymers*, 1968, **6**, 513–529.

139. J. T. Finch and A. Klug, *J. Mol. Biol.*, 1969, **46**, 597–598.

140. A. J. Adler, L. Grossman and G. D. Fasman, *Biochemistry*, 1969, **8**, 3846–3859.

141. B. Janik, R. G. Sommer and A. M. Bobst, *Biochim. Biophys. Acta*, 1972, **281**, 152–168.

142. R. Maggini, F. Secco, M. Venturini and H. Diebler, *J. Chem. Soc., Faraday Trans.*, 1994, **90**, 2359–2363.

143. W. M. Scovell, *Biopolymers*, 1978, **17**, 969–984.

144. J. Luo, M. H. Sarma, R. D. Yuan and R. H. Sarma, *FEBS Lett.*, 1992, **306**, 223–228.

145. H. Robinson and A. H. Wang., *Proc. Natl. Acad. Sci. USA*, 1993, **90**, 5224–5228.

146. Y. Wang and D. J. Patel, *J. Mol. Biol.*, 1994, **242**, 508–526.

147. C. A. Bush and H. A. Scheraga, *Biopolymers*, 1969, **7**, 395–409.

148. J. L. Alderfer and S. L. Smith, *J. Am. Chem. Soc.*, 1971, **93**, 7305–7314.

149. S. Chakraborty, S. Sharma, P. K. Maiti and Y. Krishnan, *Nucleic Acids Res.*, 2009, **37**, 2810–2817.

150. S. Saha, D. Bhatia and Y. Krishnan, *Small*, 2010, **6**, 1288–1292.

151. S. Saha, K. Chakraborty and Y. Krishnan, *Chem. Commun.*, 2012, **48**, 2513–2515.

152. C. Wang, Y. Tao, F. Pu, J. Ren and X. Qu, *Soft Matter*, 2011, **7**, 10574–10576.

153. S. Kim, J. Choi and T. Majima, *J. Phys. Chem. B.*, 2011, **115**, 15399–15405.

154. B. Yurke, A. J. Turberfield, A. P. Mills, F. C. Simmel and J. L. Neumann, *Nature*, 2000, **406**, 605–608.

155. R. P. Goodman, M. Heilemann, S. Doose, C. M. Erben, A. N. Kapanidis and A. J. Turberfield, *Nat. Nanotechnol.*, 2008, **3**, 93–96.

156. D. Sagi, T. Tlusty and J. Stavans, *Nucleic Acids Res.*, 2006, **34**, 5021–5031.

157. E. A. Lipman, B. Schuler, O. Bakajin and W. A. Eaton, *Science*, 2003, **301**, 1233–1235.

158. A. D. Keefe, S. Pai and A. Ellington., *Nat. Rev. Drug Discov.*, 2010, **9**, 537–550.

159. C. Meyer, U. Hahn and A. Rentmeister, *J. Nucleic Acids,* 904750–904768.

160. W. Tan, H. Wang, Y. Chen, X. Zhang, H. Zhu, C. Yang, R. Yang and C. Liu, *Trends Biotechnol.*, 2011, **29**, 634–640.

CHAPTER 8

Making Sense of Catalysis: The Potential of DNAzymes as Biosensors

SIMON A. McMANUS, KHA TRAM AND YINGFU LI*

Department of Biochemistry and Biomedical Sciences and Department of Chemistry and Chemical Biology, McMaster University, 1280 Main Street West, Hamilton, Ontario, Canada L8S 4K1
*Email: liying@mcmaster.ca

8.1 Introduction

8.1.1 Discovery of DNAzymes through *in vitro* Selection

DNA is well recognised as the fundamental genomic material of all life on Earth, and plays a critical role in the storage and transfer of genetic information. This molecule was first regarded as a passive player in biological systems, simply being acted upon by its own protein products, and faithfully passing the template of life from one generation to the next. With the discovery that RNA was capable of performing catalysis,[1] with efficiencies similar to protein enzymes,[2] many wondered whether DNA could also carry out additional functions. Given that single-stranded DNA shares many chemical and structural properties with RNA, it was hypothesised that when it is alleviated from its rigid double-helical form, DNA should also be capable of creating intricate tertiary structures through hydrogen bonding, charge–charge and π stacking interactions, much like its RNA counterparts.

RSC Biomolecular Sciences No. 26
DNA Conjugates and Sensors
Edited by Keith R Fox and Tom Brown
© The Royal Society of Chemistry 2012
Published by the Royal Society of Chemistry, www.rsc.org

The isolation of catalytic DNA molecules, known as DNAzymes or deoxyribozymes, was made possible through a process known as *in vitro* selection. Originally developed for isolating RNA aptamers, RNA molecules that can bind specific ligands,[3,4] *in vitro* selection enriches for functional nucleic acid molecules from pools of up to 10^{16} individual sequences through successive rounds of selection and amplification of sequences with a desired function. The power of this technique relies on its ability to separate a small number of active sequences from a vast number of inactive sequences. This allows the use of very large sequence libraries, which is not feasible with screening techniques that rely on assessment of the activity of sequences individually.

The main *in vitro* selection methods used for isolating catalytic DNA involve coupling catalysis to the addition or removal of a tag during the selection step. Catalytic molecules can then be isolated from non-reactive molecules on the basis of the size difference between the tagged and untagged sequences, or through affinity chromatography of the tag. As an example, the first *in vitro* selection for DNAzymes selected for DNA molecules that could cleave a phosphodiester linkage adjacent to a single RNA residue embedded in a DNA sequence.[5] The selection scheme is shown in Figure 8.1A. Modifying the library with a biotin moiety at the end of the substrate region before the selection step and binding the library to a streptavidin-coated column enabled sequences capable of cleavage to be eluted from the column. This allowed the separation of a small number of catalytic DNA sequences from the majority of inactive sequences that remain bound to the column. Once isolated, these sequences are amplified by polymerase chain reaction (PCR) to generate multiple copies of the initial catalytic sequences. The procedure is then repeated multiple times, enriching the pool for catalytic sequences *via* iterative rounds of selection and amplification. This technique is also tunable such that the components of the buffer used in the selection step will interact with the DNA library and may become incorporated into the active structures of the selected DNAzymes. This typically leads to DNAzymes that are dependent on the buffer components, such as metal ions, for catalysis. Over the past two decades, *in vitro* selection has been employed to isolate countless DNAzymes that can catalyse a fascinating array of chemical reactions.

8.1.2 RNA-Cleaving DNAzymes

Perhaps the best-studied DNAzymes are those that cleave RNA.[6] DNA easily catalyses this reaction, because RNA is a natural binding partner for DNA via Watson–Crick interactions. Furthermore, RNA is inherently prone to cleavage through intramolecular attack of the 2′-hydroxyl group of the ribose on the adjacent bridging phosphodiester within the RNA chain. DNA only needs to bind and position the RNA in the necessary in-line orientation for cleavage. From numerous selection experiments performed in the past 15 years, a variety of DNAzymes capable of cleaving RNA substrates, or chimeric DNA–RNA substrates, under a wide range of conditions have been isolated.[5,7–15] This

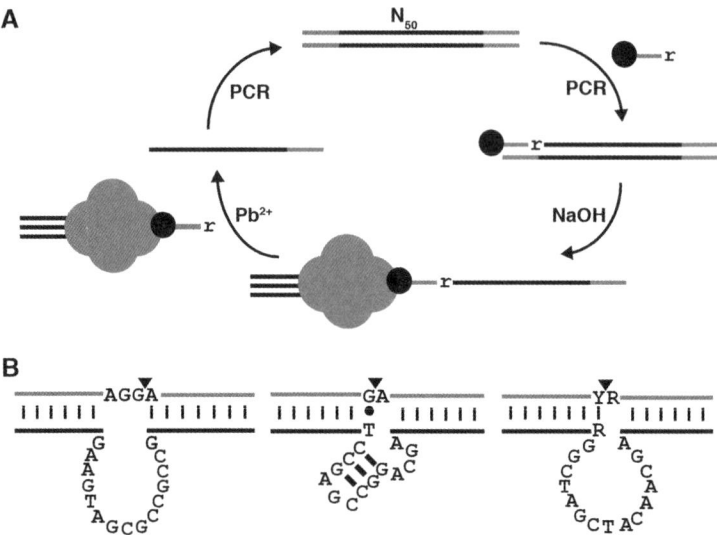

Figure 8.1   RNA-cleaving DNAzymes. (A) *In vitro* selection scheme used for the isolation of the first DNAzyme. PCR is used to incorporate a primer containing a biotin label (black circle) and a ribonucleotide (r) into a library with a 50-nucleotide random region. After binding of the library to a streptavidin column (grey complex), sequences that undergo RNA cleavage in the presence of Pb^{2+} are removed from the column and amplified by PCR. (B) Representative RNA-cleaving DNAzymes. The DNAzymes are shown in black and the substrates are shown in grey. The left structure is the first isolated Pb^{2+}-dependent DNAzyme. The center and right structures are the 8-17 and 10-23 DNAzymes, respectively. R: purine-based nucleotide; Y: pyrimidine-based nucleotide.

section will highlight the major achievements in the discovery of RNA-cleaving DNAzymes that have since been used in many downstream sensing applications.

The first DNAzyme selection was performed by the laboratory of Gerald Joyce, and led to the isolation of an RNA-cleaving DNAzyme that could cleave an RNA linkage embedded in a DNA substrate in a buffer containing Pb^{2+} and monovalent metal ions.[5] After five rounds of *in vitro* selection, a sequence that could cleave the substrate with a rate enhancement of 10^5 over the spontaneous cleavage reaction was identified. As shown in Figure 8.1B, structural analysis revealed that the DNAzyme formed two helical interactions with the substrate flanking the cleavage site, with 15 nucleotides between the helices performing catalysis. This two-binding-arm structure with a central catalytic domain has subsequently been seen in the majority of RNA-cleaving DNAzymes. The DNAzyme was also found to be able to perform bimolecular catalysis, by separation of the substrate and DNAzyme domains, with a turnover rate of $1\,min^{-1}$. These rates and enhancements over spontaneous reaction are comparable to those of many natural ribozymes. Other early selection endeavours by the Joyce[7] and Famulok[8] groups also isolated DNAzymes that can cleave an embedded RNA linkage in the presence of Mg^{2+} and Ca^{2+}, respectively.

After the initial demonstrations that DNA can cleave chimeric RNA/DNA substrates under different reaction conditions, the Joyce group went on to select for DNAzymes that could cleave an all-RNA substrate under physiological conditions.[9] DNAzymes that could function under these conditions would be useful in targeting cellular RNAs for research and therapeutic applications, and could potentially be developed into biosensors for RNA substrates. Using a similar selection protocol to their initial selection, they obtained two classes of DNAzyme that cleaved an RNA substrate at two distinct cleavage sites. These two DNAzymes were named 8-17 and 10-23 as they represented the 17th and 23rd clones isolated in the eighth and tenth round of selection, respectively. These two constructs, shown in Figure 8.1B, have ended up being the best-studied DNAzymes. Characterisation of the 10-23 DNAzyme revealed that it has a catalytic rate of $0.1\ min^{-1}$ under physiological conditions and a catalytic efficiency of $10^9\ M^{-1}\ min^{-1}$. By changing the sequences of the binding arms it was found that 10-23 could be made to cleave any RNA with a purine–pyrimidine junction at the cleavage site. Variants of 10-23 have since been used to target many messenger and viral RNAs for both biochemical studies and therapeutic applications.[16] The other DNAzyme isolated from this selection, 8-17, has also proved to be a versatile DNAzyme. Reselection and mutagenesis analysis revealed that its small catalytic core contained a small 3-base pair stem and eight essential nucleotides. While not as efficient a catalyst as 10-23, the 8-17 motif gained recognition because it was isolated multiple times in several independent *in vitro* selection processes.[8,9,15,17,18] The repeated isolation of this motif from different DNA libraries suggests that it may be the simplest structural motif for DNA-based RNA cleavage. Variations of the 8-17 motif have also been shown to use different metal ion cofactors such as Mg^{2+},[9] Ca^{2+},[8] and Zn^{2+}.[17] The versatility of 8-17 is also shown in its substrate recognition. In a study to isolate DNAzymes that could cleave all 16 possible dinucleotide junctions, DNAzymes containing the 8-17 motif were found to cleave 14 of the possible 16 junctions.[15] The small size and versatility of 8-17 has led to it being utilized in a number of different sensor schemes, described in Sections 8.2 and 8.3.

8.1.3 Other Reactions Catalysed by DNAzymes

In addition to RNA-cleaving DNAzymes, DNAzymes that catalyse many other chemical reactions have been discovered. These reactions include RNA ligation and branching,[19–21] DNA ligation,[22–25] DNA phosphorylation,[26,27] DNA adenylation,[28] DNA cleavage,[29,30] DNA hydrolysis,[31] thymine dimer repair,[32] nucleopeptide bond formation,[33] and carbon–carbon bond formation.[34] Many of these DNAzymes have attributes that are amenable to sensor development, such as low spontaneous reaction rates compared with those of RNA-cleaving DNAzymes and the requirement for many cofactors and small-molecule substrates that can be used as sensor targets. Some of these

DNAzymes also have interesting secondary structures that can be utilised in the rational design of aptazyme-based sensors (see Section 8.2.4). This section will summarise the DNAzymes that have subsequently been developed into sensors.

The catalytic attachment of two molecules can be exploited in sensing mechanisms such as fluorescent resonance energy transfer (FRET) and nanoparticle aggregation. An early study by the Szostak group selected for DNA molecules that could catalyse DNA ligation. DNA ligase DNAzymes were isolated that could mediate the ligation of one DNA strand to a second modified DNA molecule by the condensation of a 5′-hydroxyl to a 3′-phosphorimidazolide.[22] The DNAzyme was found to function in the presence of Zn^{2+} or Cu^{2+} and was capable of performing ligation in a multiple turnover fashion. The Ellington group has also isolated ligase DNAzymes that can ligate two DNA oligonucleotides containing a 3′-phosphorothioate and a 5′-iodine group.[23]

Another early DNAzyme study, by the Sen group, led to the isolation of DNA molecules that could catalyse porphyrin metalation.[35] This DNAzyme was obtained by performing a selection for aptamers that can bind an *N*-methylmesoporphyrin IX, a transition state analogue of the metalation of mesoporphyrin IX. The isolated aptamer sequence was found to be able to catalyse metalation of porphyrin. The guanine quadruplex-forming DNA molecule from this selection was also found to have affinity for haemin and to enhance its low peroxidase activity by 250 times.[36] This aptamer has been used subsequently in many sensing applications.[37–41]

The Breaker group performed a selection for DNA molecules capable of self-cleavage.[29] Given that DNA is much less labile than RNA, the selection scheme was designed to induce hydroxyl radical-based DNA cleavage. The selection was carried out in Cu^{2+} with ascorbate acting as the reducing agent. An isolated DNAzyme was later optimised to act as a restriction endonuclease that can target a variety of different single-stranded DNA molecules.[30] The low intrinsic background cleavage rate of DNA makes these DNAzymes good candidates as sensor elements.

8.2 Fluorescent DNAzyme-based Sensors

8.2.1 DNAzyme–Fluorophore Conjugates

The isolation of DNAzymes with significant rate enhancements has set the stage for exploring them as sensors. A large proportion of DNAzyme-based sensors use fluorescence as the detection platform mainly because this method offers excellent sensitivity and ease of detection, in addition to the availability of many different fluorescent dyes for DNA modification.[42,43] Several methods exist to label DNA with fluorophores. For example, fluorescent labels can be introduced into DNA either enzymatically using fluorophore-labelled dNTPs[44] or chemically during solid-phase synthesis of oligonucleotides using dye derived phosphoramidites. Doubly labelled nucleic acids capable of FRET[45] have been widely used to detect other nucleic acids,[46] as well as other biomolecules (reviewed in ref.[47]).

With the demonstration that DNA can efficiently cleave RNA under a variety of conditions, efforts were made to examine whether RNA cleavage could be coupled to fluorescence detection. The most popular utilisation of fluorescence in DNAzyme-based sensing is done *via* quenching mechanisms between the fluorophore donors and quencher acceptors. The key feature of this type of sensor is that the fluorescence signal is quenched before substrate cleavage, with subsequent DNAzyme-mediated cleavage resulting in a loss of quenching and a high level of fluorescence. This strategy requires the fluorophore and quencher to be within a few nucleotides of each other.[48] Labelling of the substrate with a fluorophore and a quencher near the cleavage site, however, can lead to a loss of DNAzyme activity because the bulky fluorophore and quencher may disrupt the interaction of the DNAzyme with the substrate. To prevent labelling from causing a loss of activity, DNAzymes and substrates have been labelled at sites distal to the cleavage site, with DNAzyme–substrate separation upon cleavage used to create an increase in fluorescence. The general scheme for this sensor design is shown in Figure 8.2A and involves placement of a fluorophore and quencher on the 5' end of the substrate and the 3' end of the DNAzyme in an extended stem region.

8.2.2 Fluorescent Sensors using Existing DNAzymes

The first sensor that uses this scheme was developed by the Lu group[17] using an 8-17 DNAzyme variant that can cleave an RNA substrate in the presence of Pb^{2+} with an observed rate of $6.5 \, min^{-1}$. To convert this DNAzyme into a fluorescent sensor, a TAMRA fluorophore (6-carboxytetramethylrhodamine) was attached to the 5' end of the substrate and a dabcyl quencher (4-(4'-dimethylaminophenylazo) benzoic acid) on the 3' end of the DNAzyme.[28] In the absence of Pb^{2+}, the substrate is bound to the DNAzyme, placing the fluorophore in proximity to the quencher, thereby quenching fluorescence. In the presence of Pb^{2+}, the substrate is cleaved and thus separated from the DNAzyme, resulting in loss of quenching and an observed enhancement of fluorescence. This sensor was found to be able to detect Pb^{2+} over a concentration range of 10 nM to 4 µM with an 80-fold specificity for Pb^{2+} over other metal ions. This sensor was improved in a follow-up study in which a second quencher moiety was used to reduce background fluorescence, resulting in a 5-fold improvement in signalling when compared with the original design.[49] The sensitivity of this sensor was further enhanced by the Tan group, who connected the substrate and DNAzyme with a polythymine linker[50] as shown in Figure 8.2B. This results in increased hybridisation between DNAzyme and substrate, which enhances the catalytic efficiency and reduces background fluorescence. This modified DNAzyme sensor exhibited a much improved detection limit of 3 nM. In a recent effort to create a fluorescent DNAzyme sensor for Pb^{2+} with high specificity, the Lu group created a

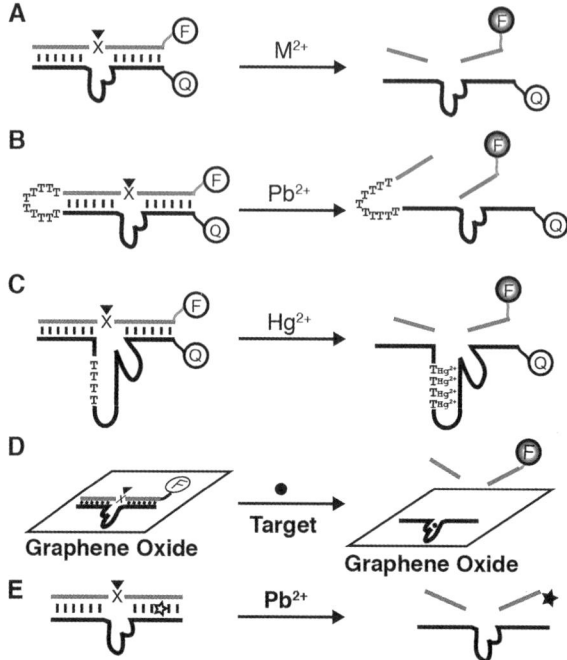

Figure 8.2   Schemes for converting existing DNAzymes into fluorescent sensors. (A) Substrate and DNAzyme labelled sensors. Substrate (light grey) and DNAzyme are modified with a fluorophore (F) and quencher (Q), respectively. In the presence of the correct metal-ion cofactor, the RNA substrate is cleaved, releasing the substrate and leading to an increase in fluorescence (dark circle). (B) Unimolecular Pb^{2+} detection. By linking the DNAzyme and substrate with a polythymine linker, hybridisation is increased and background fluorescence is reduced. (C) A DNAzyme sensor for Hg^{2+}. By placing thymine–thymine mismatches in an essential stem, an RNA-cleaving DNAzyme is deactivated. The binding of Hg^{2+} to thymines causes thymine–thymine base pairs to form, activating the DNAzyme. (D) Fluorescent detection by graphene assisted quenching. DNAzymes are bound to graphene and hybridised to RNA substrates containing a fluorophore, which is quenched by graphene. Upon cofactor (target) induced RNA cleavage, the substrate is released from the DNAzyme–graphene conjugate and an increase in fluorescence is observed. (E) Label-free fluorescent sensors. ATMND (star) binds to an abasic site in the DNAzyme–substrate complex, which quenches its fluorescence. Upon Pb^{2+}-dependent substrate cleavage, ATMND is released and its fluorescence is increased.

similarly designed sensor using the first DNAzyme ever isolated,[5] which was shown to be 40 000 times more specific than the previously designed DNA-zyme-based Pb^{2+} sensors.[51]

Sensors for other metal ions have also been developed using other DNA-zymes as the recognition elements. One sensor for uranium ions developed by the Lu group utilised *in vitro* selected DNAzymes specific for UO_2^{2+}, which were modified with a quencher and a substrate containing a second quencher and a fluorophore.[52] This sensor has a detection limit of 45 pM and a selectivity of more than 1 million-fold over other metal ions, making it one of the most sensitive uranium sensors available. Using the same DNAzyme, the Lu group[53] used a rational design approach to convert their uranium sensor into a sensor for Hg^{2+}. It has been shown that Hg^{2+} can bind pairs of thymines, creating T–T base pairs. To make the DNAzyme responsive to Hg^{2+}, T–T mismatches were first added into a stem of the DNAzyme to deactivate the DNAzyme, as shown in Figure 8.2C. When Hg^{2+} is present, the activity of the DNAzyme is restored, producing an increase of fluorescence. Using this system, it is possible to detect 4 nM of Hg^{2+}, which is below the Environmental Protection Agency limit for toxicity.

Fluorescent DNAzyme sensors have also been developed using other DNA-catalysed reactions. A fluorescent sensor for Cu^{2+} was developed using a copper-dependent DNA-cleaving DNAzyme[29] with a fluorophore and a quencher placed at the ends of an extended helical region.[54] The sensor was found to be specific for Cu^{2+} with a detection limit of 35 nM.

In addition to dyes that quench fluorescence, other types of quenching have been employed to simplify sensor design. Graphene is a two-dimensional nanomaterial consisting of a single-carbon-atom layer that is homogeneously arranged into a honeycomb crystal lattice. An interesting derivative of graphene, graphene oxide, has been used in a variety of biological applications owing to its unique surface and electronic properties.[55–58] In addition to being water-soluble and having quenching capacities, graphene oxide can interact with oligonucleotides through hydrophobic and π-stacking interactions. Interestingly, as a result of the super quenching effects of graphene oxide, the fluorescence-generating DNAzymes do not require a quenching dye partner. Selected biosensing applications that previously used fluorescence-generating DNAzymes have been converted to graphene oxide-based systems as shown in Figure 8.2D. Graphene-immobilised fluorescence-generating DNAzymes have been applied to detect metal ions such as Cu^{2+} and Pb^{2+}, with detection limits ranging between 0.3 and 2.0 nM.[59–62]

Fluorescent DNAzyme sensors that do not require substrate and DNAzyme labelling have also been developed. In one system, an abasic site is introduced into the substrate-binding region of the DNAzyme.[63] As shown in Figure 8.2E, when the DNAzyme is bound to the substrate, the abasic site is capable of binding 2-amino-5,6,7-trimethyl-1,8-naphthyridine (ATMND), which is non-fluorescent when bound. Upon substrate cleavage, the DNAzyme–substrate complex is disassembled to release the ATMND, which becomes fluorescent in its unbound state. When this system is used with a Pb^{2+}-utilising DNAzyme, it is capable of detecting as low as 4 nM Pb^{2+}, comparable with its labelled DNAzyme sensor counterparts. This method has been generalised and used to detect UO_2^{2+} and Hg^{2+}, using relevant DNAzymes mentioned above, with detection limits of 3 nM and 30 nM, respectively.[64]

8.2.3 Isolation of *de novo* Fluorescent DNAzyme Sensors through *in vitro* Selection

One drawback of the aforementioned systems is that, in order to convert the RNA-cleaving DNAzymes into fluorescent sensors, it is necessary to modify the substrate and DNAzymes with fluorescent dyes. Given that the residues near the cleavage site cannot usually be modified without a loss of enzymatic activity, the fluorophore and quencher dyes typically need to be placed distal to the active site. This can lead to issues such as loss of sensitivity and high background fluorescence. To prevent these problems, our group performed a selection for DNAzymes that can cleave substrates containing an RNA linkage flanked by a fluorophore and quencher.[65] The substrate is designed with the fluorophore and quencher separated by only one residue, resulting in strong quenching in the uncleaved substrate and low background fluorescence. Upon cleavage of the substrate, the two portions containing the fluorophore and quencher are released and separated, resulting in an increase in fluorescence. After 22 rounds, the selection yielded a DNAzyme that cleaves the fluorogenic RNA substrate at a rate of $7\,min^{-1}$, placing it amongst the fastest DNAzymes known to date. This demonstrated that placement of a fluorophore and quencher near the active site did not hinder the selection of efficient DNAzymes. This DNAzyme was found to have strong activity in the presence of Co^{2+}, Ni^{2+}, and Mn^{2+}, suggesting that this type of sensor system could be used in the detection of transition metal ions. A subsequent selection was performed to isolate fluorescence-generating DNAzymes that could cleave an RNA linkage at various ranges of pH.[66–70] By performing parallel selections in buffers with different acidity settings, DNAzymes that cleaved fluorogenic substrates at a pH of 3, 4, 5, 6 or 7 were isolated. DNAzymes from each pH class were found to perform efficient catalysis, with reaction rates around $1\,min^{-1}$.

8.2.4 Fluorescence-generating Aptazymes: Linking Target Recognition to Catalytic Activity

To use DNAzymes to sense molecules larger than metal ions, it is necessary for the sensors to have recognition elements, in addition to catalytic motifs. *In vitro* selection has been very effective in isolating DNA and RNA aptamers that can recognise and bind a target molecule with high affinity and specificity. Attempts have been made to examine whether the catalytic signalling capacity of DNAzymes could be coupled with the specific binding of DNA aptamers to generate catalytic sensors that are responsive to target binding. Early demonstrations of this type of system were made by the Breaker group using RNA aptamers appended to ribozymes. By rational design, aptamer-linked ribozymes or 'RNA aptazymes' were created in which ATP binding to an aptamer domain altered the folding of an attached hammerhead ribozyme domain, causing it to adopt an inactive conformation and inhibiting the catalytic activity of the ribozyme.[71,72] In another study, the Breaker group was able to isolate FMN-responsive aptazymes with 270-fold rate enhancements by performing an *in vitro* selection with a randomised linker region.[73]

The first DNA aptazyme was designed by the Sen group using a well-studied anti-ATP DNA aptamer[74] as a recognition element and the 10-23 DNAzyme as the catalytic reporter.[75] As shown in Figure 8.3A, their design used a DNAzyme construct with an altered substrate-binding domain and an ATP-binding aptamer domain flanked by a short region complementary to the substrate. In the absence of ATP, the rate of catalysis is reduced, because the DNAzyme cannot bind the substrate (the two shortened complementary regions on the

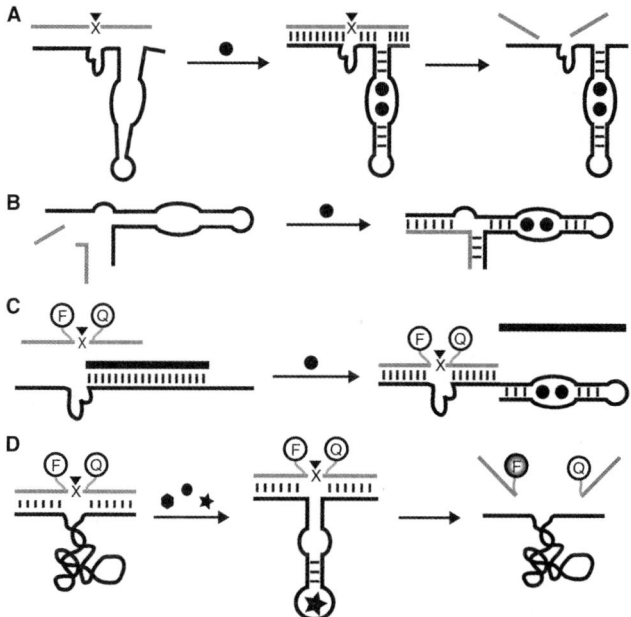

Figure 8.3 Target-responsive DNA aptazymes. (A) A DNA aptazyme constructed using an ATP aptamer and the 8-17 DNAzyme. In the absence of ATP (filled circle), the aptazyme cannot interact with the substrate (light grey) owing to a lack of base pairing. ATP binding alters the structure of the aptazyme, allowing substrate binding and RNA cleavage. (B) Aptazyme containing an ATP aptamer and a ligase DNAzyme. In the absence of ATP, the aptazyme has a conformation that does not allow it to bind its substrates. ATP binding changes the conformation, allowing substrate binding and ligation. (C) An aptazyme using a structure-switching mechanism. In the absence of ATP, a regulatory DNA binds to the aptazyme, blocking binding of a fluorogenic substrate. Upon ATP binding to the aptamer domain, the aptazyme switches from a duplex with the regulatory oligonucleotide (thick line) to a complex with ATP. This relieves the DNAzyme to bind and cleave its substrate. (D) Aptazyme sensor for *E. coli* crude extracellular mixture (CEM, represented by three different shapes). In the presence of *E. coli* CEM, the aptazyme can bind its target (star) and cleave a fluorogenic substrate.

aptazyme are separated by the unstructured ATP-binding region). When ATP is added, it binds to the aptamer domain, rearranging the aptazyme structure and permitting additional base pairing with the substrate. To show the versatility of this design, a similar system was constructed using the 8-17 DNAzyme as a reporter. The second DNA aptazyme sensor was developed by the Ellington group, and utilised the same anti-ATP DNA aptamer and a previously isolated DNA ligase,[76] as shown in Figure 8.3B. Binding of ATP to the recognition domain was found to increase the rate of ligase activity by up to 460-fold. The conformation of the aptazyme in the absence of ATP was also found to reduce background ligation levels below that of the templated reaction.

Another approach, reported by our group and modelled after a structure-switching aptamer design,[77] used the same ATP aptamer sequence linked to an RNA-cleaving DNAzyme[66] and a regulatory oligonucleotide that binds part of the substrate-binding region of the DNAzyme and part of the aptamer domain,[78] as shown Figure 8.3C. In the absence of ATP, the regulatory oligonucleotide blocks substrate binding to the DNAzyme portion. Binding of ATP to the aptamer domain disrupts the aptamer's interaction with the regulatory oligonucleotide, resulting in its release. The DNAzyme portion is now free to bind and cleave its RNA substrate. This system was shown to achieve a 30-fold rate enhancement in the presence of ATP. A unimolecular structure-switching mechanism was also developed using a fluorescence-generating RNA-cleaving DNAzyme functional at pH 6 and the ATP aptamer.[79] The sensor was designed by sequestering catalytically essential nucleotides in a base-paired stem within the ATP aptamer domain. Upon addition of ATP, the aptamer domain switches from a helical interaction with the DNAzyme to form a complex with ATP. The DNAzyme can now cleave its RNA substrate, producing a fluorescence increase.

Another method for the development of aptazymes, as opposed to rational design, is to use *in vitro* selection directly to isolate sequences that have catalytic activity that is responsive to a target of interest. This technique has potential for designing sensors for targets with no existing aptamers. Target dependence can be obtained by adding negative selection steps to the *in vitro* selection process in order to remove sequences from the DNA pool that are catalytic in the absence of target. By monitoring the level of activity in the absence of target, the stringency of the negative selection can be increased if needed, resulting in an aptazyme with very low background fluorescence. In one study, a fluorescent DNAzyme sensor was developed for the detection of bacteria.[80] As shown in Figure 8.3D, a DNA library containing a 70-nuceotide (nt) random region attached to a DNA substrate with a fluorophore and quencher flanking a ribonucleotide was used to select for a DNAzyme responsive to a crude extracellular mixture (CEM) of *Escherichia coli* (*E. coli*). Specificity was attained by performing counter-selection using CEM from *Bacillus subtilis* and discarding the population that could cleave the substrate under these conditions. After 20 rounds of selection, a sequence was isolated that could cleave the substrate and generate a fluorescence signal in the presence of *E. coli* CEM. The sensor was shown to be specific, because the cleavage and fluorescence were not seen in the presence of other Gram-negative and Gram-positive bacteria. With the addition of a culturing step, the sensor was shown to be able to detect a single seeding *E. coli* cell. It is also noteworthy that this *E. coli* sensor was developed using a complex mixture, the CEM, and was not directed for a specific target molecule. This allowed the *in vitro* selection procedure to choose the best target from the mixture. This avoids the laborious steps of finding a suitable target present in *E. coli* (or another specific bacterium of interest), but not other bacteria, and purifying the target while maintaining its native structure. Also, there is no guarantee that the purified targets will be suitable ligands in an aptamer selection. The direct aptazyme selection using a complex mixture, on the other hand, will target molecules for which the aptazyme has a great affinity.

8.2.5 Using DNAzymes to Amplify Fluorescent Signals

The sensitivity of many sensors is limited by the inability of the detection system to report low levels of recognition events. Amplification of recognition facilitates the lowering of the detection limit, creating an improved sensor. Signal amplification systems may be broken down into three modules: the sensor, activation, and amplification units. The sensor module may be viewed similarly to a molecular recognition element in which the presence of the target is recognised and acts as the starting point for initiating a series of events. The second step is the activation step, which is responsible for synthesising a particular molecule to induce amplification. The activation step is unique in that the system catalyses a particular reaction that is completely separate from the catalysed reactions seen in the amplification process. Lastly, the amplification step utilises the output product from the activation step for continuous production of fluorescent molecules.

Two examples of amplification processes that utilise DNAzymes in different capacities will be described. The first is a sensor for ATP and NAD$^+$, shown in Figure 8.4A. In the first step, these molecules act as cofactors for T4 DNA ligase, which ligates two oligonucleotides into a complete RNA-cleaving DNAzyme capable of multiple turnovers. The DNAzyme is released and cleaves multiple molecular beacon substrates, generating a large fluorescent signal.[81] In the second example (an L-histidine sensor), the roles of the fluorescence-generating DNAzymes and protein enzymes are reversed. Recognition is performed by an RNA-cleaving DNAzyme dependent on L-histidine for activity. As shown in Figure 8.4B, upon RNA cleavage a portion of the released substrate binds to a complementary region of a molecular beacon, creating a restriction site for a nicking enzyme. A nicking endonuclease specific for that site cleaves the molecular beacon strand, which separates its quencher and fluorophore, resulting in an increase in fluorescence (Figure 8.4B).[81] As the nicking enzyme does not alter the substrate fragment, it is able to bind other molecular beacons in what becomes an amplified process that greatly increases one L-histidine binding event into a large fluorescent signal.

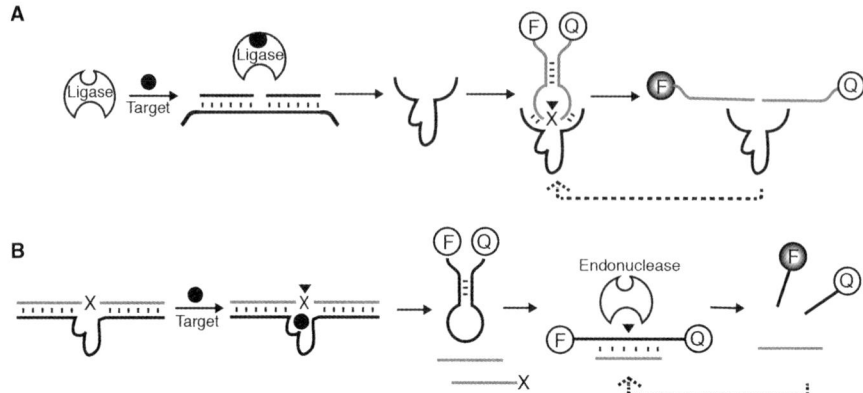

Figure 8.4   DNAzyme-based fluorescence signal amplification systems. (A) Amplification of T4 ligase-mediated ATP/NAD$^+$ detection. Upon the binding of ATP or NAD$^+$ to T4 DNA ligase, the ligase is activated and ligates two oligonucleotides into a functional RNA-cleaving DNAzyme. This DNAzyme can then cleave a fluorogenic substrate (light grey) in a multiple turnover fashion to generate an amplified signal. (B) Nicking endonuclease amplification of DNAzyme-mediated RNA cleavage. A DNAzyme cleaves an RNA substrate in the presence of a metal-ion cofactor to be detected. A portion of the released substrate hybridises to a molecular beacon, which is cleaved in the presence of a nicking endonuclease. The substrate fragment is then released and can bind to other molecular beacons, amplifying the fluorescent signal.

8.3   DNAzyme Sensors with Other Reporting Systems

8.3.1   Gold Nanoparticle-based Sensors

Colorimetric sensors offer advantages over many other detection systems because they can report the presence of a target without the need for expensive and bulky instrumentation. Gold nanoparticles in particular have gained much attention and have been used extensively for detection because they produce an intense red colour with higher extinction coefficients than most organic dyes. A change in the surface plasmon resonance upon aggregation of the nanoparticles results in a change in colour, which can be coupled to a target-responsive event to allow for sensing. Following the reports of using DNA oligonucleotide conjugated gold nanoparticles for DNA detection by the Mirkin group,[82] they have been used in an array of DNA sensing schemes.[83,84] By coupling oligonucleotide aptamers to gold nanoparticle aggregation, sensors have been made for proteins[85] and small molecules[86–90] (reviewed in ref.[91]).

   In a series of publications, the Lu group has used gold nanoparticle aggregation as a detection system for DNAzyme-based sensors. In one study, a sensor was created by designing a DNAzyme substrate that contains identical 5′ and 3′ regions and hybridises to oligonucleotides covalently bound to gold nanoparticles.[92] When the system is heated above the melting temperature of the complementary regions and cooled, the substrate forms a bridge between nanoparticles, leading to nanoparticle aggregation. Under these conditions the sensing solution appears purple. When substrate and Pb^{2+}-dependent DNAzyme are incubated with Pb^{2+} before melting and cooling, the substrate is cleaved, preventing nanoparticle aggregation. Consequently, a red colour is seen. This sensor was found to be able to detect lead ions with a dynamic range between 100 nM and 4 µM. By using a mutant DNAzyme that can bind the substrate without causing cleavage, it is possible to tune the range of the sensor by altering the ratio of active and mutant DNAzyme. In a follow-up study, the sensor was improved by changing the orientation of the nanoparticle–substrate complex and the size of the nanoparticles.[93] This allows the sensor to function at room temperature and removes the melting and cooling steps. These improvements reduced the time of detection from 2 h to 10 min. As shown in Figure 8.5A, this system has also been modified with nanoparticle disassembly, instead of assembly, being monitored in a 'light up' format.[94]

Another system for the detection of lead has been designed that utilises the properties of gold nanoparticles that aggregate in high concentrations of salt. As shown in Figure 8.5B, substrate-labelled nanoparticles are incubated with a Pb^{2+}-dependent DNAzyme and the maximum amount of salt at which the nanoparticles are still stable and not aggregated.[95] Upon the addition of Pb^{2+}, the substrate is cleaved, and a portion is released from the nanoparticles. The shortened oligonucleotides on the nanoparticles are not sufficient to stabilise the nanoparticles, resulting in aggregation and a colour change from red to purple.

Aptazyme-based colorimetric sensors using gold nanoparticles have also been developed.[96] As shown in Figure 8.5C, the aptazyme is divided into two oligonucleotides, one with an 8-17 catalytic domain and one arm of an ATP aptamer domain, and a second oligonucleotide containing the second half of the aptamer domain. Like those of the previous studies, the system contains a substrate capable of bridging oligonucleotide-modified nanoparticles, leading to nanoparticle aggregation and a purple colour. In the presence of adenosine, the aptazyme is activated, cleaving the substrate, and preventing gold particle aggregation. Gold nanoparticle aggregation has also been used as a sensing system with a DNA ligase DNAzyme. A sensor for Cu^{2+} was designed using a copper-dependent DNAzyme and two substrates capable of hybridising to oligonucleotide-modified gold nanoparticles.[97] In the presence of Cu^{2+}, the DNAzyme is active and ligates the two substrates. This ligated substrate is then capable of bridging two gold nanoparticles, leading to aggregation and a red-to-purple colour change.

8.3.2 Electrochemical-based Sensors

Surface-based detection has gained popularity in recent years. Electrochemical sensors in particular offer several advantages over solution-based systems. Electroactive labels used for detection are typically less sensitive to environmental changes than fluorescent or colorimetric dyes, and background signal from contaminants is far less an issue than that seen with optical techniques.

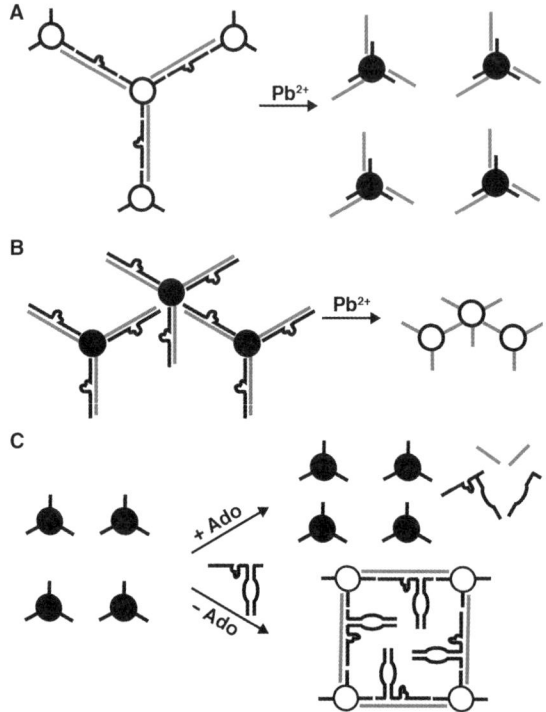

Figure 8.5   Gold nanoparticle-based DNAzyme sensors. (A) A turn-on sensor for
Pb²⁺. Oligonucleotide-modified gold nanoparticles can bind a substrate
(light grey) at its 5′ and 3′ ends, leading to nanoparticle aggregation and a
purple colour, shown with white circles. The addition of Pb²⁺ leads to
DNAzyme-mediated substrate cleavage and a loss of aggregation of the
nanoparticles leading to a colour change to red, shown by black circles.
(B) Detection of changes in salt-induced aggregation. Nanoparticles are
modified with substrate oligonucleotides and placed in salt conditions
under which they are barely stable. Upon DNAzyme-mediated substrate
cleavage, the shortened substrate fragments on the nanoparticle cannot
prevent aggregation, and a colour change from red to purple is observed.
(C) A nanoparticle-based aptazyme sensor. A substrate for an adenosine
(Ado)-responsive aptazyme is designed with complementary regions for
oligonucleotides linked to gold nanoparticles. In the absence of adenosine
the substrate can cross-link gold nanoparticles leading to aggregation and
a purple colour. When adenosine is present, the aptazyme cleaves the
substrate, which is no longer able to cross-link the nanoparticles, leading
to a red colour.

Owing to these factors, steps have been made to utilise the molecular recognition capacities of DNAzymes to develop sensitive electrochemical sensors.

The first study was carried out by the Plaxco group to develop a Pb²⁺ sensor
using a methylene blue-modified, lead-sensitive DNAzyme attached to a gold
electrode.[98] As shown in Figure 8.6A, when bound to its substrate, the double-
helical complex separates the methylene blue from the electrode. In the

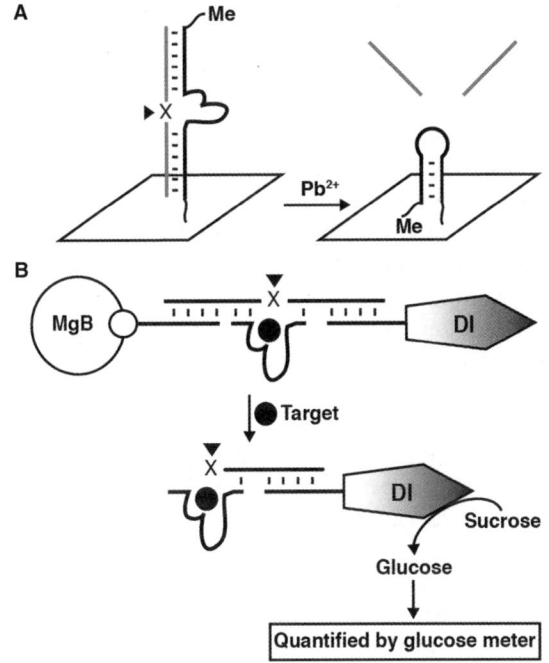

Figure 8.6 Electrochemical-based DNAzyme sensors. (A) A Pb^{2+}-dependent DNA-zyme modified with electroactive methyl blue (MB) is tethered to a gold electrode. When the DNAzyme forms a complex with the substrate, the methyl blue group is separated from the electrode, preventing electron transfer. In the presence of the Pb^{2+}, the substrate is cleaved and dissociated from the DNAzyme. The methyl blue can then interact with the electrode, producing a signal increase. (B) Linking DNAzyme activity to a personal glucose meter. A substrate is bound through complementary oligonucleotides to a magnetic bead (MgB) and an invertase enzyme (DI). In the presence of UO_2^{2+} and a UO_2^{2+}-dependent DNAzyme, the substrate is cleaved, releasing the invertase from the magnetic bead. The invertase can then hydrolyse sucrose to glucose, which is measured by a glucose meter.

presence of the lead ion, the substrate is cleaved and dissociates from the DNAzyme. In this unbound form, the DNAzyme becomes flexible and the attached methylene blue can contact the electrode, resulting in electron transfer. The sensor was shown to be able to detect as little as 300 nM Pb^{2+}. No signal was observed using 10 µM of several other divalent metal ions, showing that the sensor is very specific for lead.

Another sensor design utilises an electrode-bound Pb^{2+}-dependent DNA-zyme and $Ru(NH_3)_6^{3+}$, which binds to the negative phosphate groups of DNA, as the signal transducer.[99] Upon the addition of Pb^{2+}, the substrate is cleaved, resulting in dissociation of the DNAzyme–substrate complex and a reduction in the amount of $Ru(NH_3)_6^{3+}$ confined to the electrode surface. This system was found to be able to detect 5 nM of lead. By coupling the substrate to oligo-nucleotide-modified gold nanoparticles, the sensitivity of the sensor was increased to 1 nM Pb^{2+}. This signal amplification is accomplished by $Ru(NH_3)_6^{3+}$ being bound to multiple oligonucleotides on the nanoparticle.

Gold nanoparticles loaded with oligonucleotides have been used to amplify an electrochemical signal. In this design, gold nanoparticles are modified with Pb^{2+}-dependent DNAzymes which can hybridise to substrate DNA–RNAs.[100] In the presence of Pb^{2+}, the substrate is cleaved and released from the DNA-zyme. The solution is then added to an electrode modified with single-stranded DNA complementary to the DNAzyme sequence. The DNAzyme can hybri-dise to the electrode-bound DNA, bringing the gold nanoparticle labelled with many oligonucleotides to the electrode, resulting in a large increase in elec-trochemical signalling in the presence of $Ru(NH_3)_6^{3+}$. This sensor has been found to detect as little as 28 pM of Pb^{2+}.

A sensor with subnanomolar detection capacity has been developed by combining the benefits of electrochemical and chemiluminescence detection. The sensor utilises electrochemiluminescence, or the detection of light emitted from the excited state of an electrochemical signalling compound, to detect lead.[101] This sensor contains an electrode-bound DNAzyme and substrate labelled with tris(2,2'-bipyridine)-ruthenium(II) (TBR). Upon Pb^{2+}-induced substrate cleavage, the substrate-bound TBR is separated from the electrode, resulting in a reduction in electrochemiluminescence intensity. This system has allowed for the detection of Pb^{2+} at a concentration of 100 pM.

A recent study has elegantly coupled DNAzyme activity to a personal glu-cose meter. The personal glucose meter is an electrochemical sensor that detects glucose by monitoring the redox reaction catalysed by glucose oxidase. This platform has been optimized over the last 30 years into very efficient pocket-sized sensors. To utilise this well-established system to detect other molecules, a beta-fructofuranosidase, known as invertase, was used along with sucrose. As shown in Figure 8.6B, the invertase was conjugated to an oligonucleotide using the maleimide–thiol reaction, hybridised to a complex of a UO_2^{2+}-dependent RNA-cleaving DNAzyme and its substrate bound to a magnetic bead.[102] In the presence of UO_2^{2+}, the substrate is cleaved, releasing the DNA–invertase conjugate. Once released, the invertase can hydrolyse sucrose to glucose, which can be oxidised by glucose oxidase and detected by the meter with a detection limit of 9 nM UO_2^{2+}. This system was also shown to be versatile and was able to selectively detect cocaine, adenosine, and interferon-gamma using structure-switching aptamers.

8.4 Conclusions and Future Prospects

DNA has come a long way since its discovery as the genetic material of life. Through *in vitro* selection, DNA has been found to bind molecules, both small and large, and to catalyse a plethora of chemical reactions. With a range of aptamers and DNAzymes on hand, many researchers have carried out innovative studies that explore their potential utility as sensors. Using DNAzymes that function under different metal-ion conditions, an array of biosensors has been developed for the detection of toxic and medically relevant ions. By coupling DNAzymes to the binding abilities of DNA aptamers, aptazyme-based sensors have been developed against small molecules such as adenosine and ATP. *In vitro* selection has even allowed the isolation of aptazymes responsive to the crude extracellular matrix of *E. coli*, a complex mixture containing many proteins and small molecules. This demonstrates the power of *in vitro* selection, because the sensor does not detect other similar bacteria whose extracellular matrix contains many of the same components. The potential use of *in vitro* selection to develop a sensor, without the need for painstaking identification of a unique target in a complex mixture, offers a means of developing a sensor in a timely manner. This would be useful when a sensor is needed in a short period of time, such as following the emergence of a new pathogen.

Many of the sensors described above have detection limits and specificities that rival competing sensor techniques. Coupled with the recent advances in signal amplification and surface-based sensors that offer low background and are amenable to miniaturisation, DNAzyme-based sensors have the potential to emerge as viable options for many sensing applications. We anticipate that the exploration of DNAzymes for biosensing applications will continue to be a fruitful playground for innovative academic research in the years to come. However, the future of this field will undoubtedly depend on the development of truly unique DNAzyme systems that will offer better sensitivity, specificity and ease of use over other systems for judiciously chosen targets relevant to the sectors of health care, food and water safety, and environmental protection.

References

1. A. J. Zaug and T. R. Cech, *Science*, 1986, **231**, 470.
2. T. R. Cech, *Gene*, 1993, **135**, 33.
3. A. D. Ellington and J. W. Szostak, *Nature*, 1990, **346**, 818.
4. C. Tuerk and L. Gold, *Science*, 1990, **249**, 505.
5. R. R. Breaker and G. F. Joyce, *Chem. Biol.*, 1994, **1**, 223.
6. S. K. Silverman, *Nucleic Acids Res.*, 2005, **33**, 6151.
7. R. R. Breaker and G. F. Joyce, *Chem. Biol.*, 1995, **2**, 655.
8. D. Faulhammer and M. Famulok, *Angew. Chem. Int. Ed.*, 1996, **35**, 2837.
9. S. W. Santoro and G. F. Joyce, *Proc. Natl. Acad. Sci. USA*, 1997, **94**, 4262.
10. C. R. Geyer and D. Sen, *Chem. Biol.*, 1997, **4**, 5793.
11. A. Roth and R. R. Breaker, *Proc. Natl. Acad. Sci. USA*, 1998, **95**, 6027.
12. A. R. Feldman and D. Sen, *J. Mol. Biol.*, 2001, **313**, 283.
13. D. M. Perrin, T. Garestier and C. Helene, *J. Am. Chem. Soc.*, 2001, **123**, 1556.

14. L. Lermer, Y. Roupioz, R. Ting and D. M. Perrin, *J. Am. Chem. Soc.*, 2002, **124**, 9960.
15. R. P. Cruz, J. B. Withers and Y. Li, *Chem. Biol.*, 2004, **11**, 57.
16. Y. Isaka, *Curr. Opin. Mol. Ther.*, 2007, **9**, 132.
17. J. Li, W. Zheng, A. H. Kwon and Y. Lu, *Nucleic Acids Res.*, 2000, **28**, 481.
18. K. Schlosser and Y. Li, *Biochemistry*, 2004, **43**, 9695.
19. A. Flynn-Charlebois, Y. Wang, T. K. Prior, I. Rashid, K. A. Hoadley, R. L. Coppins, A. C. Wolf and S. K. Silverman, *J. Am. Chem. Soc.*, 2003, **125**, 2444.
20. W. E. Purtha, R. L. Coppins, M. K. Smalley and S. K. Silverman, *J. Am. Chem. Soc.*, 2005, **127**, 13124.
21. E. D. Pratico, Y. Wang and S. K. Silverman, *Nucleic Acids Res.*, 2005, **33**, 3503.
22. B. Cuenoud and J. W. Szostak, *Nature*, 1995, **375**, 611.
23. M. Levy and A. D. Ellington, *Bioorg. Med. Chem.*, 2001, **9**, 2581–2587.
24. M. Levy and A. D. Ellington, *J. Mol. Evol.*, 2002, **54**, 180.
25. A. Sreedhara, Y. Li and R. R. Breaker, *J. Am. Chem. Soc.*, 2004, **126**, 3454.
26. Y. Li and R. R. Breaker, *Proc. Natl. Acad. Sci. USA*, 1999, **96**, 2746.
27. W. Wang, L. P. Billen and Y. Li, *Chem. Biol.*, 2002, **9**, 507.
28. Y. Li, Y. Liu and R. R. Breaker, *Biochemistry*, 2000, **39**, 3106.
29. N. Carmi, L. A. Shultz and R. R. Breaker, *Chem. Biol.*, 1996, **3**, 1039.
30. N. Carmi, S. R. Balkhi and R. R. Breaker, *Proc. Natl. Acad. Sci. USA*, 1998, **95**, 2233.
31. M. Chandra, A. Sachdeva and S. K. Silverman, *Nat. Chem. Biol.*, 2009, **5**, 718.
32. D. J. Chinnapen and D. Sen, *Proc. Natl. Acad. Sci. USA*, 2004, **101**, 65.
33. P. I. Pradeepkumar, C. Hobartner, D. A. Baum and S. K. Silverman, *Angew. Chem. Int. Ed.*, 2008, **47**, 1753.
34. M. Chandra and S. K. Silverman, *J. Am. Chem. Soc.*, 2008, **130**, 2936.
35. Y. Li and D. Sen, *Nat. Struct. Biol.*, 1996, **3**, 743.
36. P. Travascio, Y. Li and D. Sen, *Chem. Biol.*, 1998, **5**, 505.
37. I. Willner, B. Shlyahovsky, M. Zayats and B. Willner, *Chem. Soc. Rev.*, 2008, **37**, 1153.
38. M. Moshe, J. Elbaz and I. Willner, *Nano Lett.*, 2009, **9**, 1196.
39. C. Teller, S. Shimron and I. Willner, *Anal. Chem.*, 2009, **81**, 9114.
40. G. Pelossof, R. Tel-Vered, J. Elbaz and I. Willner, *Anal. Chem.*, 2010, **82**, 4396.
41. X. Liu, R. Freeman, E. Golub and I. Willner, *ACS Nano*, 2011, **5**, 7648.
42. W. G. Cox and V. L. Singer, *Biotechniques*, 2004, **36**, 114.
43. R. P. van Gijlswijk, E. G. Talman, I. Peekel, J. Bloem, M. A. van Velzen, R. J. Heetebrij and H. J. Tanke, *Clin. Chem.*, 2002, **48**, 1352.
44. F. Awonusonu, S. Srinivasan, J. Strange, W. Al-Jumaily and M. C. Bruce, *Am. J. Physiol.*, 1999, **277**, L848.
45. T. Förster, *Ann. Physik.*, 1948, **2**, 55.
46. S. Tyagi and F. R. Kramer, *Nat. Biotechnol.*, 1996, **14**, 303.

47. R. Nutiu and Y. Li, *Methods*, 2005, **37**, 16.
48. W. Chiuman and Y. Li, *Nucleic Acids Res.*, 2007, **35**, 401.
49. J. Liu and Y. Lu, *Anal. Chem.*, 2003, **75**, 6666.
50. H. Wang, Y. Kim, H. Liu, Z. Zhu, S. Bamrungsap and W. Tan, *J. Am. Chem. Soc.*, 2009, **131**, 8221.
51. T. Lan, K. Furuya and Y. Lu, *Chem. Commun.*, 2010, **46**, 3896.
52. J. Liu, A. K. Brown, X. Meng, D. M. Cropek, J. D. Istok, D. B. Watson and Y. Lu, *Proc. Natl. Acad. Sci. USA*, 2007, **104**, 2056.
53. J. Liu and Y. Lu, *Angew. Chem. Int. Ed.*, 2007, **46**, 7587.
54. J. Liu and Y. Lu, *J. Am. Chem. Soc.*, 2007, **129**, 9838.
55. C. H. Lu, H. H. Yang, C. L. Zhu, X. Chen and G. N. Chen, *Angew. Chem. Int. Ed.*, 2009, **48**, 4785.
56. M. Zhou, Y. Zhai and S. Dong, *Anal. Chem.*, 2009, **81**, 5603.
57. H. Dong, W. Gao, F. Yan, H. Ji and H. Ju, *Anal. Chem.*, 2010, **82**, 5511.
58. S. He, B. Song, D. Li, C. Zhu, W. Qi, Y. Wen, L. Wang, S. Song, H. Fang and C. Fan, *Adv. Funct. Mater.*, 2010, **20**, 453.
59. X. H. Zhao, R. M. Kong, X. B. Zhang, H. M. Meng, W. N. Liu, W. Tan, G. L. Shen and R. Q. Yu, *Anal. Chem.*, 2011, **83**, 5062.
60. M. Liu, H. Zhao, S. Chen, H. Yu, Y. Zhang and X. Quan, *Biosens. Bioelectron.*, 2011, **26**, 4111.
61. Y. Wen, C. Peng, D. Li, L. Zhuo, S. He, L. Wang, Q. Huang, Q. H. Xu and C. Fan, *Chem. Commun.*, 2011, **47**, 6278.
62. M. Liu, H. Zhao, S. Chen, H. Yu, Y. Zhang and X. Quan, *Chem. Commun.*, 2011, **47**, 7749.
63. Y. Xiang, A. Tong and Y. Lu, *J. Am. Chem. Soc.*, 2009, **131**, 15352.
64. Y. Xiang, Z. Wang, H. Xing, N. Y. Wong and Y. Lu, *Anal. Chem.*, 2010, **82**, 4122.
65. S. H. Mei, Z. Liu, J. D. Brennan and Y. Li, *J. Am. Chem. Soc.*, 2003, **125**, 412.
66. Z. Liu, S. H. Mei, J. D. Brennan and Y. Li, *J. Am. Chem. Soc.*, 2003, **125**, 7539.
67. Y. Shen, J. D. Brennan and Y. Li, *Biochemistry*, 2005, **44**, 12066.
68. S. A. Kandadai, W. W. Mok, M. M. Ali and Y. Li, *Biochemistry*, 2009, **48**, 7383.
69. S. A. Kandadai and Y. Li, *Nucleic Acids Res.*, 2005, **33**, 7164.
70. M. M. Ali, S. A. Kandadai and Y. Li, *Can. J. Chem.*, 2007, **85**, 261.
71. J. Tang and R. R. Breaker, *Chem. Biol.*, 1997, **4**, 453.
72. J. Tang and R. R. Breaker, *Nucleic Acids Res.*, 1998, **26**, 4214.
73. G. A. Soukup and R. R. Breaker, *Proc. Natl. Acad. Sci. USA*, 1999, **96**, 3584.
74. D. E. Huizenga and J. W. Szostak, *Biochemistry*, 1995, **34**, 656.
75. D. Y. Wang, B. H. Lai and D. Sen, *J. Mol. Biol.*, 2002, **318**, 33.
76. M. Levy and A. D. Ellington, *Chem. Biol.*, 2002, **9**, 417.
77. R. Nutiu and Y. Li, *J. Am. Chem. Soc.*, 2003, **125**, 4771.
78. J. C. Achenbach, R. Nutiu and Y. Li, *Anal. Chim. Acta.*, 2005, **534**, 41.

79. Y. Shen, W. Chiuman, J. D. Brennan and Y. Li, *ChemBioChem*, 2006, **7**, 1343.
80. M. M. Ali, S. D. Aguirre, H. Lazim and Y. Li, *Angew. Chem. Int. Ed.*, 2011, **50**, 3751.
81. L. M. Lu, X. B. Zhang, R. M. Kong, B. Yang and W. Tan, *J. Am. Chem. Soc.*, 2011, **133**, 11686.
82. R. Elghanian, J. J. Storhoff, R. C. Mucic, R. L. Letsinger and C. A. Mirkin, *Science*, 1997, **277**, 1078.
83. T. A. Taton, C. A. Mirkin and R. L. Letsinger, *Science*, 2000, **289**, 1757.
84. K. Sato, K. Hosokawa and M. Maeda, *J. Am. Chem. Soc.*, 2003, **125**, 8102.
85. H. Wei, B. Li, J. Li, E. Wang and S. Dong, *Chem. Commun.*, 2007, 3735.
86. J. Liu and Y. Lu, *Angew. Chem. Int. Ed.*, 2005, **45**, 90.
87. J. Liu and Y. Lu, *Nat. Protoc.*, 2006, **1**, 246.
88. L. Wang, X. Liu, X. Hu, S. Song and C. Fan, *Chem. Commun.*, 2006, 3780.
89. W. Zhao, W. Chiuman, M. A. Brook and Y. Li, *ChemBioChem*, 2007, **8**, 727.
90. W. Zhao, W. Chiuman, J. C. Lam, S. A. McManus, W. Chen, Y. Cui, R. Pelton, M. A. Brook and Y. Li, *J. Am. Chem. Soc.*, 2008, **130**, 3610.
91. J. Liu and Y. Lu, *J. Fluoresc.*, 2004, **14**, 343.
92. J. Liu and Y. Lu, *J. Am. Chem. Soc.*, 2003, **125**, 6642.
93. J. Liu and Y. Lu, *J. Am. Chem. Soc.*, 2004, **126**, 12298.
94. J. Liu and Y. Lu, *J. Am. Chem. Soc.*, 2005, **127**, 12677.
95. W. Zhao, J. C. Lam, W. Chiuman, M. A. Brook and Y. Li, *Small*, 2008, **4**, 810.
96. J. Liu and Y. Lu, *Anal. Chem.*, 2004, **76**, 1627.
97. J. Liu and Y. Lu, *Chem. Commun.*, 2007, 4872.
98. Y. Xiao, A. A. Rowe and K. W. Plaxco, *J. Am. Chem. Soc.*, 2007, **129**, 262.
99. L. Shen, Z. Chen, Y. Li, S. He, S. Xie, X. Xu, Z. Liang, X. Meng, Q. Li, Z. Zhu, M. Li, X. C. Le and Y. Shao, *Anal. Chem.*, 2008, **80**, 6323.
100. X. Yang, J. Xu, X. Tang, H. Liu and D. Tian, *Chem. Commun.*, 2010, **46**, 3107.
101. X. Zhu, Z. Lin, L. Chen, B. Qiu and G. Chen, *Chem. Commun.*, 2009, 6050.
102. Y. Xiang and Y. Lu, *Nat. Chem.*, 2011, **3**, 697.

Electrochemical Techniques as Powerful Readout Methods for Aptamer-based Biosensors

BINGLING LI AND ANDREW D. ELLINGTON*

Institute for Cellular and Molecular Biology, Center for Systems and Synthetic Biology, Department of Chemistry and Biochemistry, University of Texas at Austin, Austin, TX 78712, USA
*Email: andy.ellington@mail.utexas.edu

9.1 Introduction to Aptamer-based Sensors (Aptasensors)

9.1.1 Aptamers

Aptamers are single-stranded nucleic acids that can be selected *in vitro* to bind to many different small-molecule, protein, and cellular targets. They typically fold into compact secondary and tertiary structures that allow them specifically to recognise their targets, and they can act as antagonists or agonists of biological activities.[1–3]

The technique for the production of aptamers is often referred to as the 'Systematic evolution of ligands by exponential enrichment (SELEX)'.[4–6] SELEX generally starts with the synthetic generation of a pool of random nucleic acid sequences in which a core of random sequence is flanked by constant sequences required for amplification and manipulation. Binding species are sieved from non-binding species by any of a variety of methods, such as

RSC Biomolecular Sciences No. 26
DNA Conjugates and Sensors
Edited by Keith R Fox and Tom Brown
Published by the Royal Society of Chemistry, www.rsc.org

target-mediated filter-binding or affinity chromatography, and after multiple rounds of selection (generally 6–18 rounds) the few 'fittest' nucleic acids are isolated from the original, large ($> 10^{13}$ different sequences) random sequence population. Modified nucleotides (5-naphthylmethylaminocarbonyl-deoxyuridine, 5-(1-pentynyl)-2′-deoxyuridine, 2′-*O*-methylpurines, and others) can be included in the initial nucleic acid pool and copied by the right combination of polymerases; these modifications can impart both stability and improved functionality to aptamers.[1,7,8] The selection process usually takes from weeks to months when done manually, but recent improvements of the method and automation can potentially hasten this time to only a few days.[9]

To obtain a more complete understanding of selected aptamers, it is useful not only to determine what sequences have been selected, but also to measure their affinities and determine their secondary and even tertiary structures. Functional secondary structures can be modelled using relatively accurate programs such as MFOLD or NuPack, and can be further confirmed by chemical and enzymatic probing.[10–12] Aptamers have been found to assume a number of different secondary structures (Figure 9.1), and it is likely that these different structures provide multiple options for presenting binding residues at an interface, or forming pockets that can enfold small molecules.

Some aptamer tertiary structures have been determined by nuclear magnetic resonance (NMR) or crystallography.[13,14] Aptamers against proteins present specific residues for hydrogen-bonding and stacking. For example, in an anti-MS2 coat protein aptamer (Figure 9.1B), residues A4 and A10 in a loop region make

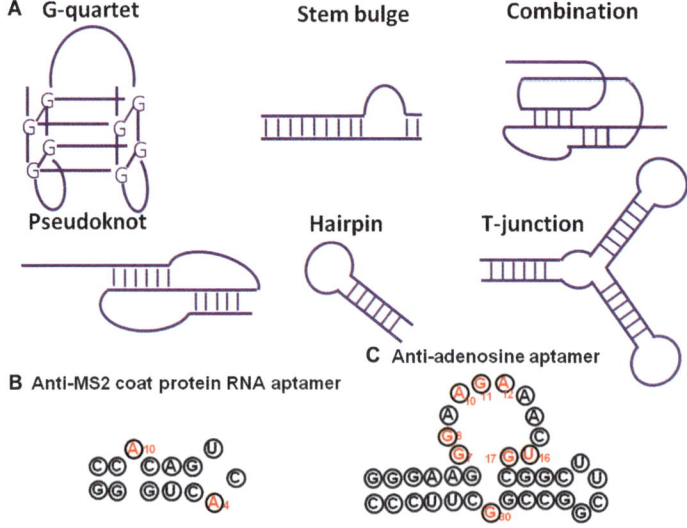

Figure 9.1 Examples of aptamer secondary structures.[1,13,14,16] A: Generic structures assumed by many different aptamers. B: The sequence and structure of an anti-MS2 coat protein aptamer. C: The sequence and structure of an anti-adenosine aptamer.

hydrogen bond contacts with specific amino acids in the protein.[13] Aptamers against small molecules can fold to form compact binding pockets similar to those found in proteins. For example, in an anti-adenosine aptamer, the internal loop folds into a compact zeta structure to make a pocket supported by two asymmetric G7–G11 and G17–G30 non-canonical base-pairs and a G8–A19–A10 U-turn. The ligand stacks stably with the pocket, and makes specific hydrogen bonds with residues G8 and A12 (Figure 9.1C).[14]

Because of the precise positioning of binding residues, aptamers against proteins can exhibit affinities for their targets, ranging from picomolar (1×10^{-12} M) to high-nanomolar (1×10^{-7} M), and from nanomolar (1×10^{-9} M) to micromolar (1×10^{-6} M) for small organics. The structured binding pockets in aptamers can even discriminate between very closely related molecules; for example, the anti-theophylline aptamer shows 10^5-fold less binding to caffeine, based on the presence of a single methyl group.[15]

9.1.2 Aptamers as Biosensors

Aptamers are viable alternatives to antibodies for bioanalysis/biosensors,[3,17–19] and their abilities to recognise their targets tightly and specifically have led to applications in fundamental research, drug screening, clinical diagnosis, and disease therapy.[20–22] In addition, aptamers have at least some advantages over antibodies, in that they are relatively easy to synthesise or modify *in vitro*, and can be appended with linkers and reporters for adaptation to various sensing strategies.

Similar to many other biosensors, a classical aptamer-based biosensor (aptasensor) contains three parts: 1) a bio-recognition system (or sensing platform), in which the aptamer is used as the key recognition element to capture the target; 2) a transducer, including a signal reporter/probe, that transfers recognition into a measurable signal; and 3) a corresponding instrument/device (measurement/readout technique) that can collect the signal from the reporter and convert it into readable data or a spectrogram. While aptasensors can be classified by different methods, they are most commonly distinguished on the basis of the measurement/readout technique. A number of sensing strategies have been reduced to practice, including adapting aptasensors to fluorescence, electrochemistry, surface plasmon resonance (SPR), Raman, quartz crystal microbalance (QCM), colorimetry, and other readouts.[3,18]

That said, fluorescent and electrochemical aptasensors predominate in the literature. Fluorescence is a classically powerful technique and has many options for signal reporters, ranging from small dyes to polymers to novel nano-/micro-fluorescent materials. These fluorescent readouts can be readily used by pioneer researchers who have developed or adapted aptamers. In comparison, electrochemical readouts are more technically challenging and less readily adapted by researchers not already versed in these methods. This may be why the first electrochemical aptasensor appeared in 2004,[23] almost 10 years after the introduction of fluorescent aptasensors. Electrochemical methods have now caught up, however, and account for up to one-third of the total publications on aptasensors (see Figure 9.2A, below). Ultimately, we can expect further growth in the development of electrochemical aptasensors because of the intrinsic advantages this approach offers, such as: 1) high sensitivity; 2) a variety of electrochemical readouts, such as cyclic voltammetry (CV), differential pulse voltammetry (DPV), square wave voltammetry (SWV), and others (see Figure 9.3, below); 3) relatively low cost; 4) amenability to miniaturisation and arraying; and 5) reusability.[24–27] In the following sections, we provide an overview of electrochemical aptasensors and then examine in greater detail the different types of electrochemical aptasensor.

9.2   Electrochemical Aptasensors

9.2.1   Introduction to Electrochemical Transduction

Electrochemistry is a powerful tool to study reduction–oxidation (redox) chemical reactions that take place at electrode–liquid interfaces. Such reactions are driven either by internal potential differences (such as in a battery discharging) or by external voltage (such as in electrolysis and battery charging, the preferred mechanism for electrochemical aptasensors). Therefore, signal reporters/indicators in electrochemical aptasensors are usually redox probes that have good electrochemical activity within a certain range of potential. The readout signals can be one of several parameters, including current (i), current density (j, i/A), impedance (Z), capacity (C), or charge passed (Q). Among these parameters, current density or current are used most frequently, because they directly reflect the electrochemical reaction activity or rate ($v = i/nFA$), and are closely related to many of the other parameters mentioned.[28] There are several factors that can affect the signal transduction, including the potential added (E), temperature, effective electrode material/area (A), the redox reporter species/concentration (C), diffusion coefficients (D), tunnelling distance or pathway between the electrode and redox probe (d, relevant to the normal reaction rate constant, $k°$), and additional surface conditions on the electrode. Aptasensors are typically designed to modulate one of these factors while holding the others constant.

The binding constant of an aptamer is probably the most important parameter for generating a good aptasensor. Below, we also examine other key variables that can be modified to affect the final detection limit, detection range, detection selectivity, and signal-to-background ratio.

1) **Electrochemical reporters.** As described above, reporters should have good redox activity with respect to the voltage on the working electrode (usually $k° \gg 0.1$ cm/s, in the case of a first-order reaction). Commonly used probes include traditional redox-active molecules such as methylene blue (MB), Ferrocene (Fc), and $Ru(NH_3)_6^{3+}$ (RuHex). Enzyme conjugates to horseradish peroxidase (HRP) and glucose oxidase have also proven useful to initiate electrochemical signalling. More recently, nanomaterials such as platinum, cadmium, and lead have been used to increase the density of aptasensors and/or electrochemical reporters, and have thereby further improved signal gains.[3]

2) **Electrodes.** The electrode is the most basic element of electro-analytical systems. The most commonly used electrode system is the classical three-electrode set, consisting of a working electrode, a counter electrode, and a reference electrode (as shown in Figures 9.2B,C). The sensing system is usually fabricated on the working electrode surface. Working electrodes are often constructed from carbon, gold (Au), indium tin oxide (ITO), or platinum (Pt) (Figure 9.2), and following aptamer immobilisation they have an effective contact or detection area ranging from sub-μm^2 to sub-cm^2. During sensing, electrochemical changes near the working electrode–solution interface will be sensitively and directly reflected in electronic parameter changes. Counter electrodes typically complete a closed circuit and the reactions taking place on the counter electrode are usually not monitored, except when an electrochemical biofuel cell is used as the readout (an example is elaborated in Section 9.2.2.2.3, Figure 9.9B).

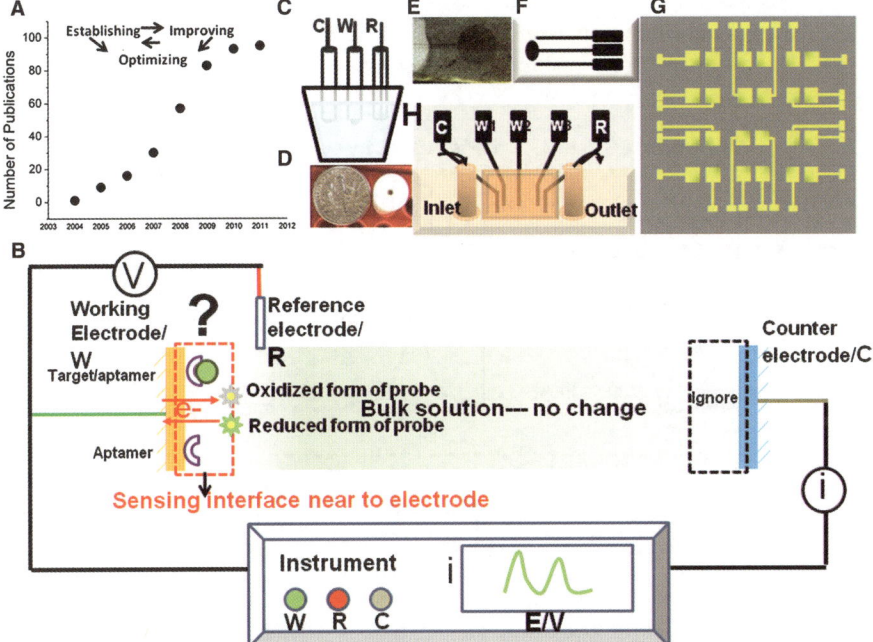

Figure 9.2 Electrochemical aptasensors. A: Number of publications on electrochemical aptasensors over time. B: Scheme of three-electrode system, sensing surface and instrument connection. C: Example of a classical three-electrode setup. D: Example of an Au disc working electrode (right, 1.2 mm in diameter, used in ref.[29]) wrapped by a white insulated polymer, shown alongside a US dime (17.91 mm in diameter) for comparison. E: Modified ITO electrode after layer-by-layer assembly.[30] F: Screen-printed three-electrode system, with carbon working electrode.[31] G: Example of gold-sputtered dual array electrodes with attached chambers.[32] H: Microfluidic chamber coupled with integrated three-electrode system, with array Au working electrodes.[33]

With the development of chemical and/or physical printing techniques, the entire three-electrode system can be arrayed onto small substrates such as glass, with concomitant opportunities for the development of multi-analysis, parallel-analysis, or micro-fluidic-assisted analysis (Figure 9.2).

3) **Electrochemical instruments.** The performance of the electrochemical instrument used for measurements is an important component of sensor sensitivity and noise. Most instruments contain both potentiostats and galvanostats, which can introduce different potentials or currents to the working electrode. Some traditional sub-techniques and corresponding parameters/readouts are listed in Figure 9.3. Many of these techniques rely on current as an output, and DPV, anodic stripping voltammetry (ASV), SWV, and alternating current voltammetry (ACV) in particular can provide high signal-to-background. CV displays less sensitivity but is a mature and simple technique to characterise the fabrication process and determine $k°$, D, effective electrode area, and other parameters. The sensitivity of CV can be significantly enhanced when coupled with an electrochemiluminence (ECL) readout. chronocoulometry (CC) is sensitive to changes in surface absorption, and electrochemical impedance

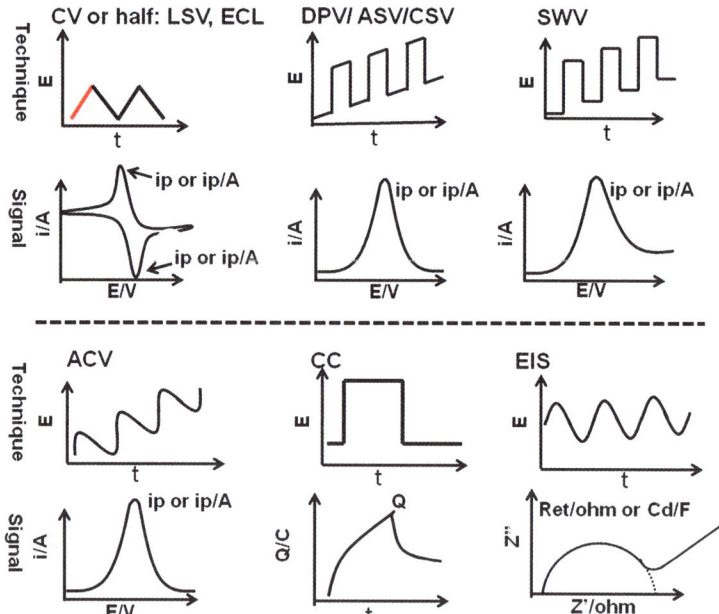

Figure 9.3   Sub-techniques and corresponding signals commonly used in electrochemical aptasensors. CV: Cyclic voltammetry. LSV: Linear sweep voltammetry. DPV: Differential pulse voltammetry. ASV or CSV: Stripping voltammetry. SWV: Square wave voltammetry. ACV: Alternating current voltammetry. CC: Chronocoulometry. EIS: Electrochemical impedance spectra. ECL: Electrochemiluminence, using the peak light (a.u.) as the readout.

spectra (EIS) is a powerful technique that reports surface conditions. Both techniques allow label-free detection.

9.2.2 Electrochemical Aptasensors

The detailed relationships among targets, aptamers, redox probes, electrodes and instruments will ultimately determine the specificities and sensitivities of aptasensors. Several distinct classes of electrochemical aptasensor are discussed, with an emphasis on how these variables can be combined.

9.2.2.1 Covalent Attachment of the Reporter to the Aptamer

9.2.2.1.1 Electrochemical Aptasensors that Fold upon Interaction with their Ligands (FOLD Strategy).
In 2005, Plaxco's group developed a reagentless electronic aptamer-based (E-AB) sensor.[34] This 'E-AB' sensor expanded on the previous development of 'E-DNA' sensors,[27] but the target was changed from DNA to a protein, α-thrombin. As shown in Figure 9.4A, a two-end-modified, 32-base single-stranded (ss)DNA probe (15 bases of which are anti-thrombin aptamer) is the sensing element. The 3'-end of the probe is modified with an electrochemically active redox reporter, methylene blue (MB). The 5' end of the probe is modified with a thiol that allows immobilisation of DNA onto a gold electrode. Non-specific interactions are reduced by subsequently blocking the electrode surface with a neutral 6-carbon thiol (1-mercaptohexanol, MCH). In the absence of target, the DNA probe is largely unstructured, and MB molecules readily interact with the electrode surface (fast-transfer kinetics, with an apparent $k°$ of $88 \, cm \, s^{-1}$). Using a potential scan from –0.4 to 0 V (ACV method), an oxidation current can be achieved with a peak current (i_p) of between –0.3 V and –0.2 V (Figure 9.4B). This aptamer is in equilibrium between its unstructured and quadruplex binding conformations, and upon addition of thrombin, the binding conformation is stabilised. The rigidity of the G-quartet increases the electron-tunnelling distance, which is correspondingly reflected in an i_p decrease (slower transfer kinetics, an apparent $k°$ of $58 \, cm \, s^{-1}$). The deviation of the i_p in the presence of thrombin is thrombin specific and dose dependent, even in the presence of diluted calf blood samples.

At around the same time, O'Sullivan's group developed a similar 'E-AB' sensor[35] in which Fc was used as the reporter. These constructs are shorter (15-mers rather than 32-mers), and upon thrombin binding, the quadruplex structure formation leads to an increase in current. Plaxco's group has further developed similar 'E-AB' aptasensors for the detection of cocaine[36] and platelet-derived growth factor (PDGF).[37–38] In these instances, aptamer structure formation should drive the reporter close to the site of aptamer attachment, and hence close to the surface of the electrode, resulting in an increase in signal in the presence of target. The same strategy has also been coupled with a microfluidic system, as shown in Figure 9.2H, allowing the 'signal-on' detection of cocaine in undiluted blood serum in real time.[33]

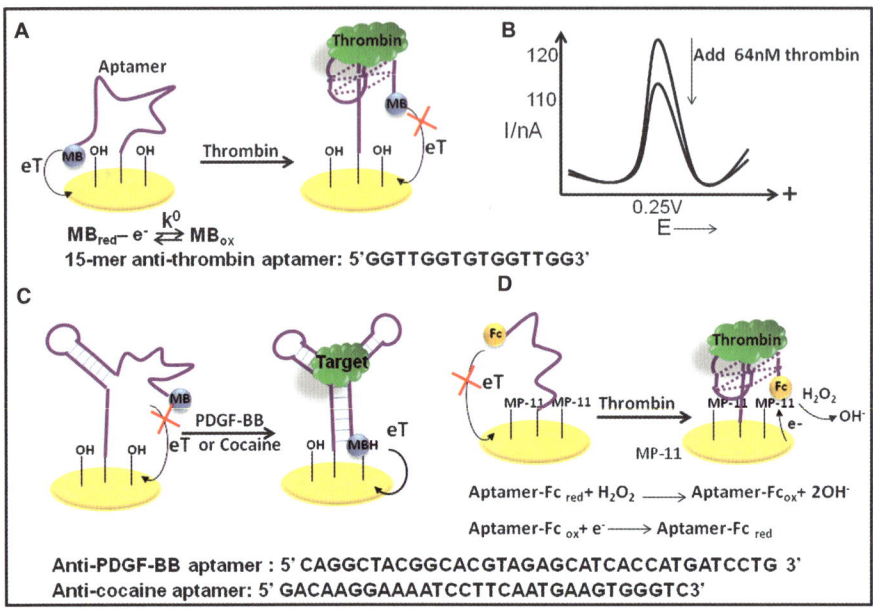

Figure 9.4   FOLD strategies for 'E-AB' sensors. A: The first signal-off E-AB sensor
containing an MB-labelled aptamer directed against thrombin.[34] B: The
mimic ACV curve of the E-AB sensor responds robustly to the thrombin
of Figure 9.4A. C: Signal-on E-AB sensor with cocaine or PDGF-BB as
the target.[36,37] D: E-AB sensor using co-immobilisation of MP-11 and
thiol-Fc labelled anti-thrombin aptamer on gold before and after thrombin interaction.[39]

The sensitivity of the E-AB folding strategy has been improved through
the use of enzyme amplification of signal (Figure 9.4D).[39] In the presence
of thrombin, a ferrocyanide (Fc) reporter on an aptamer can be brought
into proximity with a peroxidase-coated electrode. Using EIS as readout
technique, a 30 fM detection limit has been claimed, a remarkable achievement
given that the anti-thrombin aptamer has a reported K_d in the mid-nanomolar
range.[40]

**9.2.2.1.2   Electrochemical Aptasensors that Refold upon Interaction with their
Ligands (REFOLD Strategy).**   The 'E-AB' sensor is representative of a
number of electrochemical sensor designs,[25,27,41] all of which follow the same
basic principle: target binding influences structure formation, and this in turn
changes the electron-tunnelling distance or pathway of redox reporters. In
addition to stabilising a folded conformation, as described in the previous
section, it is also possible to change the equilibrium between an already folded (but inactive) aptamer and its active, folded conformation, and in so
doing change the opposition of an electrochemical reporter to the electrode
surface. In general, it is hoped that the signal amplitude and direction will be

influenced by a change in the flexibility or availability of the nucleic acid strand carrying the reporter.[42–44]

In some cases it is possible to predict roughly whether the signal will increase or decrease, because a number of techniques (including atomic force microscopy (AFM), surface plasmon resonance (SPR), and electrochemistry) have demonstrated that if nucleic acid density is high enough, nucleic acids should be oriented on the electrode to form a compact monolayer (with different angles). Because duplex DNA is about 50-fold more rigid and harder to bend than ssDNA, reporters on double-stranded (ds)DNAs in monolayers are less likely to approach the surface than reporters on ssDNAs. In addition, because aptamers are folded nucleic acids, they are more similar in nature to dsDNA than ssDNA, and thus sensing schemes can treat them like rigid barriers. For example, the anti-thrombin quadruplex immobilised on an electrode has been hybridised to an antisense oligonucleotide, leading to denaturation of the aptamer. The electrochemical reporter MB has been attached to the far end of the antisense oligonucleotide and thus, in principle, held away from the electrode owing to its the relatively rigid duplex structure (Figure 9.5A,B).[43,45]

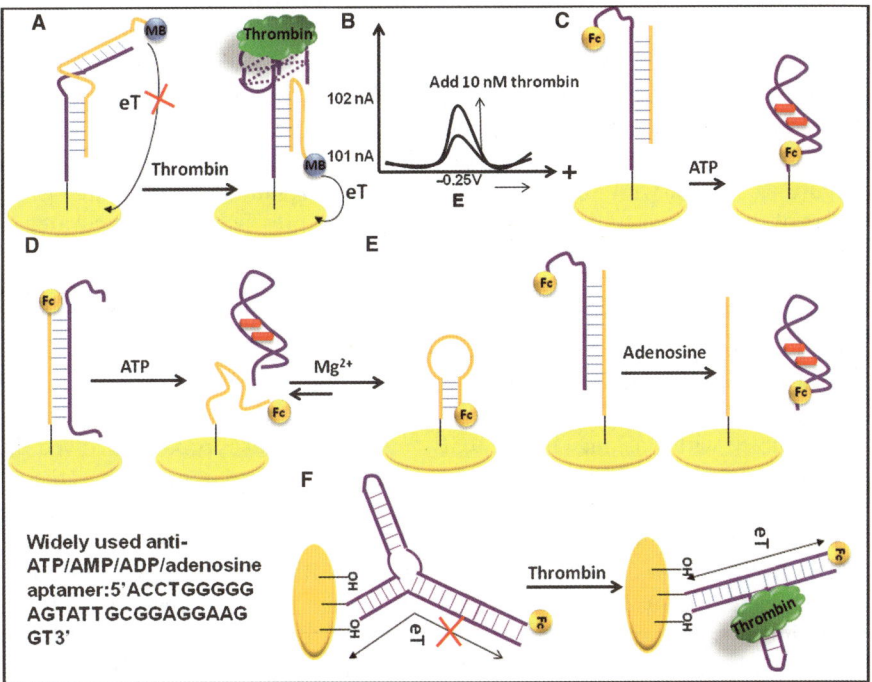

Figure 9.5 REFOLD strategies for 'E-AB' sensors. A: The first signal-on E-AB sensor based on the strand-displacement mechanism.[46] B: Mimic ACV response to thrombin. C: Reformed signal-on E-AB sensor based on the strand-displacement mechanism.[49] D: Improved strategy by stabilising the hairpin.[50] E: Reporter 'on-to-off' strategy.[51] F: Deoxyribosensor constructs on a gold chip for direct electronic detection of target.[52]

In the presence of the target, the slow equilibrium between the denatured, non-binding, antisense-bound and native, binding, antisense-unbound conformations could be perturbed.[46] In this case, the direction of signal change (by ACV, signal 'on') could be predicted because the released antisense-MB strands could diffuse to be closer to the electrode surface.[43] Interestingly, if the density of nucleic acids on a surface is too high, the aptasensor may be incapable of movement, limiting signal generation. Thus, optimisation of nucleic acid density is very important in most electrochemical DNA sensors. A second approach has been to spread out the sensors by mounting aptamers on DNA nanostructures.[47] This approach has also been shown to passivate the surface better, reducing protein binding[48] and allowing sensitive detection of thrombin (<1 pM detection limit in 50% serum) and cocaine (33 nM detection limit in complex samples such as a sugar solution, soda solution, 10% serum, and 50% serum).

Interestingly, even though a REFOLD strategy should perturb the apparent K_d of the aptamer more than just a FOLD strategy (because refolding requires dissociation of the antisense oligonucleotide, essentially creating an unstructured intermediate that can then fold into the aptamer structure), the strategy in Figure 9.5A has nonetheless yielded a modest increase in sensitivity, from 20 nM to 3 nM in detection limit, compared with the experimental design shown in Figure 9.4A. It might therefore be supposed that a reduction in apparent affinity is more than offset by an improvement in the signal-to-background ratio, which in turn may be due to the hypothesised restricted diffusion of the reporter prior to target-binding.

Various other versions of the REFOLD strategy have been reported (Figures 9.5D–F), with signal transduction again resulting from changing the access of the reporter to the electrode surface in the presence of target. Strategies very similar to those in Figure 9.5A have been employed in which antisense release leads to aptamer[49] or hairpin[50] folding, and presumably juxtaposition of the reporter to the electrode surface (Figures 9.5C,D). A detection limit of 10 nM for ATP was reported for the strategy in Figure 9.5C; 10 nM for ATP and 2 nM for thrombin were reported for the strategy in Figure 9.5D. In order to reduce background further, release of the reporter strand upon target-binding has been coupled with washing (Figure 9.5E).[51] A more novel approach involved the use of the DNA duplex itself to enable electron transfer via tunneling (Figure 9.5F).[52] A detection limit of 5 pM for thrombin was realised, and this sensor could work in the presence of diluted serum.

The E-AB sensors, whether of the FOLD or REFOLD variety, are particularly useful because they are 'reagentless', and are thus of potential interest for point-of-care applications or *in situ* sample detection. A summary of E-AB sensors and their characteristics is shown in Table 9.1. However, the sensitivities of these methods are ultimately limited by the affinities of aptamers, and these affinities are often further compromised by the signal transduction strategies themselves. Moreover, the principles that govern whether and how a change in conformation will result in a change in signalling are for the most part unknown. E-AB sensors therefore require extensive, empirical optimisation,[53–57] including adjustments of temperature and salt concentrations.[29,58]

Table 9.1 Summary of 'E-AB' sensors.

Target	Signal probe	Strategy	Detection limit	Linear range	Detection methods	Electrode	Reference
Thrombin	MB	FOLD-Off[a]	20 nM, in diluted fetal calf serum[b]	20–768 nM	ACV	Au–C[a]	34
Thrombin	MB	REFOLD-On[a]	3 nM	5–35 nM	ACV	Au–C	46
Thrombin	Fc	FOLD-On	0.5 nM		DPV	Au–C	35
PDGF-BB	MB	FOLD-On	50 pM	50 pM–tens of nM	ACV	Au–C	37
Cocaine	MB	FOLD-On	10 μM, in cutting/masking agents		ACV	Au–C	36
K$^+$	Fc	FOLD-On	0.015 mM, in mixed ion buffer	0.1–1 mM	SWV	Au–C	59
ATP	Fc	REFOLD-On	10 nM, in lysed cells		SWV	Au–C[a]	49
Adenosine	Fc	REFOLD-Off	0.02 μM	0.1 to 10 μM	ACV	Au–C	51
Thrombin	Fc, MP-11	FOLD-On[a]	30 fM		EIS	Au–C	39
Cocaine	Fc	FOLD-On	0.5 μM	1–15 μM	SWV	Au–AuNPs–C[a]	60
Theophylline	Fc	FOLD-On	0.2 μM, in diluted serum	0.2–10 μM	DPV	Au–C	61
ATP	Fc	REFOLD-On	10 nM	10–80 nM	DPV	Au–C	50
Thrombin	Fc	REFOLD-On	2 nM	2–40 nM	DPV	Au–C	50
Thrombin	Fc	REFOLD-On[c]	5 pM, in diluted serum		SWV	Au–C/on chip	52
ATP	Ru(bpy)$_2$(cbpy)	REFOLD-On	0.02 nM	0.05–10 nM	ECL	Au–C	62
Cocaine	Ru(bpy)$_3^{2+}$	FOLD-On	10 pM	10–5000 pM	ECL	PIGE–C[a]	63
Aminoglycoside antibiotics	MB	FOLD-On	–, in diluted calf serum		SWV	Arrayed Au–C	64

222

Chapter 9

Table 9.1 (*Continued*)

Target	Signal probe	Strategy	Detection limit	Linear range	Detection methods	Electrode	Reference
ATP	QDs[a]	REFOLD-Off	0.01 nM	0.01–100 nM	ASV	macroporous Au–C	65
Lysozyme	Fc	REFOLD-Off	0.1 pM	0.1 pM–1 nM	SWV	Au–AuNPs–C	66
Adenosine	Fc	REFOLD-On[d]	0.1 nM	3.74 nM–37.4 μM	DPV	Au–C	67
Interferon gamma (IFN-γ)	MB	FOLD-Off	~60 pM, in release from human leukocytes	–	SWV	Arrayed Au–C/on chips	68
Thrombin	MNPs, TMB, H_2O_2[e]	FOLD-Off	0.1 nM	1–75 nM	DPV	Au–AuNPs–C	69
Thrombin, lysozyme	Fc, anthraquinone	REFOLD-Off[f]	–;–	–;–	DPV	Au–C	70
Cocaine	MB	FOLD-On	–, in undiluted blood serum	–	ACV	Au–C/on chips	33

[a]:**FOLD-On**: FOLD strategy, signal-on; **FOLD-Off**: FOLD stratey, signal-off; **REFOLD-On**: REFOLD strategy, signal-on; **REFOLD-Off**: REFOLD strategy, signal-off; **Au-C**: Au electrode; **C** means DNA is covalently immobilised on the electrode; **AuNPs**: gold nanoparticles; [b]: the ones with real sample test are marked; [c]: deoxyribosensors; **PIGE**: paraffin-impregnated graphite electrode; **QDs**: quantum dots; [d]: enzyme participated detection; [e]: MNPs are labelled on the nucleic acids, which contain catalytic activity for H_2O_2-oxidising TMB. **TMB**: 3,3′,5,5′-tetramethylbenzidine; [f]: multianalysis; thrombin and lysozyme were detected on the same electrode.

While many of the reported detection limits are admirably low, these are for the most part obtained using pure anlaytes rather than real-world samples in complex mixtures.

9.2.2.1.3 Sandwich Assays with Electrochemical Aptasensors. A common strategy for signal transduction that also helps to ensure specificity is a sandwich assay in which a ligand is pinioned between two receptors, one of which can report the presence of the ligand (Figure 9.6A).[23] A standard system that is used by many authors employs two anti-thombin aptamers that bind to different sites on thrombin.[23,71] Another system that is widely used utilises the fact that anti-PDGF aptamers can bind jointly to the PDGF–BB dimer. Many groups do not go beyond these proofs-of-principle because the number of targets with aptamers that bind to two distinct epitopes is so far quite limited. Antibody–aptamer pairs have also been used to expand the range of aptasensors (Table 9.2).

In addition, because the reporters on the second aptamer will be distant from the electrode surface (separated by a target and another aptamer), electron tunnelling may prove to be very inefficient, limiting detection sensitivity. To overcome this issue, signal reporters that are catalytic or that can otherwise undergo amplification are usually employed. For example, Pt nanoparticles (PtNPs) have been used as a reporter by Willner's group (Figure 9.6B), with the signal being generated from the catalytic electrochemical reduction of H_2O_2 by the PtNPs.[71] The detection limit reported for thrombin was 1 nM. In another example,[72] Wang's group has employed CdS QDs as the reporter (Figure 9.6C). Bound QDs are dissolved by acid to release Cd^{2+}, which is detected with a solid-contact, Cd^{2+}-selective microelectrode. A detection limit of 0.14 nM for thrombin was realised. Obviously, the reusability of this approach would be limited. AuNPs have also been used to increase the effective electrode surface area and enrich the cadmium-based or other QDs (Figure 9.6D). Metal ions dissolved from these QDs (Cd^{2+} in this example) will further contribute to electrochemical signalling via their reduction. The stripping process will deposit the metal on the electrode and lead to an enhanced oxidation current; this enhancement has led to a further decrease in the detection limit for thrombin, to 0.55 fM.[73] As with many other electrochemical sensors, this result was obtained with pure samples. Ferrocenylhexanethiol-loaded silica nanocapsules (FcSH/SiNCs) that bound thrombin via a sandwich assay have also been stripped to create a DPV measurement.[74] Ramos cells were detected by loading multiple quantum dots onto a self-assembled DNA nanostructure,[75] and as few as 10 cells could be specifically detected with this method.

Another type of sandwich assay co-localises both aptamers on the surface. In this case, two immobilised, Fc-labelled anti-PDGF aptamers are brought together by the PDGF dimer, leading to current generation (Figure 9.6E).[76] As is the case with other proximity methods such as the proximity ligation assay,[77] the need for two binding events to occur simultaneously in order to generate signal greatly reduces background, and the reported detection limit was very good (approximately 68 fM).

9.2.2.1.4 Quaternary Structure Aptasensors. A second type of aptamer sandwich is not really a sandwich *per se*, in that it does not recognise two different epitopes on the target. Rather, a given aptamer is split into two pieces, and these pieces are reassembled in the presence of the target.[89] In this regard, these 'quaternary structure aptasensors' are much more like the FOLD aptasensors described above, and have many of the same advantages and disadvantages. In addition, however, it is not always possible to divide an aptamer such that it can be reassembled, or reassembled to form a functional binding site. Nonetheless, quaternary structure aptasensors have two major advantages. First, unlike conventional sandwich assays they can be used to develop sensors for small molecules. For example, both the anti-cocaine and anti-ATP aptamers have been successfully divided to create

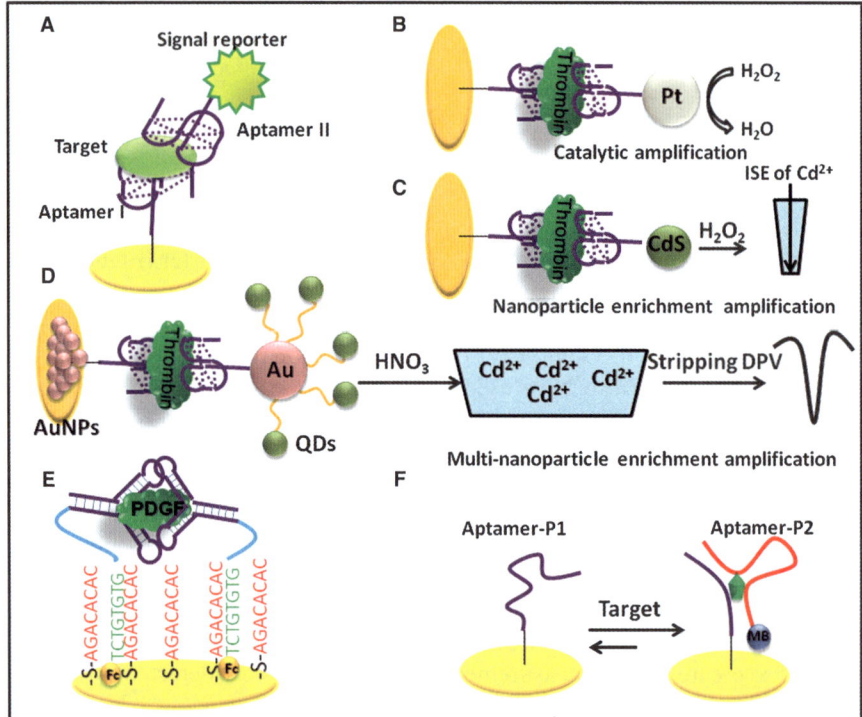

Figure 9.6 Sandwich and quaternary structure-forming aptasensors (A–D). A: A general schema for a sandwich aptasensor. B: PtNPs as catalytic signal reporters.[71] C: QDs as signal amplifiers.[72] D: Using AuNP–QD conjugates to expand the effective electrode surface.[73] E: Proximity method for the detection of PDGF.[76] F: A general scheme for quaternary structure-forming aptasensors.

Table 9.2 Sandwich aptasensors.

Target	Signal reporter labelled	Property of reporter	Detection limit	Detection methods	Electrode	Reference
Thrombin	GDH[a]	Catalyse oxidisation of glucose in presence of m-PMS electron mediator	1 μM	LSV	Au–C	23
Thrombin	(PQQ)GDH[a]	Catalyse oxidisation of glucose in presence of m-PMS electron mediator	10 nM	LSV	Au–C	78
Thrombin	PtNPs	Catalyse reduction of H_2O_2	1 nM	LSV	Au–C	71
IgE[a]	ALP[a]	Catalyse reduction of Ag^+ by ascorbic acid 2-phosphate	0.02 nM	LSV	Atu–C	79
Thrombin	CdS QDs	Release Cd^{2+} to be detected	0.14 nM	Special method	Au–C for sensing/Cd^{2+} – selective electrode detection	72
C Reactive protein	ALP	Catalyse oxidisation of a-naphthyl-phosphate to form a-naphthol as redox probe	$0.2\,mg\,L^{-1}$	DPV	Screen-printed electrodes – A[a]	80
Thrombin	ALP	Catalyse reduction of Ag^+ by ascorbic acid 2-phosphate	0.1 nM	LSV	Au–C	81
Thrombin	CdS QDs	Release Cd^{2+} to be detected	1 pM	SWSV[a]	Au–C for sensing/GCE for detecting Cd^{2+}	82
Thrombin	PbS QDs-modified AuNPs	Release Pb^{2+} to be detected	6.5 fM	DPASV[a]	Au–C for sensing/GCE for detectingCd^{2+}	83
Thrombin	ALP-modified SWCNT	Catalyse oxidisation of p-aminophenyl phosphate in presence of auxiliary enzyme DI and NAD^+ as electron mediator	8.3 fM	CV	GCE–C	84

Table 9.2 (*Continued*)

Target	Signal reporter labelled	Property of reporter	Detection limit	Detection methods	Electrode	Reference
Thrombin	CdS QDs-modified AuNPs	Release Cd^{2+} to be detected	0.55 fM	DPV	Au–C for sensing/GCE for detecting Cd^{2+}	73
Thrombin	Polystyrene microbead composite with multilayer of CdTe QDs	QD ECL	0.35 pM	CV-ECL	Au–C	85
Thrombin	Thi/Pt–AuNPs/HRP–SWCNTs	Catalyse reduction of H_2O_2 in presence of thionine as electron mediator	3.6 fM	DPV	Au–C	86
Thrombin	HPtCoNCs-haemin-GOD-Fc[a]	Catalyse oxidation of D-glucose with Fc as electron mediator	0.39 pM	DPV	Au–C	87
Thrombin	HRP	Catalyse reduction of H_2O_2 in presence of TMB as electron mediator	100 pM	CV	Au–C	48
PDGF-BB	RCA[a] coupled with ALP	Catalyse reduction of Ag^+ by ascorbic acid 2-phosphate	10 fM in fetal calf serum, human serum	ASV	Au–C	88

[a]**GDH**: glucose dehydrogenase; **(PQQ)-GDH**: pyrroloquinoline quinone glucose dehydrogenase; **M-PMS**: methoxyphenazine methosulfate; **IgE**: immunoglobulin E; **ALP**: alkaline phosphatase; **SWSV**: square wave stripping voltammetry; **DPASV**: Differential pulse anodic stripping voltammetry; **GCE**: Glass carbon electrode; **SWCNT**: single wall carbon nanotube; **HPtCoNCs**: one type of nano-composite; **Screen-printed electrodes – A**: Here the sensing surface was fabricated on magnetic beads, which was further absorbed onto the electrode surface by magnetic source; **RCA**[a]: rolling circle amplification.

electrochemical aptasensors (Figure 9.6F). Interestingly, the half aptamers (called part1, P1 and part 2, P2 here) attached to the electrode could still form a structure even in the absence of target, leading to a background current under ACV. Though such background current can be minimised by optimising the concentration of P2, the detection sensitivities and ranges for this method were also limited. Detection limits of 1 μM for both ATP and cocaine were obtained using this approach.[90] Second, as with other methods, the MB reporter can be replaced with signal generators such as PtNPs and SiNPs with $Ru(bpy)_3^{2+}$, generating a catalytic H_2O_2 reduction current and ECL signal, respectively.[91,92] Multi-analysis of ATP and cocaine has also been realised on a single sensing surface, using PbS and Cds QDs as dual reporters.[93]

9.2.2.2 Label-Free Electrochemical Aptasensors

'Label-free' electrochemical sensors represent those without covalently appended reporters. Instead, target-binding is detected on the basis of perceived changes in electrostatic adsorption/repletion, hydrogen bonding, van der Waals interactions, or other coordination interactions between the reporters and aptamers or targets. Accordingly, label-free electrochemical aptasensors can be roughly classified into three sets: 1) EIS electrochemical aptasensors that repel reporters in most cases; 2) electrochemical aptasensors that attract reporters; and 3) electrochemical aptasensors in which reporters are pre-immobilised onto the electrode. In addition, label-free electrochemical aptasensors can be developed that are based on the 'FOLD', 'REFOLD', and sandwich assays (both canonical and those involving quaternary structure assembly). However, in many instances, reporter-free detection is not dependent on any change in the conformation of the aptamer, but rather on changes in the overall electrochemical properties induced by aptamer–target binding.

9.2.2.2.1 EIS Electrochemical Aptasensors. In 2005, two EIS, label-free aptasensors were featured in the same issue of *Analytical Chemistry*.[94,95] Almost all EIS aptasensors both then and since have used the $Fe(CN)_6^{4-}$–$Fe(CN)_6^{3-}$ redox couple as a signal indicator, based on a simple hypothesis: when the electrode surface is changed, the electron transfer rate between [Fe $(CN)_6]^{4-}$ and $[Fe(CN)_6]^{3-}$ can be inhibited or accelerated, which further changes the electron transfer resistance (Ret or Rct, as shown in Figures 9.7A,B). As 'E-AB' sensors, the EIS aptasensors started from the simplest 'FOLD' mode – the signal is modulated following ligand stabilisation of the aptamer structure. In work from the O'Sullivan lab, an Au electrode has been modified with an anti-thrombin aptamer monolayer,[94] which produces a negatively charged interface and a repulsive force against anionic $[Fe(CN)_6]$, resulting in an increased Ret compared with a bare electrode. Changes in Ret can then be read out as an increase of quasi-circle diameter

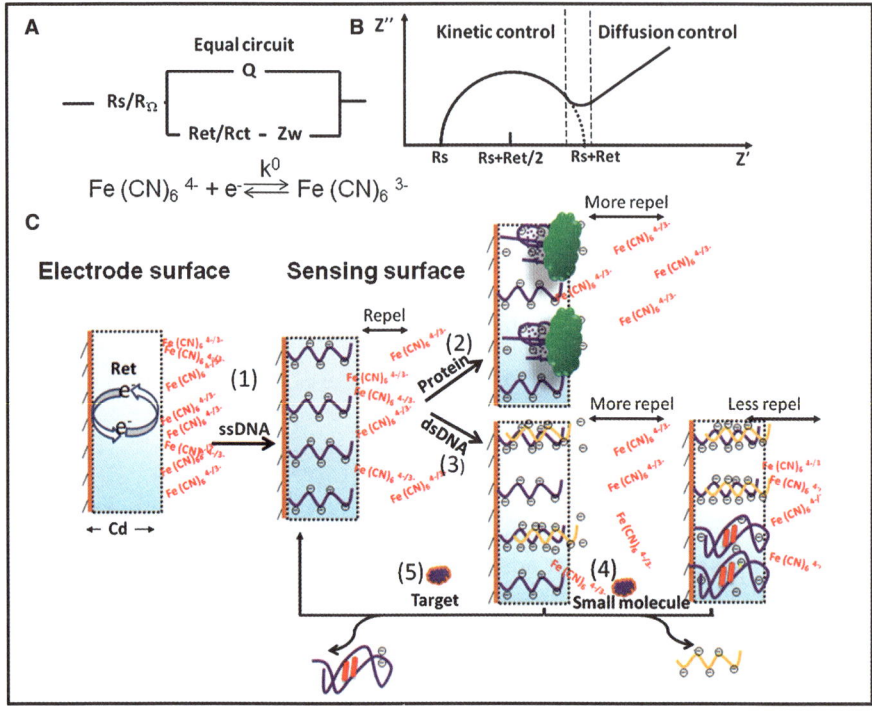

Figure 9.7   Label-free EIS aptasensors. A: Equal circuit of an aptasensor. B: Nequist plot of EIS readout. C: Difference strategies of EIS aptasensors.

in a Nequist plot, as shown in Figures 9.7B,C, process 1. In addition, while thrombin has a pI close to 7, after it binds to the aptamer on the electrode the Ret is further increased owing to the bulk protein, which forms a resistive hydrophobic layer insulating the conductive support, as shown in Figures 9.7B,C, process 2. The detection limit reported for this reagentless sensor was 5 nM.

Similarly, the Ma group showed that IgE could be detected in a similar way on an electrode array.[95] A 'signal-off' aptasensor was constructed for lysozyme, based on the fact that once bound, its net 8 positive charge could attract more [Fe(CN)$_6$] anion to the electrode.[96] Other targets that have been detected based on modulation of the electrostatic repulsion of the aptamer include 17β-estradiol, PDGF, interferon-γ, CCRF-CEM cells, and both PSMA (+) and PSMA (−) prostate cancer cells.[97–106] These targets tend to be of the same size as the aptamer, or larger.

While EIS aptasensors are versatile, they suffer from increased background due to non-specific target-binding that can also modulate the approach of the [Fe(CN)$_6$] reporter to the electrode. This is not true for the corresponding labelled aptasensors, which require a target-specific conformational transition in order to signal. Because of this, it has been wisely suggested that EIS aptasensors should be rigorously controlled and optimised.[107] That said, if the surface is sufficiently passivated, EIS aptasensors have successfully detected targets in complex mixtures.[108,109] For example, Dong's group was able specifically to detect thrombin and ATP in 1% human plasma.[110]

In order to expand the range of EIS approaches and potentially reduce background further, this strategy can be melded with the different conformational strategies previously applied to labelled aptasensors. A REFOLD strategy has been employed by Willner's group to allow small molecule detection.[111] As shown in Figure 9.7C, process 3 to 4, interactions with a target stabilise the release of an antisense strand (similar to Figure 9.5C), reducing repulsion and decreasing the Ret. Conversely, Dong's group has released the aptamer–target complex from an immobilised antisense oligonucleotide as shown in Figure 9.7C, process 3 to 5 (similar to Figure 9.5E).[29] Lysozyme has also been detected with a similar aptasensor.[112] Finally, sandwich assays have been employed to increase specificity further, and further modulate the electrostatic repulsion of the aptamer. DNA-modified AuNPs and SDS-modified AuNPs have both enhanced the sensitivity of thrombin detection,[109,113] leading to a detection limit under 0.05 nM.

9.2.2.2.2 Electrochemical Aptasensors that Attract Reporters. In EIS aptasensors, the signal comes from a change in electrode conductivity, which is reflected in a change in the electron transfer rate constant (kinetics) for the $[Fe(CN)_6]^{3-/4-}$ couple. Thus, $[Fe(CN)_6]^{3-/4-}$ is always present at high concentrations during electrochemical measurements. Alternatively, a redox reporter can be attracted to an aptamer, and background signal can be washed away. One commonly used reporter is the aromatic cationic dye methylene blue (MB), which belongs to the phenothiazine family and has the ability to bind ssDNA, dsDNA, or transfer (t)RNA via intercalation or electrostatic absorption. Kim's group was the first to use MB as a non-covalent reporter in an aptasensor for thrombin.[114] A hairpin aptamer beacon was designed to be stabilised in a quadruplex conformation by thrombin-binding; MB should intercalate into the non-thrombin-binding duplex and then be released upon addition of thrombin and assumption of the quadruplex conformation. This process can be monitored by DPV, with a peak current decrease of MB (signal-off) in the presence of thrombin. A detection limit of 11 nM was observed. Later, Dong's group used a REFOLD strategy (similar to Figure 9.7C, process 5) for ATP detection, in which both aptamer and bound MB were released from the electrode.[115] This method was further improved by adding to the aptamer DNA-coated AuNPs that could soak up additional MB molecules.[116] The DNAs on the AuNPs bind more MB molecules, resulting in an even greater negative signal and decreasing the detection limit by 100-fold, from 10 nM to 0.1 nM.

Yu's group has employed a canonical sandwich assay for signal-on detection. As shown in Figure 9.8A, an anti-thrombin aptamer was extended with a long duplex to soak up MB, and upon binding and washing yielded a 0.5 nM

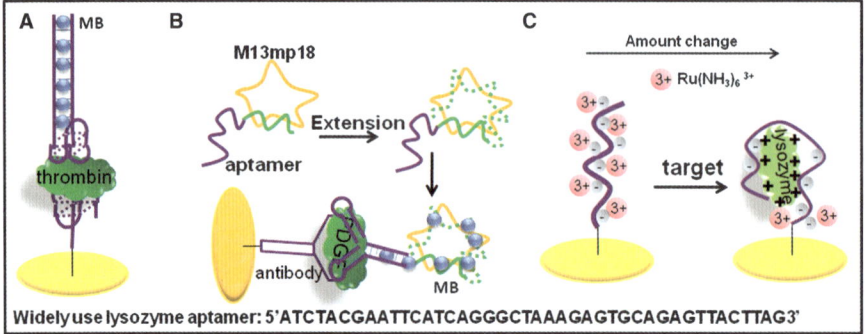

Figure 9.8   Examples of electrochemical aptasensors based on MB staining (A[116] and B[119]) and RuHex absorption (C).[120]

detection limit for thrombin.[117] As with signal-off methods, DNA-coated AuNPs have also been used as MB acceptors in sandwich assays.[118] Another method of generating DNA to soak up MB is to couple an aptamer probe to extension-based amplification (Figure 9.8B).[119]

Other aptasensors have used the positively charged $Ru(NH_2)_6^{3+}$ cation as a redox probe, rather than MB. Binding of targets such as positively charged lysozyme can potentially interfere with cations, leading to a decrease in signal.[120] Similarly, target-dependent displacement of the aptamer (REFOLD) has led to less $Ru(NH_2)_6^{3+}$ being bound adjacent to the electrode,[121] while quaternary structure-forming aptasensors have attracted more $Ru(NH_2)_6^{3+}$ to the electrode.[122]

Beyond MB and $Ru(NH2)_6^{3+}$, other absorptive redox reporters have been employed, such as $Ru(phen)_3^{3+}$, Fc-bearing cationic polythiophene, and positively charged Fc-AuNPs.[123–127] While in general these methods have the advantage of being label-free, they also have the potential for large amounts of background binding (and background signalling), which must be controlled by stringent washing procedures that are not particularly amenable to the development of diagnostic or other assays.

9.2.2.2.3 'Label-Free' Electrochemical Aptasensors with Solid-State Reporters. 'Solid-state' means that redox reporters are pre-immobilised on an electrode surface prior to the addition of aptasensors. Placing the redox probe near to the electrode guarantees a more sensitive electrochemical response. Such fabricated sensors may be directly used for point-of-care analysis, without adding and/or washing away reporters. Aptasensors that rely on solid-state reporters are very similar to EIS sensors, in that they rely on observation of changes in the sensing surface condition following target binding. When the surface is covered by poor conductors (such as nucleic acids), the electron transfer kinetics of immobilised reporters will be inhibited, and they will therefore exhibit decreased current intensity. One of the first solid-state aptasensors was based on a FOLD strategy (similar to Figure 9.7C, process 2) in which AuNPs bearing anti-thrombin aptamers are deposited on the electrode. In the presence of thrombin, the electrode surface is insulated from conductive electrolyte, leading to a peak current decrease in CV. The sensor reported a detection limit down to ~10 fM.[128] Another solid-state device from Dong's group involves an ITO electrode surface fabricated layer-by-layer (LBL) using ferrocene-appended poly(ethyleneimine) (Fc-PEI, positive), carbon nanotubes (CNTs, negative), and aptamer (negative) (Figure 9.9A). The multilayer reproducibly detects DPV background current from Fc, which is in turn sensitively inhibited by protein-binding.[129] This leads to impressive detection limits of 4 pM for thrombin and 10 pM for lysozyme. Quaternary structure assembly has also been used to detect cocaine,[30] but with AuNPs replacing CNTs. An array ITO electrode yielded reproducible signals, with a detection limit of 0.1 μM cocaine in buffer. These solid-state sensors have also yielded specific signals in samples such as 25% human plasma, serum and saliva. Further improvements in electrode performance have been obtained by fabricating a stable PEI/PSS/(Fc-PEI/PSS)$_2$/PEI/GSGHs multilayer, with the outside layer as GSGHs, a recently invented graphene–mesoporous silica–Au NP hybrid. The graphene was shown to be better for electron transfer than electrodes modified with only AuNPs, while

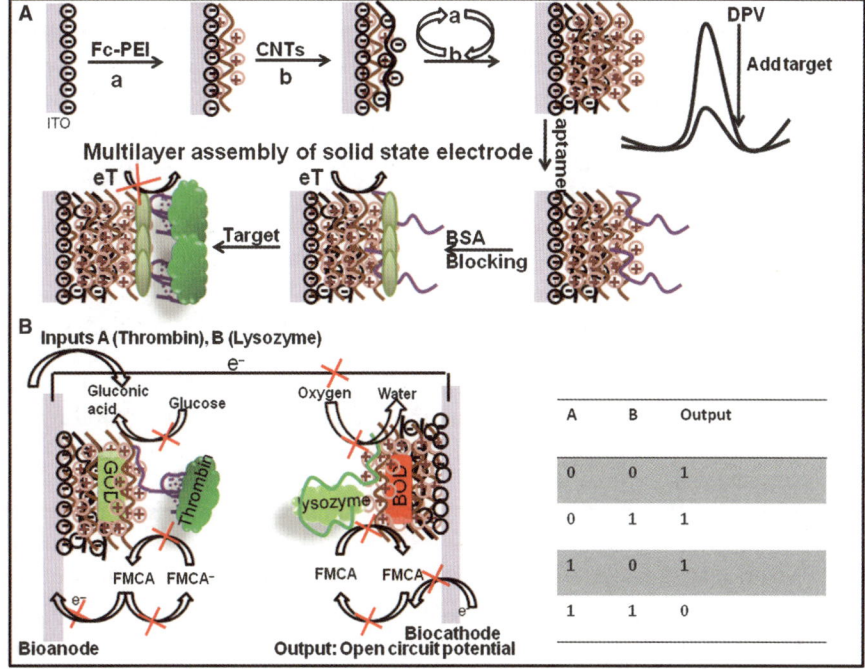

Figure 9.9 Examples of (A) solid-state-based electrochemical aptasensors,[128] (B) and aptamer-controlled power release of biofuel cell.[131]

the AuNPs help immobilise thiolated anti-D-vasopressin (D-VP) half-aptamers (similar to P1, in Figure 9.6F). Using a quaternary structure assembly approach, a detection limit of 4 nM was reported for D-VP, while greater than 100 nM L-VP did not lead to any signal change.[130]

Aptamers coupled with solid-state reporters have also been used for controlled power release in biofuel cells (BFCs).[131,132] An ITO bioanode and biocathode have been modified with solid-state reporters consisting of a (PDDA-CNTs)/GOD/(PDDA-CNTs)/anti-thrombin aptamer and a (PDDA-CNTs)/BOD/(PDDA-CNTs)/anti-lysozyme aptamer, respectively (Figure 9.9B). Either thrombin or lysozyme can inhibit the power density (open circuit potential) of the BFC.[131]

While these solid-state sensors are remarkable, fabricating stable and functional electrodes is technically difficult, and the methods to immobilise redox reporters on electrodes (listed in Table 9.3) are still being explored. Unstable immobilisation will lead to loss of reporters during sensing, and may thereby create false 'off-signals'. Solid-state sensors are also prone to issues with non-specific binding, similar to EIS sensors. Research into this class of sensors requires stringent optimisation and rigorous controls in order to generate reproducible and believable data.

9.2.2.3   *Enzyme Amplification Strategies*

While electrochemical reporters and reagentless sensors are both excellent strategies for adapting aptasensors to electrochemical platforms, the advantages inherent in the use of protein enzymes for signal amplification can lead to even further improvements in sensitivity. One obvious method by which aptasensors can utilise enzyme amplification is via a canonical sandwich assay (see Table 9.2). However, there are other strategies that rely on making more nucleic acids for sensing. For example, two aptamers can bind to thrombin; in this system, one aptamer serves as the primer for rolling circle amplification while the other aptamer serves as template. When Fc-modified dUTPs are included in the amplification, there is a large increase in label-mediated signalling (Figure 9.10A).[139] Nuclease degradation as well as polymerisation can lead to signal amplification (Figure 9.10B).[140] Shifting the equilibrium towards ochratoxin A (OTA) leads to concomitant nuclease digestion and recycling of ochratoxin A, one of the examples of 'target recycling'. As low as 2 pM OTA has been detected.

Transduction strategies can be even more complex. Dong's group has coupled a solid-state electrode (ITO/PEI/PSS/(Fc-PEI/PSS)$_2$/PEI/GSGHs (as in ref.[129]) with strand displacement-based polymerisation, and in consequence has realised ultra-sensitive ATP detection (Figure 9.10C).[141] In greater detail, an antisense strand (called a 'blocker' here) is displaced from micro-silver particles by the binding of ATP to its corresponding aptamer. After removal of the particles and transfer to a new tube, the free ('blocker') strand can bind to and open a hairpin, allowing extension by a polymerase and nucleoside

Table 9.3 Summary of solid-state 'label-free' electrochemical aptasensors.

Target	Electrode	Strategy	Detection limit	Detection method	Reference
Thrombin	SPE/DDAB-AuNPs (C)^a	FOLD-Off	1 nM	SWV	128
Thrombin	ITO/(Fc-PEI/CNTs)n-Fc-PEI (A)	FOLD-Off	0.14 ng mL^{-1}	DPV	129
Lysozyme	ITO/(Fc-PEI/CNTs)n-Fc-PEI (A)	FOLD-Off	0.17 ng mL^{-1}	DPV	129
D-vasopressin	ITO/PEI/PSS/(Fc-PEI/PSS)$_2$/PEI/GSGHs(C)	QS-Off^a	4 nM	DPV	130
cocaine	ITO/(Fc-PEI/AuNPs)n-AuNPs array (C)	QS-Off	0.1 µM indiluted bio-samples	DPV	30
lysozyme	ITO/(Fc-PEI/AuNPs)n-AuNPs array (C)	FOLD-Off	0.1 ng mL^{-1}	DPV	30
Thrombin	Au-Pt/MB/Nafion@MWCNTs/GCE (C)	REFOLD-On	3 pM	DPV	133
L-selectin	JUG-thio monolayer (A)^a	FOLD-Off	1000 cells mL^{-1}	DPV	134
Thrombin	Au(nano-Au/Thi +)n (C)	FOLD-Off	40 pM	CV	135
Thrombin	H$_2$O$_2$-Pt-AuNPs/HRP/Pt-AuNPs/NiHCFNPs/AuNPs/Au (C)^a	FOLD-Off	6.3 pM	CV	136
Thrombin	AuNPs/MB/nafion@graphene/Au (C)	FOLD-Off	6 pM	CV	137
Daunomycin	GCE/AuNPs/polyTTBA/PS-aptamer/AuNPs (C)^a	FOLD-On	32.3 pM in real urine	DPV	138

^{a}The signal reporter is underlined; **QS-off**: 'quarterternary structure' sandwich manner, signal-off; **SPE**: Screen-printed electrode; **DDAB**: didodecyldimethylammonium bromide; **JUG-thio**: 5-hydroxy-3-hexanedithiol-1,4-naphthoquinone; **NiHCFNPs**: nickel hexacyanoferrates nanoparticles; **PolyTTBA**: [2,2_:5_,2_-terthiophene-3_-(p-benzoic acid)]; C means that the nucleic acids were covalently immobilised, and A means that the nucleic acids were adsorbed on the electrode.

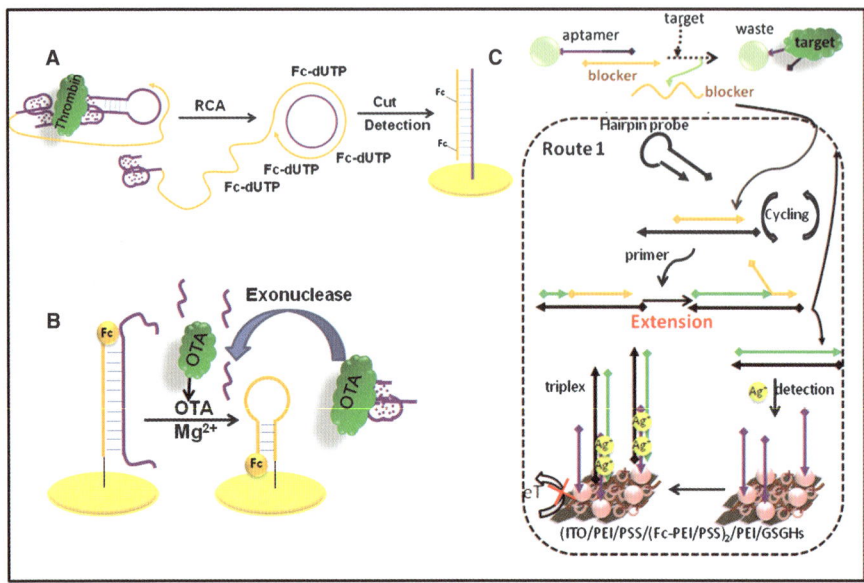

Figure 9.10   Examples of enzyme-participated amplification though circling reactions.[138–140]

triphosphates, which in turn frees the blocker to initiate an additional round of hairpin opening and extension. After double-stranded DNAs have been synthesised, they are transferred to an electrode and captured on a modified surface via silver-stabilised triplex interactions. The final signal is generated via the inhibition of Fc electrochemistry on the solid surface. While this scheme is complex and requires multiple, somewhat impractical transfers, the polymerase-mediated amplification and capture led to a detection limit as low as 23 pM, which is 300-fold above the background in the absence of signal amplification. Similar work has been carried out by Zhang's group, with $Ru(NH_2)_6^{3+}$ as reporter and Ramos cells as targets.[142] In another approach, enzymatic amplifications have been coupled to the deposition of DNA–QD conjugates on surfaces, yielding 'signal-on' ECL detection. As few as 210 Ramos cells mL^{-1} could be reliably detected in this manner.[143]

9.3   Conclusions and Future Prospects

Aptasensors are an especially useful class of electrochemical sensors because of their ability to undergo programmed conformational changes with respect to ligands, and hence to alter the apposition of reporters relative to the electrode surface, or properties of the electrode surface itself. In particular, E-AB sensors with FOLD strategies, and recently developed solid-state sensors should make electrochemical aptasensors a viable option for a variety of real-world applications, especially those involving real-time detection and on-site use.

When coupled with improvements in surface fabricating techniques such as printing multiple electrodes onto small and cheap surfaces that also integrate micro-fluidic control, the benefits of device minimisation and high-throughput production should make electrochemical aptasensors commercially viable. At the start of what is likely to be an exponential progression in development, preliminary logic assays, parallel assays, and multi-assays based on either multi-reporters or multi-addresses have been developed.[38,70,110,121,131,132,144,145]

Nonetheless, from another vantage point, the development of aptasensors is still at a purely academic stage. Researchers have so far mostly relied upon a set number of well-studied aptamers (*i.e.* anti-thrombin, anti-ATP, anti-PDGF, anti-lysozyme, anti-cocaine) for proofs of principle. As research groups become more diversified and interdisciplinary, aptamer selection and engineering should occur side by side with analytical method development. While some publications have demonstrated the quantitation of analytes in unmodified physical samples,[33,88] most of the experiments performed to date have been carried out with purified analytes or seriously diluted samples. In addition, those strategies that rely on multiple separation or washing steps may bring less reproducibility and more onerous operation, increasing the difficulties in developing commercial devices. Overall, analytical results are not enough in and of themselves; the systems that are being developed must become simpler and more technically reliable.

Acknowledgements

This review chapter was funded by National Institute of Health (NIH Plaxco-1R01EB0007689), the National Security Science and Engineering Faculty Fellowship (NSSEFF-FA9550-10-1-0169), as well as the National Institute of Health (NIH-TR01- 5R01AI092839-02). The published material represents the position of the author(s) and not necessarily that of the sponsors.

References

1. S. E. Osborne and A. D. Ellington, *Chem. Rev.*, 1997, **97**, 349.
2. M. Famulok, J. S. Hartig and G. Mayer, *Chem. Rev.*, 2007, **107**, 3715.
3. J. Liu, Z. Cao and Y. Lu, *Chem. Rev.*, 2009, **109**, 1948.
4. A. D. Ellington and J. W. Szostak, *Nature*, 1990, **346**, 818.
5. C. Tuerk and L. Gold, *Science*, 1990, **249**, 505.
6. S. M. Shamah, J. M. Healy and S. T. Cload, *Acc. Chem. Res.*, 2008, **41**, 130.
7. L. Gold, D. Ayers, J. Bertino, C. Bock, A. Bock, E. N. Brody, J. Carter, A. B. Dalby, B. E. Eaton, T. Fitzwater, D. Flather, A. Forbes, T. Foreman, C. Fowler, B. Gawande, M. Goss, M. Gunn, S. Gupta, D. Halladay, J. Heil, J. Heilig, B. Hicke, G. Husar, N. Janjic, T. Jarvis, S. Jennings, E. Katilius, T. R. Keeney, N. Kim, T. H. Koch, S. Kraemer, L. Kroiss, N. Le, D. Levine, W. Lindsey, B. Lollo, W. Mayfield, M. Mehan,

R. Mehler, S. K. Nelson, M. Nelson, D. Nieuwlandt, M. Nikrad, U. Ochsner, R. M. Ostroff, M. Otis, T. Parker, S. Pietrasiewicz, D. I. Resnicow, J. Rohloff, G. Sanders, S. Sattin, D. Schneider, B. Singer, M. Stanton, A. Sterkel, A. Stewart, S. Stratford, J. D. Vaught, M. Vrkljan, J. J. Walker, M. Watrobka, S. Waugh, A. Weiss, S. K. Wilcox, A. Wolfson, S. K. Wolk, C. Zhang and D. Zichi, *PLoS ONE*, 2010, **5**, e15004.

8. K. S. Schmidt, S. Borkowski, J. Kurreck, A. W. Stephens, R. Bald, M. Hecht, M. Friebe, L. Dinkelborg and V. A. Erdmann, *Nucleic Acids Res.*, 2004, **32**, 5757.

9. J. C. Cox, P. Rudolph and A. D. Ellington, *Biotechnol. Prog.*, 1998, **14**, 845.

10. N. Li, Y. Wang, A. Pothukuchy, A. Syrett, N. Husain, S. Gopalakrisha, P. Kosaraju and A. D. Ellington, *Nucleic Acids Res.*, 2008, **36**, 6739.

11. N. Li, H. H. Nguyen, M. Byrom and A. D. Ellington, *PLoS ONE*, 2011, **6**, e20299.

12. A. T. Bayrac, K. Sefah, P. Parekh, C. Bayrac, B. Gulbakan, H. A. Oktem and W. Tan, *ACS Chemical Neuroscience*, 2011, **2**, 175.

13. D. H. J. Bunka and P. G. Stockley, *Nat. Rev. Microbiol.*, 2006, **4**, 588.

14. J. Feigon, T. Dieckmann and F. W. Smith, *Chem. Biol.*, 1996, **3**, 611.

15. R. D. Jenison, S. C. Gill, A. Pardi and B. Polisky, *Science*, 1994, **263**, 1425.

16. N. de-los-Santos-Alvarez, M. Jesus Lobo-Castanon, A. J. Miranda-Ordieres and P. Tunon-Blanco, *Trends Anal. Chem.*, 2008, **27**, 437.

17. E. J. Cho, J.-W. Lee and A. D. Ellington, *Annu. Rev. Anal. Chem.*, 2009, **2**, 241.

18. A. B. Iliuk, L. Hu and W. A. Tao, *Anal. Chem.*, 2011, **83**, 4440.

19. Y. Zhang, H. Hong and W. Cai, *Curr. Med. Chem.*, 2011, **18**, 4185.

20. A. D. Keefe, S. Pai and A. Ellington, *Nat. Rev. Drug Discovery*, 2010, **9**, 537.

21. S. M. Nimjee and B. A. Sullenger, *Drugs Future*, 2009, **34**, 897.

22. K. L. Rialon and R. R. White, *Anti-Cancer Agents Med. Chem.*, 2011, **11**, 434.

23. K. Ikebukuro, C. Kiyohara and K. Sode, *Analy. Lett.*, 2004, **37**, 2901.

24. T. Hianik and J. Wang, *Electroanalysis*, 2009, **21**, 1223.

25. D. Li, S. Song and C. Fan, *Acc. Chem. Res.*, 2010, **43**, 631.

26. I. Willner and M. Zayats, *Angew. Chem. Int. Ed. Engl.*, 2007, **46**, 6408.

27. A. A. Lubin and K. W. Plaxco, *Acc. Chem. Res.*, 2010, **43**, 496.

28. A. J. Bard and L. R. Faulkner, *Electrochemical Methods, 2nd ed.*, John Wiley & Sons: New York, 2001.

29. B. Li, Y. Du, H. Wei and S. Dong, *Chem. Commun.*, 2007, 3780.

30. Y. Du, C. Chen, J. Yin, B. Li, M. Zhou, S. Dong and E. Wang, *Anal. Chem.*, 2010, **82**, 1556.

31. S. Centi, S. Tombelli, M. Minunni and M. Mascini, *Anal. Chem.*, 2007, **79**, 1466.

32. E. Komarova, K. Reber, M. Aldissi and A. Bogomolova, *Biosens. Bioelectron*, 2010, **25**, 1389.

33. J. S. Swensen, Y. Xiao, B. S. Ferguson, A. A. Lubin, R. Y. Lai, A. J. Heeger, K. W. Plaxco and H. T. Soh, *J. Am. Chem. Soc.*, 2009, **131**, 4262.
34. Y. Xiao, A. A. Lubin, A. J. Heeger and K. W. Plaxco, *Angew. Chem. Int. Ed. Engl.*, 2005, **44**, 5456.
35. A. E. Radi, J. L. A. Sanchez, E. Baldrich and C. K. O'Sullivan, *J. Am. Chem. Soc.*, 2006, **128**, 117.
36. B. R. Baker, R. Y. Lai and M. S. Wood, E. H. Doctor, A. J. Heeger and K. W. Plaxco, *J. Am. Chem. Soc.*, 2006, **128**, 3138.
37. R. Y. Lai, K. W. Plaxco and A. J. Heeger, *Anal. Chem.*, 2007, **79**, 229.
38. F. Xia, X. Zuo, R. Yang, R. J. White, Y. Xiao, D. Kang, X. Gong, A. A. Lubin, A. Vallee-Belisle, J. D. Yuen, B. Y. B. Hsu and K. W. Plaxco, *J. Am. Chem. Soc.*, 2010, **132**, 8557.
39. M. Mir, A. T. A. Jenkins and I. Katakis, *Electrochem. Commun.*, 2008, **10**, 1533.
40. N. Hamaguchi, A. Ellington and M. Stanton, *Anal. Biochem.*, 2001, **294**, 126.
41. Y. Xiao and K. W. Plaxco, Electrochemical aptamer sensors. In: Functional nucleic acids for sensing and other analytical applications, Y. Lu and Y. Li, Eds. Springer, New York. 2009, pp. 179–198.
42. T. Uzawa, R. R. Cheng, R. J. White, D. E. Makarov and K. W. Plaxco, *J. Am. Chem. Soc.*, 2010, **132**, 16120.
43. A. Anne, A. Bouchardon and J. Moiroux, *J. Am. Chem. Soc.*, 2003, **125**, 1112.
44. T. M. Herne and M. J. Tarlov, *J. Am. Chem. Soc.*, 1997, **119**, 8916.
45. C. Bustamante, Z. Bryant and S. B. Smith, *Nature*, 2003, **421**, 423.
46. Y. Xiao, B. D. Piorek, K. W. Plaxco and A. J. Heeger, *J. Am. Chem. Soc.*, 2005, **127**, 17990.
47. Y. Wen, H. Pei, Y. Wan, Y. Su, Q. Huang, S. Song and C. Fan, *Anal. Chem.*, 2011, **83**, 7418.
48. H. Pei, N. Lu, Y. Wen, S. Song, Y. Liu, H. Yan and C. Fan, *Adv. Mater.*, 2010, **22**, 4754.
49. X. Zuo, S. Song, J. Zhang, D. Pan, L. Wang and C. Fan, *J. Am. Chem. Soc.*, 2007, **129**, 1042.
50. Y. Lu, X. Li, L. Zhang, P. Yu, L. Su and L. Mao, *Anal. Chem.*, 2008, **80**, 1883.
51. Z.-S. Wu, M.-M. Guo, S.-B. Zhang, C.-R. Chen, J.-H. Jiang, G.-L. Shen and R.-Q. Yu, *Anal. Chem.*, 2007, **79**, 2933.
52. Y. C. Huang, B. Ge, D. Sen and H.-Z. Yu, *J. Am. Chem. Soc.*, 2008, **130**, 8023.
53. R. J. White and K. W. Plaxco, *Anal. Chem.*, 2010, **82**, 73.
54. Y. Xiao, T. Uzawa, R. J. White, D. DeMartini and K. W. Plaxco, *Electroanalysis*, 2009, **21**, 1267.
55. R. J. White, N. Phares, A. A. Lubin, Y. Xiao and K. W. Plaxco, *Langmuir*, 2008, **24**, 10513.

56. R. J. White, A. A. Rowe and K. W. Plaxco, *Analyst*, 2010, **135**, 589.
57. J. Lluis Acero Sanchez, E. Baldrich, A. El-Gawad Radi, S. Dondapati, P. Lozano Sanchez, I. Katakis and C. K. O'Sullivan, *Electroanalysis*, 2006, **18**, 1957.
58. Y. Xiao, R. Y. Lai and K. W. Plaxco, *Nat. Protoc.*, 2007, **2**, 2875.
59. A.-E. Radi and C. K. O'Sullivan, *Chem. Commun.*, 2006, 3432.
60. X. Li, H. Qi, L. Shen, Q. Gao and C. Zhang, *Electroanalysis*, 2008, **20**, 1475.
61. E. E. Ferapontova, E. M. Olsen and K. V. Gothelf, *J. Am. Chem. Soc.*, 2008, **130**, 4256.
62. W. Yao, L. Wang, H. Wang, X. Zhang and L. Li, *Biosens. Bioelectron*, 2009, **24**, 3269.
63. B. Sun, H. Qi, F. Ma, Q. Gao, C. Zhang and W. Miao, *Anal. Chem.*, 2010, **82**, 5046.
64. A. A. Rowe, E. A. Miller and K. W. Plaxco, *Anal. Chem.*, 2010, **82**, 7090.
65. J. Zhou, H. Huang, J. Xuan, J. Zhang and J.-J. Zhu, *Biosens. Bioelectron*, 2010, **26**, 834.
66. L.-D. Li, Z.-B. Chen, H.-T. Zhao, L. Guo and X. Mu, *Sensors Actuat B-Chem.*, 2010, **149**, 110.
67. S. Zhang, R. Hu, P. Hu, Z.-S. Wu, G.-L. Shen and R.-Q. Yu, *Nucleic Acids Res.*, 2010, **38**.
68. Y. Liu, N. Tuleouva, E. Ramanculov and A. Revzin, *Anal. Chem.*, 2010, **82**, 8131.
69. S. Zhang, G. Zhou, X. Xu, L. Cao, G. Liang, H. Chen, B. Liu and J. Kong, *Electrochem. Commun.*, 2011, **13**, 928.
70. E. Hayashi, T. Takada, M. Nakamura and K. Yamana, *Chem. Lett.*, 2010, **39**, 454.
71. R. Polsky, R. Gill, L. Kaganovsky and I. Willner, *Anal. Chem.*, 2006, **78**, 2268.
72. A. Numnuam, K. Y. Chumbimuni-Torres, Y. Xiang, R. Bash, P. Thavarungkul, P. Kanatharana, E. Pretsch, J. Wang and E. Bakker, *Anal. Chem.*, 2008, **80**, 707.
73. C. Ding, Y. Ge and J.-M. Lin, *Biosens. Bioelectron*, 2010, **25**, 1290.
74. Y. Wang, X. He, K. Wang, X. Ni, J. Su and Z. Chen, *Biosens. Bioelectron*, 2011, **26**, 3536.
75. H. Zhong, Q. Zhang and S. Zhang, *Chem. Eur. J.*, 2011, **17**, 8388.
76. Y.-L. Zhang, Y. Huang, J.-H. Jiang, G.-L. Shen and R.-Q. Yu, *J. Am. Chem. Soc.*, 2007, **129**, 15448.
77. A. Blokzijl, M. Friedman, F. Ponten and U. Landegren, *J. Intern. Med.*, 2010, **268**, 232.
78. K. Ikebukuro, C. Kiyohara and K. Sode, *Biosens. Bioelectron*, 2005, **20**, 2168.
79. K. Feng, Y. Kang, J.-J. Zhao, Y.-L. Liu, J.-H. Jiang, G.-L. Shen and R.-Q. Yu, *Anal. Biochem.*, 2008, **378**, 38.
80. S. Centi, G. Messina, S. Tombelli, I. Palchetti and M. Mascini, *Biosens. Bioelectron*, 2008, **23**, 1602.

81. T. H. Degefa, S. Hwang, D. Kwon, J. H. Park and J. Kwak, *ElectroChim. Acta*, 2009, **54**, 6788.
82. H. Yang, J. Ji, Y. Liu, J. Kong and B. Liu, *Electrochem. Commun.*, 2009, **11**, 38.
83. X. Zhang, B. Qi, Y. Li and S. Zhang, *Biosens. Bioelectron*, 2009, **25**, 259.
84. Y. Xiang, Y. Zhang, X. Qian, Y. Chai, J. Wang and R. Yuan, *Biosens. Bioelectron*, 2010, **25**, 2539.
85. Y. Chen, B. Jiang, Y. Xiang, Y. Chai and R. Yuan, *Chem. Commun.*, 2011, **47**, 7758.
86. L. Bai, R. Yuan, Y. Chai, Y. Yuan, L. Mao and Y. Wang, *Anal. Chim. Acta*, 2011, **698**, 14.
87. L. Bai, R. Yuan, Y. Chai, Y. Yuan, Y. Zhuo and L. Mao, *Biosens. Bioelectron*, 2011, **26**, 4331.
88. L. Zhou, L.-J. Ou, X. Chu, G.-L. Shen and R.-Q. Yu, *Anal. Chem.*, 2007, **79**, 7492.
89. M. N. Stojanovic, P. de Prada and D. W. Landry, *J. Am. Chem. Soc.*, 2000, **122**, 11547.
90. X. Zuo, Y. Xiao and K. W. Plaxco, *J. Am. Chem. Soc.*, 2009, **131**, 6944.
91. E. Golub, G. Pelossof, R. Freeman, H. Zhang and I. Willner, *Anal. Chem.*, 2009, **81**, 9291.
92. Q. Cai, L. Chen, F. Luo, B. Qiu, Z. Lin and G. Chen, *Anal. Bioanal. Chem.*, 2011, **400**, 289.
93. H. Zhang, B. Jiang, Y. Xiang, Y. Zhang, Y. Chai and R. Yuan, *Anal. Chim. Acta*, 2011, **688**, 99.
94. A. E. Radi, J. L. A. Sanchez, E. Baldrich and C. K. O'Sullivan, *Anal. Chem.*, 2005, **77**, 6320.
95. D. K. Xu, D. W. Xu, X. B. Yu, Z. H. Liu, W. He and Z. Q. Ma, *Anal. Chem.*, 2005, **77**, 5107.
96. M. C. Rodriguez, A. N. Kawde and J. Wang, *Chem. Commun.*, 2005, 4267.
97. T. H. Degefa and J. Kwak, *Anal. Chim. Acta*, 2008, **613**, 163.
98. H. Cai, T. M. H. Lee and I. M. Hsing, *Sensors Actuat B-Chem.*, 2006, **114**, 433.
99. K. Min, M. Cho, S.-Y. Han, Y.-B. Shim, J. Ku and C. Ban, *Biosens. Bioelectron*, 2008, **23**, 1819.
100. C. Pan, M. Guo, Z. Nie, X. Xiao and S. Yao, *Electroanalysis*, 2009, **21**, 1321.
101. Z. Zhang, W. Yang, J. Wang, C. Yang, F. Yang and X. Yang, *Talanta*, 2009, **78**, 1240.
102. Y.-J. Kim, Y. S. Kim, J. H. Niazi and M. B. Gu, *Bioprocess Biosyst. Eng.*, 2010, **33**, 31.
103. K. Min, K.-M. Song, M. Cho, Y.-S. Chun, Y.-B. Shim, J. K. Ku and C. Ban, *Chem. Commun.*, 2010, **46**, 5566.
104. E. Gonzalez-Fernandez, N. de-los-Santos-Alvarez, M. Jesus Lobo-Castanon, A. Jose Miranda-Ordieres and P. Tunon-Blanco, *Biosens. Bioelectron*, 2011, **26**, 2354.

105. A. Periyakaruppan, P. U. Arumugam, M. Meyyappan and J. E. Koehne, *Biosens. Bioelectron*, 2011, **28**, 428.

106. Y. S. Kim, H. S. Jung, T. Matsuura, H. Y. Lee, T. Kawai and M. B. Gu, *Biosens. Bioelectron*, 2007, **22**, 2525.

107. A. Bogomolova, E. Komarova, K. Reber, T. Gerasimov, O. Yavuz, S. Bhatt and M. Aldissi, *Anal. Chem.*, 2009, **81**, 3944.

108. Y. Xu, L. Yang, X. Ye, P. He and Y. Fang, *Electroanalysis*, 2006, **18**, 1449.

109. B. Li, Y. Wang and H. Wei, andS. Dong, *Biosens. Bioelectron*, 2008, **23**, 965.

110. Y. Du, B. Li, H. Wei, Y. Wang and E. Wang, *Anal. Chem.*, 2008, **80**, 5110.

111. M. Zayats, Y. Huang, R. Gill, C.-a. Ma and I. Willner, *J. Am. Chem. Soc.*, 2006, **128**, 13666.

112. Y. Peng, D. Zhang, Y. Li, H. Qi, Q. Gao and C. Zhang, *Biosens. Bioelectron*, 2009, **25**, 94.

113. C. Deng, J. Chen, Z. Nie, M. Wang, X. Chu, X. Chen, X. Xiao, C. Lei and S. Yao, *Anal. Chem.*, 2009, **81**, 739.

114. G. S. Bang, S. Cho and B. G. Kim, *Biosens. Bioelectron*, 2005, **21**, 863.

115. J. Wang, F. Wang and S. Dong, *J. Electroanal. Chem.*, 2009, **626**, 1.

116. Y. Du, B. Li, F. Wang and S. Dong, *Biosens. Bioelectron*, 2009, **24**, 1979.

117. Y. Kang, K.-J. Feng, J.-W. Chen, J.-H. Jiang, G.-L. Shen and R.-Q. Yu, *Bioelectrochemistry*, 2008, **73**, 76.

118. J. Wang, A. Munir, Z. Li and H. S. Zhou, *Talanta*, 2010, **81**, 63.

119. Y. Huang, X.-M. Nie, S.-L. Gan, J.-H. Jiang, G.-L. Shen and R.-Q. Yu, *Anal. Biochem.*, 2008, **382**, 16.

120. A. K. H. Cheng, B. Ge and H.-Z. Yu, *Anal. Chem.*, 2007, **79**, 5158.

121. L. Shen, Z. Chen, Y. Li, P. Jing, S. Xie, S. He, P. He and Y. Shao, *Chem. Commun.*, 2007, 2169.

122. Y. Du, C. Chen, M. Zhou, S. Dong and E. Wang, *Anal. Chem.*, 2011, **83**, 1523.

123. Z. Liu, W. Zhang, L. Hu, H. Li, S. Zhu and G. Xu, *Chem. Eur. J.*, 2010, **16**, 13356.

124. F. Le Floch, H. A. Ho and M. Leclerc, *Anal. Chem.*, 2006, **78**, 4727.

125. D. Kwon, H. Jeong and B. H. Chung, *Biosens. Bioelectron*, 2011, **28**, 454.

126. B. Shen, Q. Wang, D. Zhu, J. Luo, G. Cheng, P. He and Y. Fang, *Electroanalysis*, 2010, **22**, 2985.

127. A. Qureshi, Y. Gurbuz, S. Kallempudi and J. H. Niazi, *Phys. Chem. Chem. Phys.*, 2010, **12**, 9176.

128. E. Suprun, V. Shumyantseva, T. Bulko, S. Rachmetova, S. Rad'ko, N. Bodoev and A. Archakov, *Biosens. Bioelectron*, 2008, **24**, 831.

129. Y. Du, C. Chen, B. Li, M. Zhou, E. Wang and S. Dong, *Biosens. Bioelectron*, 2010, **25**, 1902.

130. Y. Du, S. Guo, H. Qin, S. Dong and E. Wang, *Chem. Commun.*, 2011, **48**, 799.

131. M. Zhou, Y. Du, C. Chen, B. Li, D. Wen, S. Dong and E. Wang, *J. Am. Chem. Soc.*, 2010, **132**, 2172.

132. M. Zhou, C. Chen, Y. Du, B. Li, D. Wen, S. Dong and E. Wang, *Lab Chip*, 2010, **10**, 2932.

133. Y. Yuan, R. Yuan, Y. Chai, Y. Zhuo, L. Bai and Y. Liao, *Biosens. Bioelectron*, 2010, **26**, 881.

134. Z. Shao, Y. Li, Q. Yang, J. Wang and G. Li, *Anal. Bioanal. Chem.*, 2010, **398**, 2963.

135. Y. Yuan, R. Yuan, Y. Chai, Y. Zhuo, Z. Liu, L. Mao, S. Guan and X. Qian, *Anal.Chim. Acta*, 2010, **668**, 171.

136. L. Bai, R. Yuan, Y. Chai, Y. Yuan, L. Mao and Y. Zhuo, *Analyst*, 2011, **136**, 1840.

137. T. Sun, L. Wang, N. Li and X. Gan, *Bioprocess Biosyst. Eng.*, 2011, **34**, 1081.

138. P. Chandra, H.-B. Noh, M.-S. Won and Y.-B. Shim, *Biosens. Bioelectron*, 2011, **26**, 4442.

139. D. A. Di Giusto, W. A. Wlassoff, J. J. Gooding, B. A. Messerle and G. C. King, *Nucleic Acids Res.*, 2005, **33**.

140. P. Tong, L. Zhang, J.-J. Xu and H.-Y. Chen, *Biosens. Bioelectron*, 2011, **29**, 97.

141. S. Guo, Y. Du, X. Yang, S. Dong and E. Wang, *Anal. Chem.*, 2011, **83**, 8035.

142. R. Ren, C. Leng and S. Zhang, *Chem. Commun.*, 2010, **46**, 5758.

143. G. Jie, L. Wang, J. Yuan and S. Zhang, *Anal. Chem.*, 2011, **83**, 3873.

144. J. A. Hansen, J. Wang, A. N. Kawde, Y. Xiang, K. V. Gothelf and G. Collins, *J. Am. Chem. Soc.*, 2006, **128**, 2228.

145. H. Zhang, C. Fang and S. Zhang, *Chem. Eur. J.*, 2011, **17**, 7531.

CHAPTER 10

Oligonucleotide Conjugates for Detection of Specific Nucleic Acid Sequences

HIROMU KASHIDA AND HIROYUKI ASANUMA*

Graduate School of Engineering, Nagoya University, Furo-cho, Chikusa-ku, Nagoya 464-8603, Japan
*Email: asanuma@mol.nagoya-u.ac.jp

10.1 Introduction

In the genomic and post-genomic era, there is an ever-increasing demand for the detection of DNA and RNA targets with high sequence specificity and sensitivity. Comprehensive analysis of the human genome revealed that there are a huge number of genetic variations such as single nucleotide polymorphisms (SNPs) (Figure 10.1). In addition, recent studies have revealed that non-SNP alterations such as insertion/deletion polymorphisms account for about 70% of all variant bases.[1] Detection of such variants is required in order to assess the disease risks and drug responses of individuals. Oligonucleotides spontaneously form a double-helical structure with their complementary strands with extremely high sequence specificity, and subtle sequence changes trigger destabilisation of the duplex and/or distortion of local structure. Accordingly, naturally occurring oligodeoxyribonucleotides (ODNs) themselves are excellent tools for recognition of DNA and RNA sequences specifically. However, they are passive, involve only the four naturally occurring nucleotides, and do not send a signal with acceptable intensity in response to

RSC Biomolecular Sciences No. 26
DNA Conjugates and Sensors
Edited by Keith R Fox and Tom Brown
© The Royal Society of Chemistry 2012
Published by the Royal Society of Chemistry, www.rsc.org

```
Wild type  5'-...TATC A GCAA...-3'
           3'-...ATAG T CGTT...-5'

SNPs       5'-...TATC G GCAA...-3'
           3'-...ATAG C CGTT...-5'

Deletion   5'-...TATCGCAA...-3'
           3'-...ATAGCGTT...-5'

Insertion  5'-...TATC AA GCAA...-3'
           3'-...ATAG TT CGTT...-5'
```

Figure 10.1   Examples of genetic polymorphisms.

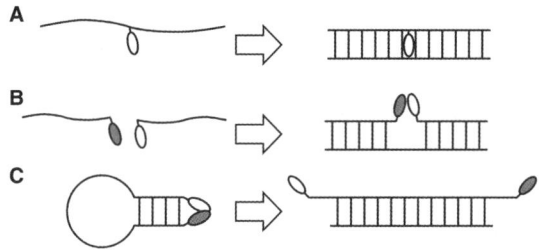

Figure 10.2   Schematic illustrations of (A) linear probe, (B) binary probe (BP), and (C) molecular beacon (MB).

binding of the target sequence. Hence, natural ODNs must be conjugated with non-natural active molecules in order to create efficient signalling probes that detect specific sequences of DNA and RNA.

Fluorescent probes are widely used as active signalling molecules in biotechnology owing to their high sensitivity and ease of handling. Phosphoramidite monomers tethered to a variety of fluorophores are now commercially available. These monomers can be used during DNA or RNA synthesis on commercial synthesisers to incorporate fluorophores into the desired positions of ODNs. Simple fluorophore–ODN conjugates are widely used for cell imaging and in DNA microarray applications. In this case, the unbound probe emits significant fluorescence and the free probe must be removed prior to analysis to eliminate this background noise. An ideal fluorescent probe should satisfy the following prerequisites: the probe should be 'silent' before hybridisation (*i.e.* it should emit little or no signal in the absence of target) or the unbound probe should emit a different signal from that after hybridisation with the target, and the signal upon target binding should be strong. In this chapter, we describe DNA-based functional probes for detection of target DNA and/or RNA that satisfy the above prerequisites. Locked nucleic acid (LNA)-based probes will be discussed in another chapter. We categorise DNA-based functional fluorescent probes into three types: 1) linear probes, 2) binary probes, and 3) molecular beacons (Figure 10.2).

10.2   Linear Probes

We first describe the design of a 'linear probe' that is composed of natural deoxynucleotides and a fluorophore or fluorophores that change in fluorescent signal upon target binding. Given that a linear probe does not form secondary structures such as hairpins, the hybridisation rate of the ODN–fluorophore conjugate should be almost the same as that of a natural ODN of the same length. Several strategies have been proposed to change fluorescence intensity and/or emission wavelength upon duplex formation. For example, French *et al.* reported a design they call the HyBeacon, a DNA probe that contains a fluorophore (FAM, HEX, or TET) at the 5-positions of uracil bases (Figure 10.3A).[2] In the single-stranded state, the emission is quenched, probably by stacking of the fluorophores with the nucleobases. Bright fluorescence is observed in the presence of the target DNA because the dye is flipped-out into a groove. Thus, duplex formation can be detected by monitoring the emission intensity of the dye.

Yamana *et al.* reported fluorescent probes that detect mismatches by excimer emission (Figure 10.3B).[3] In their design, two pyrene chromophores are introduced at the terminus of an ODN through a propanediol linker. In the single-stranded state, the probe shows monomer emission of pyrene, probably due to the interaction between pyrene and nucleobases. However, when the complementary strand is added, two pyrenes form a dimer, and excimer emission from pyrene is observed because its emission wavelength changes upon dimer formation (excimer emission). In addition, the excimer emission is highly sensitive to mismatches at the neighbouring base pair: only weak excimer emission is observed when a mismatch is introduced adjacent to the pyrene.

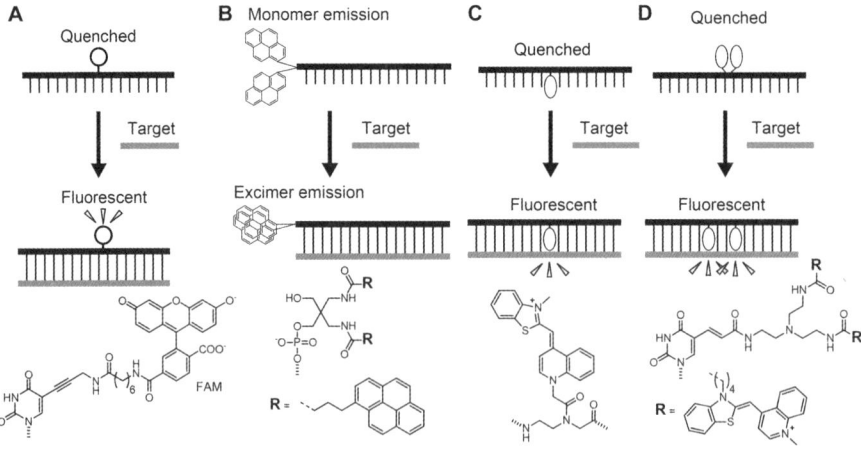

Figure 10.3   Linear probe detection schemes developed by (A) French *et al.*,[2] (B) Yamana *et al.*,[3] (C) Seitz *et al.*[4] and (D) Okamoto's group.[8,9] Examples of chemical structures of fluorophores are also shown.

Thus, mismatches can be detected by monitoring the ratio of monomer to excimer emission.

Seitz's group reported a single-stranded Peptide Nucleic Acid (PNA) probe, which they termed a forced intercalation (FIT) probe, with a thiazole orange at its centre (Figure 10.3C).[4–6] Fluorescence of thiazole orange is enhanced upon intercalation between base pairs, owing to planarisation of its structure. Hence, fluorescence of the FIT probe is enhanced upon hybridisation with the target. The FIT probe strategy has been applied to the monitoring of mRNA in living cells.[7] Okamoto's group reported the exciton-controlled hybridisation-sensitive fluorescent oligonucleotide (ECHO) design with two thiazole orange derivatives (Figure 10.3D).[8–11] In the single-stranded state, two dyes excitonically interact to form H-aggregates, which are usually non-fluorescent. When the probe hybridises with the target, the two chromophores are intercalated between natural base pairs. Strong emission from the intercalated dyes is observed because coherent interaction between dyes is suppressed by the intervening base pairs. Recently, tag technology for RNA imaging in a living cell has been developed based on ECHO probes.[12]

Okamoto and Saito *et al.* reported several base-discriminating fluorosides (BDFs) that change their emission depending on counterbases in duplexed states. In their design, modified bases are introduced into the corresponding position of an SNP site in ODNs. The BDFs change emission depending on the counterbase and, therefore, SNPs can be detected by monitoring the fluorescence emission. For example, they synthesised an adenine-discriminating base, [Py]U, by tethering pyrene via the 5-position of uracil (Figure 10.4).[13] When the counterbase of [Py]U is adenine, [Py]U forms a base pair and pyrene is flipped-out from the duplex. Consequently, strong emission from pyrene is observed because pyrene-1-carboxaldehyde derivatives show strong emission in a polar environment. On the other hand, pyrene is stacked within the duplex in the case

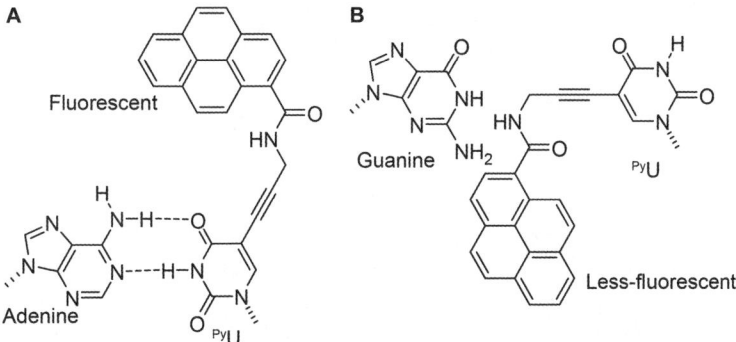

Figure 10.4 BDF detection scheme developed by Okamoto *et al.*[13] (A) When the counterbase of [Py]U is adenine, pyrene is flipped out from the duplex and strong fluorescence is observed. (B) In the case of other counterbases, pyrene is located inside the duplex and its emission is quenched.

of other counterbases (C, G and T). As a result, the emission is lower because of the presence of the hydrophobic environment around pyrene. Pyrene-modified adenine, cytosine and uracil derivatives have also been synthesised for the detection of SNPs.[13–16] The same group has also synthesised several size-expanded base analogues that change in emission intensity depending on the counterbase.[17,18]

Our group has reported the synthesis of fluorescent linear probes tethered to perylene that are able to detect deletion polymorphisms.[19,20] Perylene shows excimer or exciplex emission when it forms a dimer or heterodimer, respectively. In addition, the quantum yield of perylene is relatively high even in a DNA duplex. In our design, two perylenes are introduced via D-threoninol into the middle of the probe on both sides of the target nucleotides that are missing in the polymorphism (Figure 10.5). Nuclear magnetic resonance (NMR) analyses have revealed that a dye introduced via D-threoninol is intercalated between base pairs.[21,22] Hence, when the probe hybridises with wild-type DNA, each of the two perylenes is intercalated between the base pairs and yields monomer emission, because base pairing interrupts the interaction between the perylenes. In contrast, when the probe hybridises with a deletion mutant, a bulge-like structure is formed and the two perylenes are in close proximity. Accordingly, monomer emission from each perylene is quenched and excimer emission with a longer wavelength appears.

By using this design, we have prepared a fluorescent probe able to detect three-base deletion polymorphisms.[20] The target is the cystic fibrosis transmembrane conductance regulator (CFTR) gene (Figure 10.6); a specific three-base deletion is responsible for many cases of cystic fibrosis.[23] Here, we utilised perylene derivatives tethered to ethynyl or phenylethynyl groups as fluorophores in order to ensure complex formation in the five-base bulge.

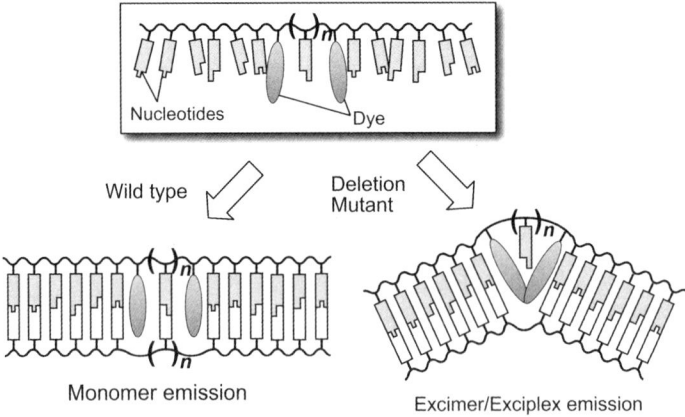

Figure 10.5 Schematic illustration of the detection of deletion polymorphisms (*n* represents the number of bases deleted).

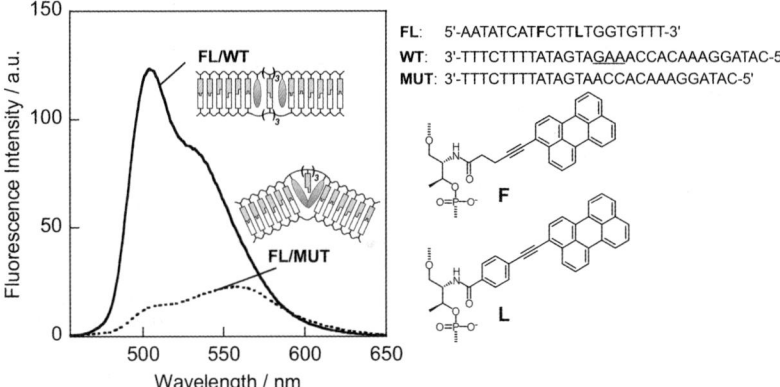

Figure 10.6 Fluorescence emission spectra of linear fluorescent probe **FL** bound to wild type (**WT**) or three-base deletion mutant (**MUT**). Sequences of the probe and targets are also shown. The concentration of probe was 1.0 μM, the target concentration was 1.2 μM, and the solution was 100 mM NaCl, 10 mM phosphate buffer (pH 7).

Figure 10.6 shows fluorescence emission spectra of the probe (**FL**) with the wild type (**WT**) gene or the deletion mutant (**MUT**). When **FL** is hybridised with **WT**, strong monomer emission of the phenylethynyl perylene moiety (**L**) is observed only at 507 nm. The emission from the ethynyl perylene conjugate (**F**) at 476 nm is not observed because of highly efficient fluorescent resonance energy transfer (FRET) from **F** moiety to **L** moiety. In contrast, when **FL** is hybridised with the three-base deletion mutant, monomer emission of **L** decreases and a new band concurrently appears at 556 nm corresponding to the exciplex emission between **F** and **L** moieties. Given that the emission colours of **FL** with **WT** and **MUT** are green and orange, respectively, deletion polymorphisms are detectable with the naked eye. We have also demonstrated that one- and two-base deletion polymorphisms can be detected using ODN probes tethered to pyrene or perylene.[19,24]

10.3 Binary (Split) Probes

A binary probe (BP) is composed of two ODNs with functional groups attached at their termini (Figure 10.2B).[25] These functional groups change fluorescent signals when two ODNs hybridise with the target. One of the advantages of binary probes is the accuracy in mismatch detection. Because a BP is usually shorter than a linear probe, incorporation of a single mismatch severely destabilises complex formation. Thus, mismatches can be detected more accurately. On the other hand, the hybridisation kinetics of BPs may be slow because of relatively large losses in entropy: three molecules, including the two probes, have to form one hybrid.[26] Here, we highlight BPs based on fluorescence detection, although there are many reports of ligation-based and

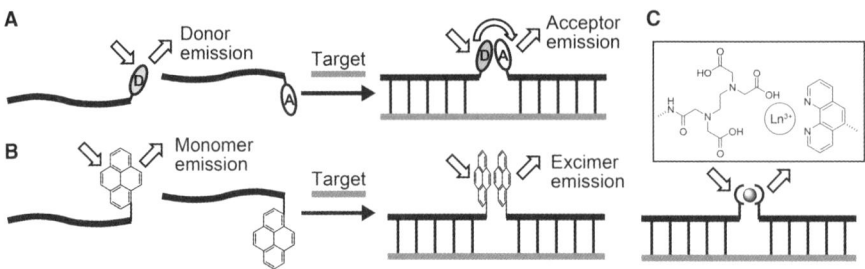

Figure 10.7 Schematic illustration of binary probes that employ (A) FRET, (B) excimer emission, and (C) metal complex formation.

deoxyribozyme-based BPs.[27] The most frequently used strategy, which makes use of FRET, was reported first by Cardullo *et al.* (Figure 10.7A).[28] In FRET BPs, one strand of the BP is conjugated with an energy donor while the other strand is attached to an acceptor. In the presence of the target, the donor and acceptor are located in close proximity and energy transfer from the donor to the acceptor occurs efficiently. As a result, acceptor emission is observed when the donor is excited. Accordingly, a target can be detected by monitoring the ratio between the donor and acceptor emissions. Any fluorophores can be used as FRET pairs as long as the emission band of the donor overlaps with the absorption band of the acceptor. These FRET-based BPs have been utilised for real-time PCR, detection of gene translocation, and mRNA visualisation in living cells.[29,30]

With FRET-based BPs, it is difficult to eliminate the background emission in the absence of the target because of the direct excitation of the acceptor.[25] Therefore, a number of other methodologies have been developed that use molecules which change in fluorescence upon complex formation. BPs utilising excimer emission of pyrene were independently reported by the laboratories of Masuko and Kool (Figure 10.7B).[31,32] In these BPs, a pyrene moiety is conjugated at each terminus of the BP. In the absence of the target, there is no interaction between pyrene moieties, and monomer emission is observed. In contrast, two pyrene moieties come into close proximity and excimer emission is observed in the presence of the target. Bichenkova *et al.* utilised exciplex emission between pyrene and *N,N*-dialkylnaphthylamine moieties for the signalling of BPs.[33] Ihara *et al.* reported another strategy that utilised lanthanide ions (Figure 10.7C).[34,35] This group incorporated EDTA and 1,10-phenanthroline moieties as chelators into the termini of a BP and added lanthanide ion into the solution. When these chelators are located in close proximity, they bind lanthanide ion cooperatively. As a result, the phenanthroline moiety functions as a sensitiser, and strong emission from the lanthanide ion is observed.

Fluorescence changes can also be triggered by chemical reactions. Kool's group reported quencher autoligation (QUAL) probes that utilise the reaction between phosphorothioate and dimethylamino-azobenzenesulfonyl (dabsyl) groups (Figure 10.8A).[36,37] In the QUAL probe, one strand is conjugated with

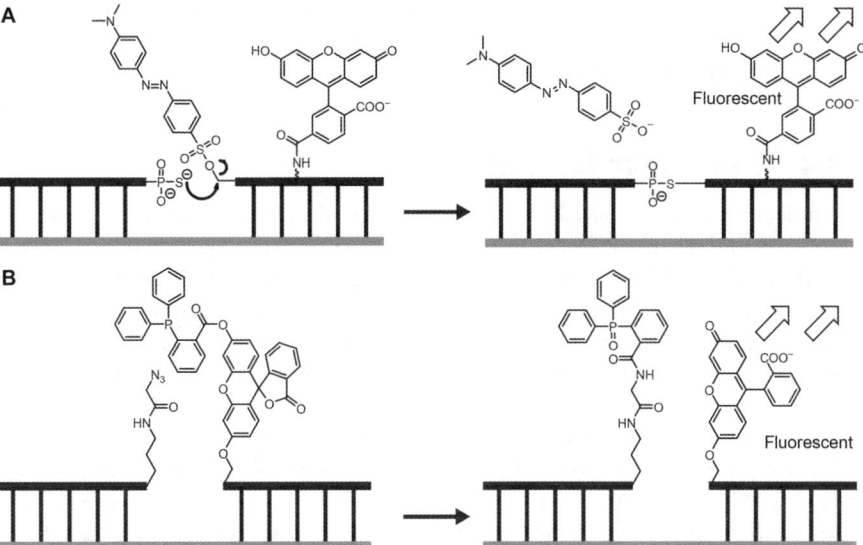

Figure 10.8 Examples of binary probes utilising chemical reactions developed by (A) Kool's group[36] and (B) Taylor's group.[44]

both a quencher (dabsyl) and a fluorophore (*e.g.* FAM, Alexa 350, Cy5) while the other has a phosphorothioate group at its terminus. In the absence of the target, the fluorophore is quenched by the dabsyl group. When the probe is hybridised with the target, however, the phosphorothioate group reacts with the dabsyl group. Consequently, the quencher is released and the emission is recovered. Although the reported QUAL probe detected the target successfully, the ligated product prevents turnover because of the high affinity of the ligated product for the target. Recently, the same authors reported a novel ligation-free strategy utilising the Staudinger reaction for the release of the quencher.[38,39] Grossmann and Seitz reported PNA-based BPs that utilise a quencher transfer reaction from cysteine to isocysteine attached to the termini of BP.[40,41]

Other researchers have reported BP designs that employ chemical conversion from non-fluorescent precursors to the fluorescent species mediated by catalysis or by a reactant attached to the other strand of the BP. Taylor and his group incorporated an imidazole group at one terminus as a catalyst and a less fluorescent coumarin-ester at the other.[42,43] In the presence of the target, imidazole catalyses the hydrolysis of the coumarin ester and releases fluorescent 7-hydroxycoumarin into the solution. They also reported PNA-based BPs in which the Staudinger reaction converts fluorescein ester (non-fluorescent) to fluorescein (Figure 10.8B).[44] Abe *et al.* introduced triphenylphosphine and azidomethyl fluorescein into BPs.[45] The azido-masked fluorescein is converted to fluorescein through the Staudinger reaction, and strong fluorescence is emitted. They successfully used this probe to monitor rRNA and mRNA in living cells. There are several other reports of BPs designed to take advantage of the Staudinger reaction from non-fluorescent precursor to emissive fluorophores (coumarin, rhodamine, *etc.*).[46–48] One of the advantages of these BPs is the signal amplification that results from template turnover.

10.4 Molecular Beacons

A molecular beacon (MB) is a hairpin DNA probe, originally reported by Tyagi and Kramer (see Figure 10.2C).[49] The basic design of an MB is as follows: the MB is typically 20–35 nucleotides in length and is able to form a 5–8 base pair stem portion (Figure 10.9). One terminus of the MB is conjugated to a fluorophore while the other is conjugated to a quencher. The loop portion is complementary to the target. In some designs, one arm of the stem participates in target binding (shared-stem design).[50] In the absence of a target DNA or RNA, the hairpin forms and fluorescence of the MB is quenched because of the close proximity of the fluorophore and the quencher. Upon binding with target DNA or RNA, the hairpin opens, and emission is recovered. MBs usually show higher sequence specificity than linear probes because of the equilibrium between a random coil and a hairpin structure.[51]

A wide variety of fluorophores, including coumarin, fluorescein, tetramethylrhodamine and boron-dipyrromethene (BODIPY) derivatives, have been used as fluorophores in MB designs.[52] 4-(Dimethylaminoazo)benzene-4-carboxylic acid (dabcyl) and Black-Hole Quenchers (BHQ1 and BHQ2) are most often used as quenchers.[53] Recently, inorganic molecules, including metal complexes and quantum dots, have also been utilised as a fluorophores and/or quenchers.[54] The simplicity and selectivity of MBs have led to their application as biochemical tools in real-time PCR, multiplex genetic analysis, analysis of protein–DNA interactions and cell imaging.[54–56] Recently, MBs have been immobilised on a solid surface for the preparation of DNA microarrays. Because these microarrays do not require labelling of sample DNAs or washing, they can be utilised as high-throughput bio-sensors of DNA/RNA sequences. However, the sensitivity of MBs on a solid surface is frequently lower than that in solution.[57]

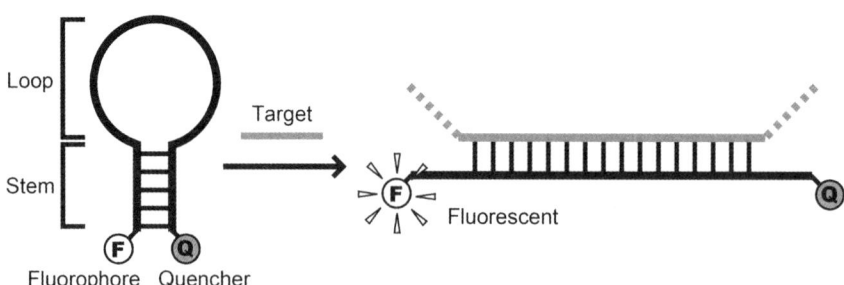

Figure 10.9   General scheme of molecular beacon for the detection of target DNA or RNA.

Several factors are required for the practical application of MBs: 1) low background emission in the closed form, 2) high signal emission in the open state, 3) rapid response and 4) high sequence selectivity. Lowering background emission has a significant impact because the background emission severely decreases the signal-to-background ratio (S/B ratio). Ideally, an MB should not show any fluorescence in the absence of the target. However, weak background emission is actually observed even in the close state because of the small amount of open MB caused by the breathing effect. In general, MBs with longer stems have less background emission than those with shorter stems and also have higher sequence specificity. Unfortunately, longer stems also result in slower hybridisation rates. The addition of even one base pair into the stem sometimes drastically decreases the hybridisation rate.[58] Thus, optimisation of loop and stem lengths is required to increase the detection sensitivity.

Several strategies have been proposed to decrease the background emission. For instance, Dubertret *et al.* utilised a gold nanoparticle as a quencher; these particles quench fluorophores such as fluorescein, rhodamine 6G, Texas red, and Cy5.[59] Although dabsyl cannot efficiently quench fluorophores that have emission maxima at longer wavelengths, gold nanoparticles effectively quench these dyes. Tan *et al.* incorporated multiple quenchers into the terminus of an MB.[60] The S/B ratio increases monotonically as the number of quenchers increases, and the S/B ratio of MB with three quenchers is 320. Häner *et al.* utilised a pyrene–perylenediimide pair as a fluorophore–quencher pair (Figure 10.10A).[61] In their design, two pairs were incorporated into the stem portion. In the absence of the target, the excimer emission from pyrene is effectively quenched by physical separation of two pyrene moieties. When the target is added, strong excimer emission from pyrene is recovered. Consequently, the S/B ratio of excimer emission is as high as 434.

In addition to the on–off type of MBs described above, several other strategies have been proposed for the ratiometric detection of the target. The Inouye group incorporated pyrene moieties into the termini of MB (Figure 10.10B).[62] In the absence of the target, excimer emission from pyrene is observed. Upon addition of target, the MB shows monomer emission. This response is the opposite of that of excimer-based binary probe, which shows excimer emission in the presence of the target. Several researchers have utilised FRET in their MB designs by incorporating energy donors and acceptors at the termini. In the absence of the target, the donor and the acceptor are in close proximity so that efficient FRET occurs. Thus, ratiometric detection of a target is possible by monitoring the donor:acceptor emission ratio. For example, Tan's group reported FRET MBs with coumarin and fluorescein as donor and acceptor, respectively.[63] Holzhauser and Wagenknecht reported an MB showing a distinct colour change upon target binding, by using FRET, from thiazole orange to thiazole red.[64] A FRET MB tethering three dyes was reported by Turro *et al.*[26] An apparent Stokes' shift of the probe is successfully enhanced by the successive energy transfer. Grossmann and Seitz reported a stem-less PNA beacon that had an energy donor (thiazole orange) at its centre and an energy acceptor (NIR667) at its terminus (Figure 10.10C).[41] Although

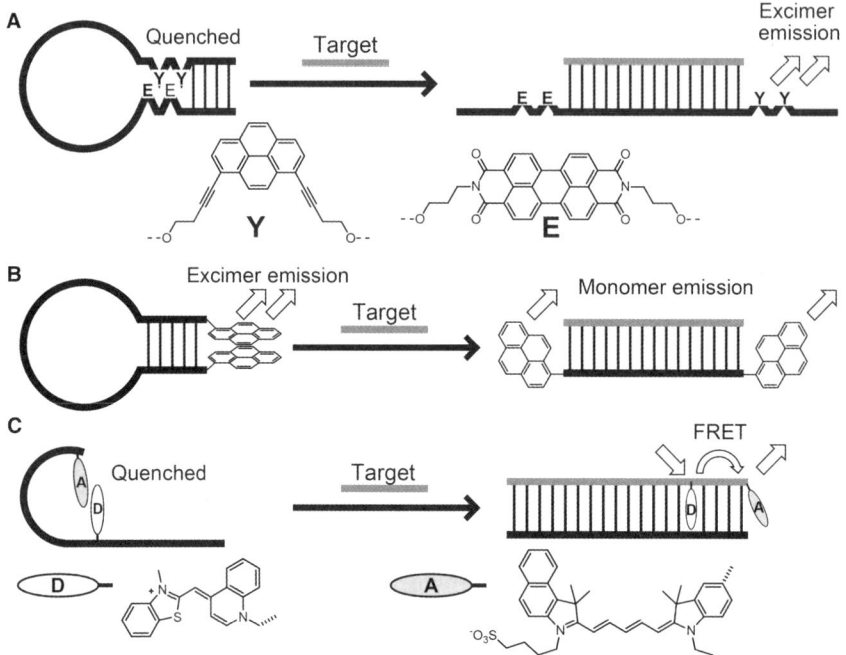

Figure 10.10 Examples of MBs developed by (A) Häner *et al.*,[61] (B) Inouye *et al.*,[62] and (C) Grossmann and Seitz.[41]

this 'beacon' does not have a hairpin structure, the fluorescence of the acceptor is quenched by the donor and/or nucleobases in the single-stranded state. In contrast, FRET occurs from the donor to the acceptor in the presence of the target. Accordingly, an S/B ratio of 108 was realised.

Our group has proposed a novel molecular beacon that we call an in-stem molecular beacon (ISMB), which has a fluorophore and a quencher in its stem.[65,66] We introduced perylene (**E**) and anthraquinone (**Q**) as a fluorophore–quencher pair into the middle of the stem (Figure 10.11). When dyes are introduced into adjacent base-pairing positions via D-threoninol, the dyes are stacked in an anti-parallel manner and strongly interact with each other.[67,68] Thus, the background emission from perylene in the closed state should be suppressed, owing to the strong interaction with anthraquinone. In addition, incorporation of multiple perylene–anthraquinone (**E–Q**) pairs is possible because the interactions between fluorescent dyes, which often cause quenching, are suppressed by the intervening base pairs. Therefore, the emission intensity in the open state is enhanced as the number of pairs is increased. Lower background emission and stronger emission upon target binding, relative to those of the classic MB, can be achieved by using the ISMB design. We synthesised ISMBs tethering multiple (one to four) perylene–anthraquinone pairs (**MB₁** to **MB₄** in Figure 10.11). The target sequence (**Surv**) is the survivin gene, which is highly expressed in breast cancer cells. The cationic polymer

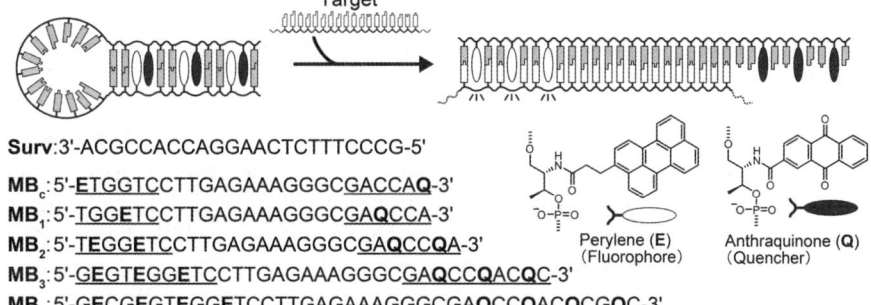

Surv:3'-ACGCCACCAGGAACTCTTTCCCG-5'

MB_c: 5'-<u>ETGGTC</u>CTTGAGAAAGGGC<u>GACCAQ</u>-3'
MB₁: 5'-<u>TGGETC</u>CTTGAGAAAGGGC<u>GAQCCA</u>-3'
MB₂: 5'-<u>TEGGETC</u>CTTGAGAAAGGGC<u>GAQCCQA</u>-3'
MB₃: 5'-<u>GEGTEGGETC</u>CTTGAGAAAGGGC<u>GAQCCQACQC</u>-3'
MB₄: 5'-<u>GECGEGTEGGETC</u>CTTGAGAAAGGGC<u>GAQCCQACQCGQC</u>-3'

Figure 10.11 Schematic illustration of in-stem molecular beacon (MB). Sequences of MBs and the target (Surv) are also shown.

Table 10.1 Signal/background (S/B) ratio of each MB.

Sequence	Fluorescence intensity/a.u.[a]		S/B ratio
	Without target	*With target*	
MB_c	32	292	9
MB₁	19	330	17
MB₂	8.2	530	65
MB₃	1.2	679	571
MB₄	0.9	155	168

[a] Solution conditions: $0.2\,\mu M$ molecular beacon (MB), $0.8\,\mu M$ Surv, $100\,mM$ NaCL, $10\,mM$ phosphate buffer (pH 7), N/P ([cationic group]$_{copolymer}$/[phosphate]$_{DNA}$ charge) of 2, $20\,^\circ C$.

(poly(L-lysine)-*graft*-dextran), previously reported by Maruyama's group,[69] was added to accelerate the hybridisation with the target. Emission intensities of ISMB with and without **Surv** are shown in Table 10.1. **MB₁**, which has one E–Q pair inside the stem, showed lower emission in the absence of the target than the MB with dyes at the termini (**MB_c**), because the in-stem strategy results in effective suppression of background emission. The S/B ratio of **MB₁** (17) was significantly higher than that of **MB_c** (9). Background emission without the target was monotonically lowered as the number of E–Q pairs increased, because multiple quenchers contribute to the quenching of the perylene emission. In contrast, as the number of E–Q pairs increased, the signal emission in the presence of target increased with all designs except **MB₄**, which contained four E–Q pairs. The emission intensity from **MB₄/Surv** was lower than that of **MB₃/Surv** because the longer stem of **MB₄** slowed hybridisation with the target. Consequently, the highest S/B ratio was 571, for **MB₃** with three perylenes. This high S/B ratio of **MB₃** allows detection of 1/1000 equivalents of the target. This in-stem technique is applicable to conventional fluorescent dyes such as Cy3 and thiazole orange[70] that are efficiently quenched by dabcyl analogues *via* excitonic interactions.

10.5   Conclusion

In this chapter, we have summarised the designs of fluorophore-modified nucleic acid probes used for the detection of specific sequences of DNA or RNA. These probes successfully detect target DNA/RNA with high-sensitivity, and some of these probes have been used in real-time PCR and SNP typing. Sequence-specific detection of nucleic acids has become quite important not only for genotyping but also in mRNA imaging and identification of micro-organisms. For example, localisation of mRNA in a living cell has been widely studied because mRNA localisation is related to the spatial and temporal control of gene expression.[71] Improvements in the sensitivity, selectivity and nuclease resistance of these probes will lead to more widespread applications in chemical biology, biotechnology and medicine.

References

1. S. Levy, G. Sutton, P. C. Ng, L. Feuk, A. L. Halpern, B. P. Walenz, N. Axelrod, J. Huang, E. F. Kirkness, G. Denisov, Y. Lin, J. R. MacDonald, A. W. C. Pang, M. Shago, T. B. Stockwell, A. Tsiamouri, V. Bafna, V. Bansal, S. A. Kravitz, D. A. Busam, K. Y. Beeson, T. C. McIntosh, K. A. Remington, J. F. Abril, J. Gill, J. Borman, Y.-H. Rogers, M. E. Frazier, S. W. Scherer, R. L. Strausberg and J. C. Venter, *PLoS Biol.*, 2007, **5**, 2113–2144.
2. D. J. French, C. L. Archard, T. Brown and D. G. McDowell, *Mol. Cell. Probes.*, 2001, **15**, 363–374.
3. K. Yamana, T. Iwai, Y. Ohtani, S. Sato, M. Nakamura and H. Nakano, *Bioconjugate Chem.*, 2002, **13**, 1266–1273.
4. O. Köler, D. V. Jarikote and O. Seitz, *ChemBioChem*, 2005, **6**, 69–77.
5. E. Socher, D. V. Jarikote, A. Knoll, L. Röglin, J. Burmeister and O. Seitz, *Anal. Biochem.*, 2008, **375**, 318–330.
6. L. Bethge, D. V. Jarikote and O. Seitz, *Bioorg. Med. Chem.*, 2008, **16**, 114–125.
7. S. Kummer, A. Knoll, E. Socher, L. Bethge, A. Herrmann and O. Seitz, *Angew. Chem. Int. Ed.*, 2011, **50**, 1931–1934.
8. S. Ikeda and A. Okamoto, *Chem. Asian J.*, 2008, **3**, 958–968.
9. A. Okamoto, *Chem. Soc. Rev.*, 2011, **40**, 5815–5828.
10. T. Kubota, S. Ikeda, H. Yanagisawa, M. Yuki and A. Okamoto, *Bioconjugate Chem.*, 2009, **20**, 1256–1261.
11. K. Sugizaki and A. Okamoto, *Bioconjugate Chem.*, 2010, **21**, 2276–2281.
12. T. Kubota, S. Ikeda, H. Yanagisawa, M. Yuki and A. Okamoto, *PLoS ONE*, 2010, **5**, e13003.
13. A. Okamoto, K. Kanatani and I. Saito, *J. Am. Chem. Soc.*, 2004, **126**, 4820–4827.
14. Y. Saito, Y. Miyauchi, A. Okamoto and I. Saito, *Chem. Commun.*, 2004, 1704–1705.
15. C. Dohno and I. Saito, *ChemBioChem*, 2005, **6**, 1075–1081.

16. Y. Saito, Y. Miyauchi, A. Okamoto and I. Saito, *Tetrahedron Lett.*, 2004, **45**, 7827–7831.
17. A. Okamoto, K. Tainaka and I. Saito, *J. Am. Chem. Soc.*, 2003, **125**, 4972–4973.
18. A. Okamoto, K. Tanaka, T. Fukuta and I. Saito, *J. Am. Chem. Soc.*, 2003, **125**, 9296–9297.
19. H. Kashida, T. Takatsu and H. Asanuma, *Tetrahedron Lett.*, 2007, **48**, 6759–6762.
20. H. Kashida, N. Kondo, K. Sekiguchi and H. Asanuma, *Chem. Commun.*, 2011, **47**, 6404–6406.
21. X. G. Liang, H. Asanuma, H. Kashida, A. Takasu, T. Sakamoto, G. Kawai and M. Komiyama, *J. Am. Chem. Soc.*, 2003, **125**, 16408–16415.
22. H. Kashida, X. G. Liang and H. Asanuma, *Curr. Org. Chem.*, 2009, **13**, 1065–1084.
23. B. Eshaque and B. Dixon, *Biotechnol. Adv.*, 2006, **24**, 86–93.
24. H. Kashida, H. Asanuma and M. Komiyama, *Chem. Commun.*, 2006, 2768–2770.
25. D. M. Kolpashchikov, *Chem. Rev.*, 2010, **110**, 4709–4723.
26. A. A. Martí, S. Jockusch, N. Stevens, J. Ju and N. J. Turro, *Acc. Chem. Res.*, 2007, **40**, 402–409.
27. A. P. Silverman and E. T. Kool, *Chem. Rev.*, 2006, **106**, 3775–3789.
28. R. A. Cardullo, S. Agrawal, C. Flores, P. C. Zamecnik and D. E. Wolf, *Proc. Natl. Acad. Sci. USA*, 1988, **85**, 8790–8794.
29. A. Tsuji, H. Koshimoto, Y. Sato, M. Hirano, Y. Sei-Iida, S. Kondo and K. Ishibashi, *Biophys. J.*, 2000, **78**, 3260–3274.
30. J.-L. Mergny, A. S. Boutorine, T. Garestier, F. Belloc, M. Rougée, N. V. Bulychev, A. A. Koshkin, J. Bourson, A. V. Lebedev, B. Valeur, N. T. Thuong and C. Hélène, *Nucleic Acids Res.*, 1994, **22**, 920–928.
31. K. Ebata, M. Masuko, H. Ohtani and M. Kashiwasake-Jibu, *Photochem. Photobiol.*, 1995, **62**, 836–839.
32. P. L. Paris, J. M. Langenhan and E. T. Kool, *Nucleic Acids Res.*, 1998, **26**, 3789–3793.
33. E. V. Bichenkova, H. E. Savage, A. R. Sardarian and K. T. Douglas, *Biochem. Biophys. Res. Commun.*, 2005, **332**, 956–964.
34. Y. Kitamura, T. Ihara, Y. Tsujimura, M. Tazaki and A. Jyo, *Chem. Lett.*, 2005, **34**, 1606–1607.
35. Y. Kitamura, T. Ihara, Y. Tsujimura, Y. Osawa, M. Tazaki and A. Jyo, *Anal. Biochem.*, 2006, **359**, 259–261.
36. S. Sando and E. T. Kool, *J. Am. Chem. Soc.*, 2002, **124**, 2096–2097.
37. S. Sando, H. Abe and E. T. Kool, *J. Am. Chem. Soc.*, 2004, **126**, 1081–1087.
38. R. M. Franzini and E. T. Kool, *J. Am. Chem. Soc.*, 2009, **131**, 16021–16023.
39. R. M. Franzini and E. T. Kool, *Chem. Eur. J.*, 2011, **17**, 2168–2175.
40. T. N. Grossmann and O. Seitz, *J. Am. Chem. Soc.*, 2006, **128**, 15596–15597.
41. T. N. Grossmann and O. Seitz, *Chem. Eur. J.*, 2009, **15**, 6723–6730.

42. Z. Ma and J.-S. Taylor, *Bioorg. Med. Chem.*, 2001, **9**, 2501–2510.

43. J. Cai, X. Li and J. S. Taylor, *Org. Lett.*, 2005, **7**, 751–754.

44. J. Cai, X. Li, X. Yue and J. S. Taylor, *J. Am. Chem. Soc.*, 2004, **126**, 16324–16325.

45. K. Furukawa, H. Abe, K. Hibino, Y. Sako, S. Tsuneda and Y. Ito, *Bioconjugate Chem.*, 2009, **20**, 1026–1036.

46. H. Abe, J. Wang, K. Furukawa, K. Oki, M. Uda, S. Tsuneda and Y. Ito, *Bioconjugate Chem.*, 2008, **19**, 1219–1226.

47. Z. L. Pianowski and N. Winssinger, *Chem. Commun.*, 2007, 3820–3822.

48. R. M. Franzini and E. T. Kool, *ChemBioChem*, 2008, **9**, 2981–2988.

49. S. Tyagi and F. R. Kramer, *Nat. Biotechnol.*, 1996, **14**, 303–308.

50. A. Tsourkas, M. A. Behlke and G. Bao, *Nucleic Acids Res.*, 2002, **30**, 4208–4215.

51. G. Bonnet, S. Tyagi, A. Libchaber and F. R. Kramer, *Proc. Natl. Acad. Sci. USA*, 1999, **96**, 6171–6176.

52. S. Tyagi, D. P. Bratu and F. R. Kramer, *Nat. Biotechnol.*, 1998, **16**, 49–53.

53. S. A. E. Marras, F. R. Kramer and S. Tyagi, *Nucleic Acids Res.*, 2002, **30**, e122.

54. K. Huang and A. Martí, *Anal. Bioanal. Chem.*, 2012, **402**, 3091–3102.

55. K. Wang, Z. Tang, C. J. Yang, Y. Kim, X. Fang, W. Li, Y. Wu, C. D. Medley, Z. Cao, J. Li, P. Colon, H. Lin and W. Tan, *Angew. Chem. Int. Ed.*, 2009, **48**, 856–870.

56. Y. Li, X. Zhou and D. Ye, *Biochem. Biophys. Res. Commun.*, 2008, **373**, 457–461.

57. A. Sassolas, B. D. Leca-Bouvier and L. J. Blum, *Chem. Rev.*, 2008, **108**, 109–139.

58. A. Tsourkas, M. A. Behlke, S. D. Rose and G. Bao, *Nucleic Acids Res.*, 2003, **31**, 1319–1330.

59. B. Dubertret, M. Calame and A. J. Libchaber, *Nat. Biotechnol.*, 2001, **19**, 365–370.

60. C. J. Yang, H. Lin and W. Tan, *J. Am. Chem. Soc.*, 2005, **127**, 12772–12773.

61. R. Häner, S. M. Biner, S. M. Langenegger, T. Meng and V. L. Malinovskii, *Angew. Chem. Int. Ed.*, 2010, **49**, 1227–1230.

62. K. Fujimoto, H. Shimizu and M. Inouye, *J. Org. Chem.*, 2004, **69**, 3271–3275.

63. P. Zhang, T. Beck and W. Tan, *Angew. Chem. Int. Ed.*, 2001, **40**, 402–405.

64. C. Holzhauser and H.-A. Wagenknecht, *Angew. Chem. Int. Ed.*, 2011, **50**, 7268–7272.

65. H. Kashida, T. Takatsu, T. Fujii, K. Sekiguchi, X. Liang, K. Niwa, T. Takase, Y. Yoshida and H. Asanuma, *Angew. Chem. Int. Ed.*, 2009, **48**, 7044–7047.

66. H. Asanuma, T. Osawa, H. Kashida, T. Fujii, X. Liang, K. Niwa, Y. Yoshida, N. Shimada and A. Maruyama, *Chem. Commun.*, 2012, **48**, 1760–1762.

67. H. Kashida, T. Fujii and H. Asanuma, *Org. Biomol. Chem.*, 2008, **6**, 2892–2899.
68. T. Fujii, H. Kashida and H. Asanuma, *Chem. Eur. J.*, 2009, **15**, 10092–10102.
69. L. Wu, N. Shimada, A. Kano and A. Maruyama, *Soft Matter*, 2008, **4**, 744–747.
70. Y. Hara, T. Fujii, H. Kashida, K. Sekiguchi, X. Liang, K. Niwa, T. Takase, Y. Yoshida and H. Asanuma, *Angew. Chem. Int. Ed.*, 2010, **49**, 5502–5506.
71. K. C. Martin and A. Ephrussi, *Cell*, 2009, **136**, 719–730.

CHAPTER 11

Nucleic Acid–Nanoparticle Conjugate Sensors for Use with Surface Enhanced Resonance Raman Scattering (SERRS)

NATALIE CLARK, KAREN FAULDS AND
DUNCAN GRAHAM*

Centre for Molecular Nanometrology, Department of Pure and Applied
Chemistry, University of Strathclyde, 295 Cathedral Street,
Glasgow G1 1XL, UK
*Email: duncan.graham@strath.ac.uk

11.1 Introduction

The rapidly expanding field of molecular diagnostics is centred on the detection of particular DNA sequences.[1] One of the primary reasons for DNA detection is in the identification and treatment of diseases. The detection of specific DNA sequences has also been successfully introduced to forensic applications for the identification of suspects in legal proceedings.[2]

The majority of detection techniques currently employed use fluorescence spectroscopy alongside DNA amplification techniques, for example in a real time polymerase chain reaction (PCR).[3] There are a number of factors introduced by the use of PCR that will have an impact on the sensitivity of the detection technique. First, owing to the intrinsic limit of detection of fluorescence, a number

RSC Biomolecular Sciences No. 26
DNA Conjugates and Sensors
Edited by Keith R Fox and Tom Brown
© The Royal Society of Chemistry 2012
Published by the Royal Society of Chemistry, www.rsc.org

of PCR cycles are required in order to generate adequate target for detection. The efficiency of the PCR cycling has a direct relationship to the number of copies of DNA produced. As a result, inefficient PCR cycling will impede detection because the number of DNA strands available for detection is minimised. Furthermore, background fluorescence associated with some of the quenching techniques employed in popular detection strategies, such as Taqman probes, can also result in a loss of sensitivity. Likewise, the selectivity of these techniques is also affected by the use of PCR. The co-amplification of undesired sequences present in the original sample reduces the selectivity of the system. Finally, owing to the broad nature of the emission spectra associated with fluorescence, multiplexing of this technique can prove problematic as a result of the potential overlapping of emission profiles.

In the case of disease diagnosis, it is beneficial to improve both the sensitivity and multiplexing capability of the detection method. An alternative to the use of fluorescence is surface enhanced (resonance) Raman scattering [SE(R)RS]. Although single molecule detection has been reported using both fluorescence[4–6] and SERS,[7–11] it has been shown that in some cases SE(R)RS detection limits can be as much as four orders of magnitude lower than those obtained by fluorescence.[12] SE(R)RS examines the vibrational output of the molecule, which relates directly to the structure. As such, it can be seen that whilst fluorescence affords a broad emission band, SE(R)RS produces distinctive fingerprint spectra. These sharp, distinctive bands allow for discrimination among mixed analytes.[13] Consequently, SE(R)RS can offer a substantial advantage over fluorescent techniques in terms of the ability to detect multiple species.

11.1.1 SERS

Surface enhanced Raman scattering (SERS) is a form of vibrational spectroscopy that gives rise to sharp bands relating directly to the molecular structure of the analyte. SERS can only occur when the analyte in question is in close proximity to a roughened metal surface.

Fleischman *et al.* initially observed SERS in 1974 when examining pyridine in an aqueous solution in the presence of a silver electrode.[14] An increased Raman signal was obtained due to the absorption of the pyridine onto the roughened silver surface of the electrode. It was not known what caused this enhancement effect and later two groups offered two differing theories. Jeanmaire and Van Duyne proposed an electromagnetic effect,[15] whilst Albrecht and Creighton proposed a charge transfer effect.[16] However, both found that an enhancement was obtained routinely in the order of 10^6 and concluded that the increased scattering was due to the surface plasmon of the roughened silver electrode.

The electromagnetic theory proposes that the enhancement is a result of an increase in the electromagnetic field experienced by a molecule on a roughened metal surface.[17] In order for an enhancement to be observed, it is necessary for the laser to be tuned to the surface plasmon. The surface plasmon can be thought of as the collective oscillation of conduction band electrons at the metal surface. The careful tuning of the laser to the frequency of the surface plasmon results in a larger number of scattered photons and, consequently, an enhancement effect is observed. The molecule does not need to bind directly to the surface; however, it must be in close proximity to experience this effect.[18,19] The charge transfer theory hypothesises that the enhancement is based upon the formation of new electronic states due to interactions between the substrate and the analyte bonded to it.[20] It is believed that the enhancement relates to a charge transfer intermediate state, which occurs owing to the powerful electron coupling between the analyte and the metal surface.[21,22] It has been concluded that both mechanisms contribute to the enhancement seen in SERS, however the electromagnetic effect appears to be the much more dominant contributor.[22,23]

The SERS enhancement is highly dependent upon the proximity of the analyte to the roughened metal surface being used as the SERS substrate. The SERS enhancement is dependent upon the distance d between the analyte in question and the metal surface, and has been calculated to be proportional to d^{-12}. Consequently, the molecule being analysed must be adsorbed directly onto a roughened metal surface for optimal SERS to be achieved. The degree of surface enhancement observed upon addition of the analyte to the metal surface has also been shown to be influenced by the degree of roughness of the metal surface used as the SERS substrate.

11.1.2   SERRS

Surface enhanced resonance Raman scattering (SERRS) is a synergic combination of resonance Raman scattering (RRS) and surface enhanced Raman scattering (SERS). This technique combines the enhancement provided by both scattering techniques, resulting in a potential increase in sensitivity by up to 10^{10} over normal Raman.[7,8,10]

For SERRS to occur the analyte must also contain a chromophore. In order for enhancement to occur, the laser excitation frequency used must be coincident or close to an electronic transition of the chromophore. This enhancement is maximised if the surface plasmon of the metal is also coincident with this frequency.

It is believed that the principles behind the enhancement effect seen in SERRS are the same as those for SERS enhancement. However; it has been revealed that the contribution proposed in the charge transfer theory is significantly smaller when the laser line is in-resonance, *i.e.* in SERRS, than out of resonance, *i.e.* in SERS.[24–26]

In SERRS, as in SERS, the analyte of choice must be adsorbed onto, or held in close proximity to, a suitable roughened metal surface. The surface commonly chosen to suit this purpose is a colloidal suspension of metallic nanoparticles. Given that the degree of surface enhancement varies with the amount of roughening of the metal surface, for metallic nanoparticles it is dependent upon the size of the nanoparticles, as well as the metal chosen and the area of the surface.[27]

11.1.3 Suitable SE(R)RS Surfaces

A series of different metals have been used to provide the surface enhancement effect, such as silver, gold and copper, as well as a selection of other transition metals.[28–35] Silver and gold are most commonly used owing to their stability and because they have surface plasmons within the visible region, however silver tends to give the largest enhancement factor.[36] The roughened surface can be made available in a variety of forms, most commonly aggregated colloidal suspensions, electrodes and cold deposited metal films. Colloidal nanoparticles are often selected as the SE(R)RS substrate of choice because they offer a number of advantages over a bulk metal surface.

The use of a suspension of nanoparticles allows for a large number of particles to be interrogated by the laser beam at once, therefore the resultant signal observed is that of an average of the signals detected from each particle. This makes quantitative analysis possible. Furthermore, different batches of nanoparticles can be mixed to result in a large stock of colloidal SE(R)RS substrate. Consequently, the signal detected using nanoparticles is often more reproducible and stable. Nanoparticles are subject to Brownian motion, a phenomenon that proves advantageous for SE(R)RS analysis, in comparison to a stationary surface. The continuous motion of the nanoparticles in the solution help to eliminate problems that arise as a result of sample drying, sample heating and photodecomposition.

11.1.4 Nanoparticles

The preparation of metallic nanoparticles is simple: a metal salt is reduced in the presence of a stabiliser molecule. The agents most commonly selected to reduce the metal ions include tri-sodium citrate,[31] sodium borohydride[37] and ethylenediaminetetraacetic acid (EDTA).[38] These methods typically afford nanoparticles in the region of 5–40 nm, depending upon the protocol followed and the metal used. Generally, the optical properties of the colloidal solution are determined by the nanoparticle size, shape and chemical composition. To get the best SE(R)RS signal, the size of the nanoparticles must be carefully selected, alongside the excitation wavelength used to excite the surface plasmon on the nanoparticles. In order for quantitative SE(R)RS analysis to be possible, the nanoparticle solutions should have a narrow size distribution, because this will result in similar SE(R)RS intensities.

11.1.4.1 Functionalised Nanoparticles

As a consequence of the successful implementation of metallic nanoparticles as SE(R)RS substrates, the field has evolved further to include the functionalisation of nanoparticles for use in SE(R)RS assays, and to act as SE(R)RS tags.

Two main types of nanoparticle functionalisation can be performed, depending upon the function of the nanoparticle in the assay. First, if the nanoparticle is to act as a tag that can be easily visualised by the use of

SE(R)RS, the nanoparticle surface is encoded with SE(R)RS active small molecules, providing an enhanced signal corresponding to the immobilised molecule.

Second, if the nanoparticle is to bind to a target biomolecule, the surface can be modified by the immobilisation of a molecule that will interact with the target molecule. For instance, the nanoparticle surface could be functionalised with an antibody that interacts with an antigen whose presence indicates the presence of a disease, thereby providing diagnostic information for the sample. The immobilisation of DNA on the nanoparticle surface is often performed in order to detect the presence of target DNA with a specific DNA sequence. This approach is highly favourable because probes can be designed to give a high level of specificity to the nanoparticle binding process. Single base mismatches in DNA can be detected easily using this approach.

Furthermore, the two functionalisation approaches can be combined, resulting in a SE(R)RS encoded nanoparticle that can bind to a molecule of interest, indicating its presence. This approach is frequently adopted in multi-plex assays where a variety of different labelled nanoparticles can be made, each indicating the presence of a different target molecule.

This section will focus first on the immobilisation of small molecules onto the nanoparticle surface to produce SE(R)RS active particles, followed by the functionalisation of nanoparticles with DNA to produce DNA conjugates for SE(R)RS detection.

11.1.4.2 Encoded Nanoparticles

Owing to the distance-dependent nature of SE(R)RS enhancement, the proximity of an analyte to the roughened metal surface of choice is of vital importance. As a result of their robust nature, and surface plasmon in the visible region, metallic nanoparticles are often used alongside an analyte as a labelling system in assays involving SE(R)RS analysis. However, steric hindrance can limit the availability of the analyte to the nanoparticle surface, thereby resulting in lowered signal enhancement, and in turn reducing the sensitivity of the assay. In recent years, the field of nanoparticle functionalisation has grown rapidly, with researchers working towards the synthesis of SE(R)RS active nano-particles which can be successfully incorporated as a labelling technique in a number of assay formats.

In 2003, Mulvaney et al. published details of the glass encapsulation of analyte tagged nanoparticles.[39] Gold nanoparticles were encoded with trans-1,2-bis(4-pyridyl)-ethylene (BPE) and 4-mercaptopyridine (MP), before glass encapsulation. The spectra observed from both sets of tagged nanoparticles were examined and it was noted this work shows potential for use in a multi-plexed manner due to the distinct spectral peaks observed for both reporters at the same excitation wavelength.

Later that year, Nie et al. detailed the synthesis of dye-coded metallic nanoparticles, encapsulated in silica.[40] Gold nanoparticles were embedded with a series of sulphur-containing dye molecules: 3,3'-diethylthiadicarbocyanine iodide (DTDC), malachite green isothiocyanate (MGITC), tetra-methylrhodamine-5-isothiocyanate (TRITC) and rhodamine − 5-(and 6)-iso-thiocyanate (XRITC). Three of the four dyes used contained the isothiocyanate group, whilst the fourth contained two sulfur atoms as part of the ring structure. Both of these functional groups act as affinity tags, binding the dye molecule to the gold metal surface. It was shown that the enhancement factors achieved are significantly large to allow potentially for single-particle spectroscopy to be carried out.

Irudayaraj and co-workers demonstrated in 2007 the possibility of using nanoparticle tags in multiplexed DNA detection assays.[41] The development of a series of nanoparticle tags containing both Raman reporter molecules and probe oligonucleotides allowed the successful detection of eight probes simultaneously.

Eight non-fluorescent Raman tags were selected, all containing either sulfur or nitrogen as an affinity tag to attach the reporter molecule to the gold nanoparticle surface. As shown in Figure 11.1, the nanoparticle tags were co-functionalised with thiol-modified oligonucleotides, allowing for hybridisation of the nanoparticle tag conjugates with a target DNA molecule of choice.

In 2009, Schlücker and co-workers published details on multiplexing SERS labels immobilised on gold nanoparticles by SAM formation.[42] First, three different nanoparticle reporter conjugates were made, each with a one-component self-assembled monolayer of a different Raman label. The labels selected were 5,5′-dithiobis(2-nitrobenzioc acid) (DTNB), 2-bromo-4-mercap-tobenzoic acid (BMBA), and 4-mercaptobenzioc acid (MBA). Two-component

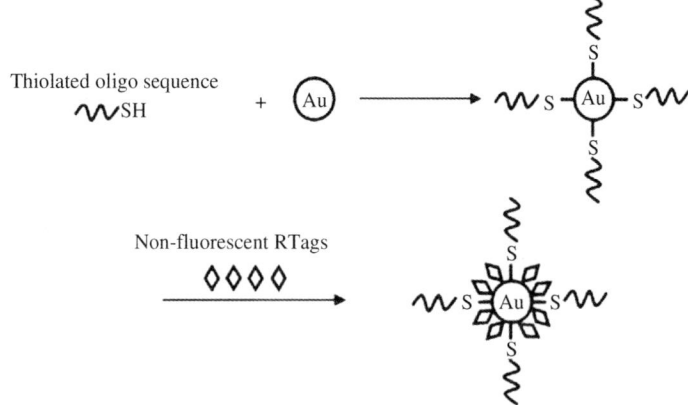

Figure 11.1 Schematic representation of the synthesis of the nanoparticle tags. Gold nanoparticles are functionalised with thiol-DNA strands to produce DNA–nanoparticle conjugates. Non-fluorescent Raman tags are then introduced to the nanoparticle surface, resulting in a labelled DNA–nanoparticle conjugate.
(Reprinted with permission from ref. 41. Copyright 2007 American Chemical Society.)

monolayer Raman tags were then synthesised. Three different combinations of Raman labels were analysed, with each combination showing clearly distinguishable peaks for the two labels present. The particles labelled with the two-component SAM could not be distinguished from a mixture of the two batches of nanoparticles labelled with the one-component SAM containing the same two Raman labels. This was then extended to a three-component SAM containing all three labels. Peaks corresponding to each individual Raman label could be distinguished from the SERS signal obtained. It was also seen that the contribution from each label to the SERS spectrum collected, could be tuned by altering the ratio of the three components.

A method for the functionalisation of nanoparticles with oligonucleotides, as well as Raman reporter molecules, without the need for formation of a mixed layer on the metal surface, was recently demonstrated by Wrzesien and Graham.[43] A series of linker molecules containing three principal groups were synthesised for the conjugation of biomolecules to metallic nanoparticles. Each linker contained a Raman tag, a surface-complexing group and a functional group for bioconjugation. The linkers contained a short-chain polyethylene glycol (PEG) to afford stability to the conjugates, and also to prevent non-specific binding of biomolecules to the nanoparticle surface. The surface-complexing group selected was thioctic acid, a cyclic disulfide. The linker molecule was capped with a carboxylic acid group, designed for the bio-conjugation of proteins or amino-functionalised oligonucleotides. A selection of dyes was incorporated into the linker molecules: fluorescein, 6-amino-fluorescein and tetramethylrhodamine (TAMRA). Both gold and silver nanoparticles were successfully functionalised with the three linker molecules produced. The successful conjugation of amino-functionalised oligonucleotides was demonstrated.

11.1.4.3   DNA–Nanoparticle Conjugates

Gold and silver nanoparticles can be functionalised with a number of different small molecules by their immobilisation onto the nanoparticle surface *via* a thiol group. Sulfur atoms are capable of covalently bonding to the gold or silver surface, providing a robust anchor for molecular labelling.

Oligonucleotide–gold nanoparticles, and oligonucleotide–silver nano-particles provide highly stable, functional nanoparticle labels. These conjugates are functionalised by oligonucleotides on the nanoparticle surface, usually by means of a thiol linkage. These nanoparticle conjugates can be used in two different ways. First, the nanoparticles themselves can be used directly as detection labels to indicate the presence of target DNA. Additionally, the conjugates can act as probes which, in the presence of target DNA, become aggregated, allowing for the novel optical properties of aggregated nano-particles in comparison to those of unaggregated nanoparticles to be taken advantage of.

11.2 DNA Detection Methods Based on SE(R)RS

The detection of DNA is an exceptionally important field of research for medical diagnostics. Many researchers in recent years have been focusing on the advent of DNA detection systems using SE(R)RS as a detection method. As such, metallic nanoparticles are often incorporated into these diagnostic assays. The use of DNA conjugates is becoming more frequent because of the unique properties they can offer these detection systems in terms of both selectivity and sensitivity. This section focuses on these detection systems incorporating DNA nanoparticle conjugates for the successful identification of a specific sequence of DNA by SE(R)RS.

11.2.1 Colloid-based Detection Techniques

11.2.1.1 Molecular Sentinels

In 2005, Vo-Dinh *et al.* detailed the use of molecular sentinels for the successful detection of DNA relating to the *gag* gene sequence of the human immuno-deficiency virus type 1 (HIV-1).[44] The probes were used to detect PCR amplicons of the HIV gene.

The molecular sentinel molecule consists of a metallic nanoparticle and a dye-labelled stem–loop DNA strand, as seen in Figure 11.2. The DNA sequence consists of a central section that is complementary to the target DNA sequence, and two flanking complementary sequences, which give rise to the hairpin loop structure. In the absence of target DNA, the stem–loop config-uration remains in the 'closed' position, with the Raman label held in close

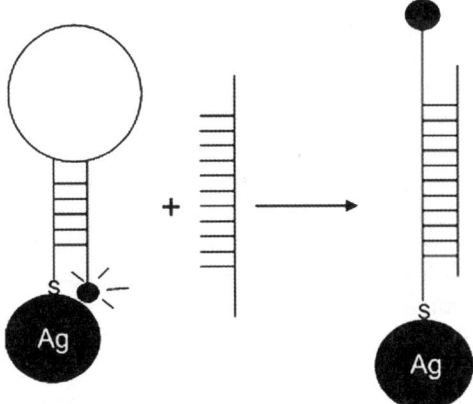

Figure 11.2 Schematic of molecular sentinel concept. When in the closed hairpin configuration, a SERS signal corresponding to the dye can be seen. When in the open conformation, the signal is reduced.
(Reprinted with permission from ref. 44. Copyright 2005 American Chemical Society.)

proximity to the nanoparticle surface. Consequently, upon interrogation with a laser of suitable wavelength, a strong SERS signal is observed. In the presence of target DNA, the stem–loop configuration is changed, resulting in the Raman reporter being removed from the nanoparticle surface. In this case, the SERS enhancement is dependent upon the distance d between the Raman reporter and the nanoparticle surface. Accordingly, the SERS effect is dramatically reduced owing to the increase in d, resulting in the related reporter signal being reduced.

Later, the same group reported the first proof of concept that the molecular sentinel approach could be multiplexed in a homogeneous solution assay format without the need for separation or washing steps.[45] Two nanoprobes containing different Raman reporter molecules were designed to target two separate genes, *erbB-2* and *ki-67*. The two genes selected are used clinically as diagnostic biomarkers for breast cancer. Multiplex detection was performed in the presence of both corresponding target DNA sequences, as well as in the presence of individual DNA targets. When both target molecules were present, the SERS signal associated with both reporters was reduced. However, in the presence of one target sequence, only the spectrum associated with the reporter from the corresponding nanoprobe relating to the target present was reduced.

11.2.1.2   SERRS Beacon

In 2005, Faulds *et al.* released details of a new approach to DNA detection by SERRS.[46] A thiolated DNA strand was coupled *via* Michael addition to dye labels, which were specifically designed for attachment to silver surfaces. This resulted in a 'SERRS Beacon' being formed that could be used for both fluorescence and SERS detection.

The assay was based around the concepts used for fluorescence detection of molecular beacons.[47] Molecular beacons are a dual labelled, single strand of DNA, with complementary bases at either end, resulting in the strand being closed over and held in a hairpin conformation. The loop section of the hairpin is the probe sequence, which is complementary to the target DNA sequence being detected. One end of the probe strand is functionalised with a fluorescent dye, whilst the other is typically labelled with a quencher molecule. In the 'closed' conformation, the fluorescent dye is held in close proximity to the quencher molecule, resulting in no fluorescence being observed. In the presence of target DNA, however, the hairpin is opened, removing the fluorescent dye from the proximity of the quencher molecule, thereby allowing for fluorescence to be detected.

The SERRS Beacon is a dual-labelled probe with a fluorophore at the 3′ end of the oligonucleotide strand, and a second dye at the other. In this case, benzotriazole dye was coupled to the beacon DNA strand at the 5′ end. This was carried out in order to facilitate attachment of the oligonucleotide to the surface of a silver nanoparticle *via* complexation of the benzotriazole moiety with the silver surface. 5-(and 6)-Carboxyfluorescein (FAM) was selected as the fluorophore label present at the 3′ end. The resultant beacon configuration displayed 98% fluorescence quenching when immobilised on the nanoparticle surface.

Whilst in the 'closed' conformation, the signal observed by the SERRS Beacon was dominated by FAM. This is a result of FAM giving a more intense signal than the benzotriazole dye by approximately two orders of magnitude. The fluorescence of the beacon in this arrangement was also recorded, resulting in no signal being detected. This is a direct result of the highly efficient fluorescence quenching from the silver nanoparticle. Upon addition of target DNA, a fluorescence signal could then be detected. Upon analysis by SERRS, the spectrum collected had changed to reflect the distancing of the FAM from the nanoparticle surface.

Further work was carried out using a complementary sequence containing overhanging bases to mimic hybridisation to a target sequence within a longer strand of DNA. Discrimination could clearly be seen between the signal observed in the absence of target DNA, where sharp bands relating to the spectrum of FAM could be seen, and in the presence of target DNA, where a significant fluorescent background could be observed. This fluorescent background was a result of the FAM dye moving further from the nanoparticle surface, and, consequently, the fluorescence from the dye being quenched to a lesser degree by the nanoparticle.

11.2.1.3 Assembly Induced Aggregation-based SERRS Assays

In 2008, Graham *et al.* published details of the successful synthesis of oligo-nucleotide–silver nanoparticles (OSNs), and their subsequent use in a sandwich assay format for the detection of DNA.[48] Oligonucleotide probes were immobilised on the nanoparticle surface *via* a terminal alkyl thiol group situated on either the 3′ or 5′ end of the strand. The method of synthesis for the silver nanoparticle conjugates is similar to that of their gold counterparts.[49] A spacer group was introduced to the oligonucleotide for two reasons. First, to prevent attachment of the reactive functional groups, present in the oligonucleotide bases, to the nanoparticle surface. Second, the spacer group aids the DNA hybridisation process by decreasing steric hindrance caused by other immobilised oligonucleotide strands. It was found that, subsequent to OSN synthesis, the conjugates were stable for up to 3 months at room temperature.

Hybridisation of OSN conjugates was performed in a sandwich assay format, with two different OSN conjugates being made. Each set of conjugates was functionalised with a sequence of DNA that was half complementary to the sequence of the desired target being detected. Upon hybridisation of both probes with the target DNA, the nanoparticles were brought into close proximity with one another, resulting in aggregation occurring. This aggregation process was observed by both scanning electron microscopy (SEM) and ultraviolet–visible (UV-vis) spectroscopy. The UV-vis spectrum showed a broadening of the surface plasmon peak associated with the silver nanoparticle conjugates, with an absorbance maximum at 540 nm. The ability of the system to detect single base mismatches was also tested in comparison with oligonucleotide–gold nanoparticles (OGNs). The OSN conjugates gave a minimum detectable target oligonucleotide concentration 50 times lower than that of their gold counterparts.

This initial work was later expanded upon by introducing the use of SERRS for sequence-specific DNA detection.[50] A Raman dye was introduced to the surface of the OSN detailed previously, resulting in SERRS active conjugates. The assay concept focuses on the enhancement of SERRS signals when nanoparticles are brought in close proximity to each other, creating 'hot spots' in the interstices between particles, as shown in Figure 11.3 (A). Two sets of OSN conjugates were synthesised, each with the same dye but with different probe sequences. The conjugates prepared were stable in 0.3M sodium chloride and 10 mM phosphate buffer at pH 7 for over 6 months. Upon addition of target DNA with a sequence complementary to that of the two probe sequences, a yellow to green–blue colour change was observed. As indicated in Figure 11.3 (B), this colour change could also be observed in the extinction spectrum. This change is indicative of aggregation of the nanoparticle conjugates having occurred.

When this assay construction was analysed by monitoring the SERRS spectral intensity of the oligonucleotide conjugates, a dramatic increase in signal was observed upon addition of target DNA, as shown in Figure 11.3 (C). It was observed that upon heating the assay sample, the SERRS signal was reduced to a minimal level. This is a direct consequence of the heat inducing the denaturation of the duplex, resulting in the nanoparticles becoming

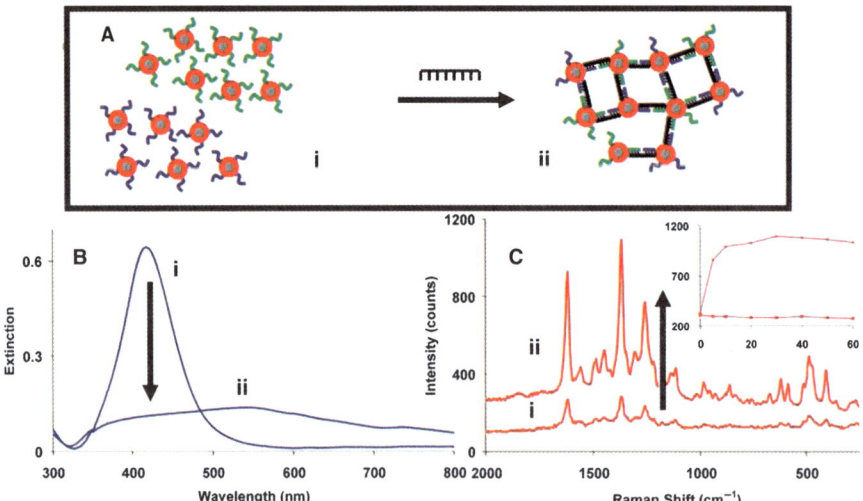

Figure 11.3   (A) A diagrammatic representation of the controlled aggregation of OSNs; (B) A UV-vis spectrum (i) pre- and (ii) post target addition, indicating a red shift in surface plasmon alongside a drop in peak intensity; (C) SERRS spectra before (i) and after (ii) target addition showing the increase in the SERRS intensity. The inset graph shows the increase in SERRS intensity over time.
(Reprinted by permission from Macmillan Publishers Ltd: Nature Nanotechnology, copyright 2008.)

monodispersed. To test the specificity of this approach, three separate batches of nanoparticle conjugates were functionalised with different DNA sequences and dyes. It was shown that, in the presence of the three different conjugates, SERRS signals are only observed for the dyes from the conjugates with a probe complementary to the targets present. It was later shown that in the presence of a single base mismatch in the target sequence, there was no increase in SERRS intensity [51]

This work has been expanded to include the use of mixed metal nanoparticle assemblies containing a mixed network of silver and gold nanoparticles.[52] It has also been shown that DNA triplexes containing LNA bases can be used successfully to direct nanoparticle assembly in this approach.[53]

11.2.1.4 Magnetic Capture Assay

In 2011, Johnson and co-workers published details of the successful detection of DNA corresponding to the genome of the West Nile virus by SERS.[54] The assay format adopted a split-probe configuration, with two probes for detection of a specific sequence of DNA. An oligonucleotide functionalised-paramagnetic nanoparticle was designed to act as a capture probe, immobilising the target–probe duplex upon successful hybridisation. The second probe utilised within the assay was a gold nanoparticle functionalised with a Raman reporter and a reporter oligonucleotide. In the presence of target DNA, the probes will successfully bind to the target sequence, resulting in a duplex labelled with both a SERS active nanoparticle and a magnetic nanoparticle. This duplex could then be immobilised by application of a magnetic field, resulting in the hybridised SERS active nanoparticle probes being removed from the solution and compacted into a pellet, before interrogation with a laser. The Raman reporter molecule selected was 5,5'-dithiobis(succinimidy-2-nitrobenzoate), DSNB.

In the presence of target DNA, the magnetically pulled-down assay gave a large SERS signal, relating to the SERS peaks of DSNB. However, in the absence of target DNA, the nanoparticle conjugates were not immobilised by the magnetic pull-down process, thereby resulting in a small SERS response being observed. The origin of this minimal response was investigated with a negative control containing only the reporter nanoparticles and the magnetic capture nanoparticles. This control also displayed the same small SERS response, indicating that this minimal signal was an artefact of non-specific binding of the reporter and capture nanoparticles. Furthermore, it was shown that hybridisation mixtures that were not concentrated into a pellet by the application of a magnetic field displayed SERS signals similar to those seen for the negative assay.

The specificity of the assay was tested by the introduction of a non-complementary 50 base oligonucleotide, in place of the target DNA strand. The same minimal response was seen in the presence of non-complementary DNA as was observed for the negative control samples. It was shown that the assay in the absence of target DNA (or the presence of nonsense DNA) gave a 200-fold decrease in SERRS response when compared with the signal observed in the presence of target DNA.

Further work was described, which investigated the limit of detection of the assay system, with varying concentrations of target DNA being introduced to the probe mixture. It was observed that, upon dilution of the target DNA, there was a progressive decrease in the SERS signal observed. The limit of detection was found to be 10 pM.

11.2.1.5 Nanoparticle–Aptamer Conjugates

Nucleic acid aptamers are synthetic DNA sequences, normally 20–80 nucleotides long, capable of binding with high selectivity to a target molecule. They have been shown to be capable of binding successfully to a range of analytes, including small molecules,[55] proteins[56] and peptides. The properties of aptamers are commonly compared to those of antibodies because of their molecular recognition and specificity. However, aptamers have been shown to exhibit a number of advantages over antibodies, including their increased stability at ambient temperatures, their chemical stability in a range of environments and their ease of synthesis. As such, aptamers have been introduced into detection assays as a substitute for antibodies.[57]

The thrombin aptamer is one of the most highly researched nucleic acid aptamers, and also one of the earliest published.[58] In 2004, Willner *et al.* published details of the optical detection of thrombin on glass surfaces, using gold nanoparticle–aptamer conjugates.[59] The thrombin aptamer was covalently bound to the glass surface *via* a maleimide-functionalised siloxane monolayer. The aptamer-functionalised nanoparticles were then bound to the thrombin, and the gold nanoparticle interface enlarged in a growth solution containing $HAuCl_4$. The higher the concentration of thrombin present, the more gold nanoparticle seeds are present for enlargement. Consequently, a higher absorbance spectrum was observed in the presence of higher levels of thrombin. This selective binding of aptamer functionalised nanoparticles to thrombin enabled a detection system with a sensitivity limit of 2 nM.

In 2011, Graham *et al.* facilitated this aptamer conjugate–thrombin interaction to demonstrate the use of the thrombin aptamer, conjugated to silver nanoparticles *via* a thiol linkage, for the successful detection of thrombin by SERRS.[60] The aptamer conjugates were labelled with a Raman reporter, in order to enable the monitoring of the interaction between the aptamer and thrombin by SERRS. Upon SERRS analysis, it was shown that in the presence of thrombin, the thrombin aptamer on the nanoparticles would bind to the free thrombin in solution, thereby resulting in the nanoparticles being brought into close proximity to each other. A subtle change in peak intensities was demonstrated, indicating that nanoparticle aggregation had occurred. Subsequently, it was shown that the SERRS response observed varied depending upon the concentration of thrombin present.

Irudayaraj *et al.* have taken advantage of the specificity of aptamers for the detection of a pathogenic species of bacterium, *Salmonella typhimurium*.[61] Gold nanoparticles were functionalised with thiolated anti-*S. typhimurium* aptamers, followed by labelling of the conjugates with 4-mercapto-benzoic acid (MBA). In the presence of *S. typhimurium*, after filtration through a filter membrane, a SERS signal corresponding to MBA could be observed. In the absence of the target pathogen, only background SERS signal could be observed. It was also demonstrated that the aptamer conjugates could be utilised in a multiplex pathogen detection assay, alongside antibody nanoconjugates. The limit of detection reported for this technique is in the range of approximately 10^2 colony-forming units (CFU) mL^{-1}.

Whilst a vast array of research into the use of DNA conjugates for DNA detection has been done in solution, a number of research groups have had success developing techniques utilising DNA–nanoparticle conjugates with a bulk substrate. These developments are discussed in the following section.

11.2.2 Bulk Substrate-based Detection Techniques

11.2.2.1 Scanometric Detection

In 2000, Mirkin *et al.* detailed the use of DNA–nanoparticle probes in a scanometric array format to detect successfully a specific DNA sequence.[62] The detection utilised the light scattering properties of metallic nanoparticles, in conjunction with a conventional flatbed scanner.

The assay consisted of a three-component sandwich assay immobilised on a glass slide, shown in Figure 11.4. Gold nanoparticles were modified with oligonucleotides complementary to a section of the target DNA being detected. A glass microscope slide was functionalised by attachment of 3′ thiol-modified capture oligonucleotides to the slide surface. Upon addition of target DNA, the capture probe and nanoparticle probe hybridise to the target sequence, resulting in immobilisation of the gold nanoparticles on the glass surface. A signal amplification process was introduced to allow the detection of lower levels of DNA. The immobilised gold nanoparticles were subjected to staining with silver *via* reduction of silver ions on the gold surface by hydroquinone. The process increased the scattering intensity by as much as a factor of 10^5.

Further to the publication of the above findings in 2000, a two colour scanometric assay was developed for the selective detection of two different DNA targets.[63] The assay incorporated two different sizes of gold nanoparticle, as the basis for the oligonucleotide–nanoparticle probes, in the assay system detailed previously.[62] Owing to the differing sizes of the nanoparticles, the surface plasmon of each had a different λ_{max} value. This property was exploited, with each nanoparticle probe thereby giving a different colour of scattered light. In the presence of both targets, signals for both respective nanoparticle probes were observed. Selective independent detection of each of the targeted DNA sequences was also performed, resulting in only the signal for the nanoparticle probe corresponding to the target present being seen.

Expanding further on the concept of multiplexed DNA detection, Mirkin *et al.* detailed the use of Raman reporters in tandem with DNA–nanoparticle

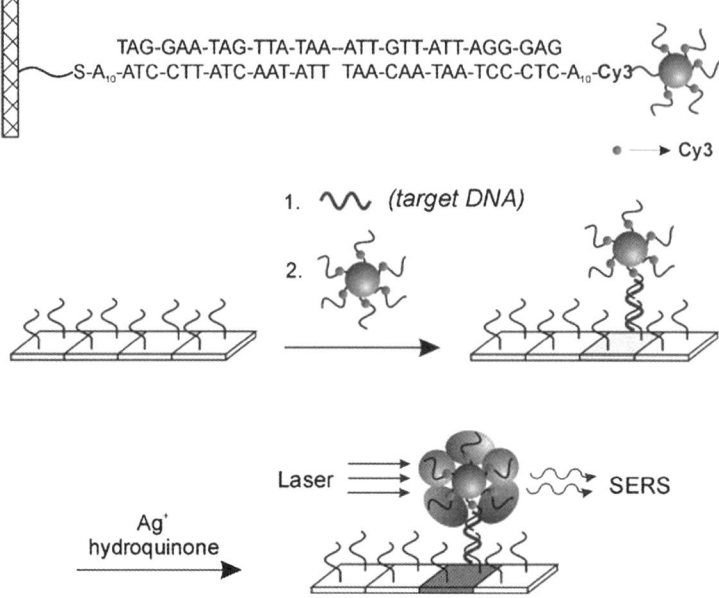

Figure 11.4   (Top) SERS active nanoparticles are immobilised on the surface of the
glass slide in the presence of target DNA by formation of a target-probe
sandwich, (bottom) target DNA is immobilised by the capture oligonucleotides bound to the slide surface, followed by the hybridisation of the
dye-labelled nanoparticle probes. The gold nanoparticles are subjected to
shelling by silver ions in the presence of hydroquinone, resulting in
amplification of the SERS signal.
(From Taton, T. A., Mirkin, C. A., Letsinger, R. L., *Science* 2000, *289*
(5485), 1757–1760. Reprinted with permission from AAAS.)

probes. These probes were designed for parallel detection of six different
nanoparticle probes, each relating to a different target sequence, using SERS as
a detection technique.[64] Gold nanoparticles were functionalised with Raman
dye-labelled oligonucleotides. Six target DNA sequences were selected, with a
Raman dye designated to a complementary probe for each target. Successful
detection of all six probes simultaneously was reported, as well as varying
combinations of the six probes. Furthermore, it was shown that two targets
could be differentiated in a semi-quantitative manner by monitoring the ratios
of the major peaks from the two dyes relating to the targets present.

11.2.2.2   *DNA Detection on Smooth Metal Films*

Recently, Braun *et al.* published details of the utilisation of a smooth metal film
in conjunction with nanoparticles for the detection of a specific DNA sequence
by SERS.[65] The assay was based upon the prediction that the junctions between
nanoparticles and smooth metallic surfaces serve as SERS hot spots.

Silver nanoparticles were functionalised with a probe DNA sequence corresponding to the complement of one half of the target DNA being detected. A smooth, non-SERS active silver film was functionalised with a DNA probe corresponding to the complement of the other half of the target DNA strand, illustrated in Figure 11.5. The silver film was capped with 6-mercaptohexanol (MCH) to prevent non-specific binding and aid the DNA hybridisation. The film was then functionalised with a thiol-modified SERS dye, 5-((2-(and 3)-S-(acetylmercapto)succinoyl)amino) fluorescein (SAMSA fluorescein). The functionalised film was incubated with the target DNA strand, followed by the oligonucleotide–nanoparticle probes. As shown in Figure 11.5, the target DNA was captured onto the silver film surface, followed by binding of the silver nanoparticle to the target DNA. This resulted in the nanoparticles becoming bound, in close proximity, to the surface of the silver film. The surface was washed to remove any nanoparticles that had not been immobilised on the surface.

In the presence of target DNA, a strong SERS signal was observed corresponding to surface bound SERS dye, indicating the immobilisation of the silver nanoparticles on the silver film. Control experiments demonstrated that, in the absence of the functionalised nanoparticles, the captured target DNA does not give any SERS signal when immobilised on the dye-modified silver film. This is due to the lack of roughness of the surface; roughness is required in order for SERS signals to be observed. No observable SERS signal was observed when the functionalised silver film was incubated with

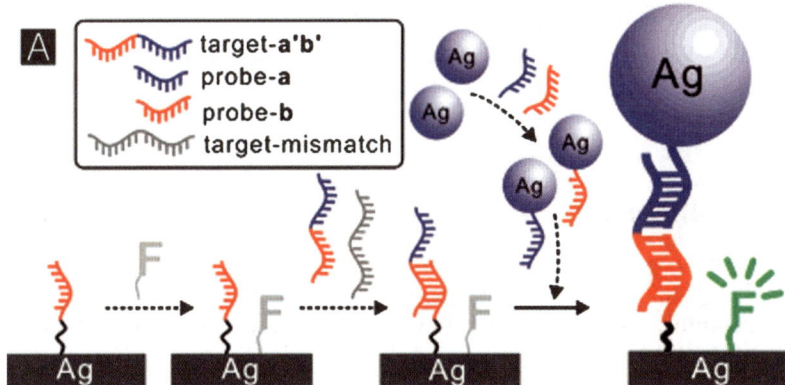

Figure 11.5 Schematic illustration for the detection of single-stranded DNA by SERS. The target DNA strand (a'b') is immobilised by the capture probe (b), and the nanoparticle probe (a) is in turn hybridised to the target strand. This results in the production of SERS hot spots between the nanoparticle and the silver surface. The presence of target DNA is indicated by the collection of SERS spectra corresponding to the surface-bound SERS dye (F).
(Reprinted with permission from ref. 65. Copyright 2007 American Chemical Society.)

the nanoparticle probes in the absence of target DNA, or when non-complementary DNA was used. This is due to the inability of the nanoparticle probes to bind with the flat surface, resulting in them being removed during washing.

Whilst the groups mentioned previously have had success using nanoparticle conjugates alongside a smooth, non-SE(R)RS active surface, research has also been conducted into the use of a bulk substrate as the roughened metal surface, functionalised with oligonucleotide probe sequences. In this work an oligonucleotide functionalised bulk substrate replaces the use of oligonucleotide–nanoparticle conjugates to provide detection of target DNA.

11.2.2.3   DNA Detection on a Gene Chip

In 2002, Vo-Dinh and co-workers published a SERS hybridisation assay for the detection of a breast cancer susceptibility gene, *BRCA1*.[66] A glass surface was etched with hydrogen fluoride, creating micro-wells, then coated with a 9 nm thick coating of silver. A monolayer of alkylmercaptans was added to the silver film, before coupling of an amino-modified capture oligonucleotide to the carboxylic acid group of the mercaptoundeconoic acid. A detection probe was utilised, labelled with Rhodamine B. The sum of the two probes was the exact complement of the target *BRCA1* DNA sequence. In the presence of target DNA, the dye-labelled detection probe was successfully immobilised to the silver surface via hybridisation of the capture and detection probes to the target DNA. Therefore, a clear SERS spectrum relating to the peaks of Rhodamine B was observed. In the absence of target DNA, however, no significant SERS peaks were seen. It was calculated that the SERS enhancement for this protocol is capable of observing in the region of 10^8 oligonucleotides in the 100 µm diameter focal area.

11.3   Conclusion

This chapter has demonstrated the vast array of research being conducted in the field of DNA detection by SE(R)RS using oligonucleotide conjugates. The most commonly used approaches have been discussed and should provide the reader with the highlights and point them in the direction of further, more detailed reading. This is a rapidly growing research area, which has seen a huge number of developments in recent years. Detection limits which could only be achieved by the use of a sample amplification technique can now be achieved by many of these spectroscopic assays. Coupled with this, there is the potential for high order multiplexing, owing to the vibrational nature of the spectra, where several sequences can be detected simultaneously without separation steps. Further research in this area is on-going, with major developments and improvements being made every day which will contribute to the sustainability of the field.

11.4 Forward Outlook

DNA–nanoparticle conjugates are at the forefront of DNA detection technology, with numerous new methods for selective target DNA detection being published annually. The limits of detection reported for many of these detection strategies make them competitive in terms of sensitivity with current detection protocols. Generally, these techniques could be more desirable than current protocols because they are capable of detecting DNA at comparable levels, without complex handling steps. This offers the opportunity for new methods of analysis to be developed that make use of the technique in combination with sophisticated sample manipulation approaches.

With high research volume in this area, it is important now to integrate this research into 'real world' applications. In order for this to be possible, it is important for the detection assays being developed to be able to detect, with a high level of sensitivity, in complex biofluids, as well as in idealised laboratory conditions. Biological samples will contain a number of interferrants and non-target DNA sequences. If these detection systems are to be used as diagnostic tests, it is vital that they can retain their selectivity and sensitivity in these challenging conditions. That said, the use of SERRS as a diagnostic technique is starting to emerge as a realistic option for use in clinical applications and we fully expect to see significant advances in this direction in the near future.

References

1. S. Yang and R. E. Rothman, *Lancet Infect. Dis.*, 2004, **4**, 337–348.
2. A. Linacre and D. Graham, *Expert Rev. Molec. Diag.*, 2002, **2**, 346–353.
3. R. Higuchi, G. Dollinger, P. S. Walsh and R. Griffith, *Bio-Technol.*, 1992, **10**, 413–417.
4. H. Li, L. Ying, J. J. Green, S. Balasubramanian and D. Klenerman, *Anal. Chem.*, 2003, **75**, 1664–1670.
5. R. A. Keller, W. P. Ambrose, P. M. Goodwin, J. H. Jett, J. C. Martin and M. Wu, *Appl. Spect.*, 1996, **50**, 12A–32A.
6. W. E. Moerner and D. P. Fromm, *Rev. Sci. Instr.* 2003, **74**, 3597–3619.
7. K. Kneipp, Y. Wang, H. Kneipp, L. T. Perelman, I. Itzkan, R. Dasari and M. S. Feld, *Phys. Rev. Lett.*, 1997, **78**, 1667–1670.
8. S. Nie and S. R. Emory, *Science*, 1997, **275**, 1102–1106.
9. H. X. Xu, E. J. Bjerneld, M. Kall and L. Borjesson, *Phys. Rev. Lett.*, 1999, **83**, 4357–4360.
10. J. A. Dieringer, R. B. Lettan, K. A. Scheidt and R. P. Van Duyne, *J. Am. Chem. Soc.*, 2007, **129**, 16249–16256.
11. E. C. Le Ru, M. Meyer and P. G. Etchegoin, *The Journal of Physical Chemistry B*, 2006, **110**, 1944–1948.
12. K. Faulds, R. P. Barbagallo, J. T. Keer, W. E. Smith and D. Graham, *Analyst*, 2004, **129**, 567–568.
13. J. A. Dougan and K. Faulds, *Analyst*, 2012, **137**, 545–554.

14. M. Fleischman, P. J. Hendra and A. J. McQuilla, *Chem. Phys. Lett.*, 1974, **26**, 163–166.

15. D. L. Jeanmaire and R. P. Van Duyne, *J.f Electroanal. Chem.*, 1977, **84**, 1–20.

16. G. M. Albrecht and A. J. Creighton, *J. Am. Chem. Soc.*, 1977, **99**, 5215–5217.

17. W. E. Smith, *Chem. Soc. Rev.*, 2008, **37**, 955–964.

18. A. R. Tao and P. D. Yang, *J. Phys. Chem. B.*, 2005, **109**, 15687–15690.

19. K. Kneipp, H. Kneipp, I. Itzkan, R. R. Dasari and M. S. Feld, *Chem. Rev.*, 1999, **99**, 2957.

20. B. N. J. Persson, *Chem. Phys. Lett.*, 1981, **82**, 561–565.

21. L. Brus, *Acc. Chem. Res.*, 2008, **41**, 1742–1749.

22. L. X. Chen and J. B. Choo, *Electrophoresis.*, 2008, **29**, 1815–1828.

23. P. Kambhampati, O. K. Song and A. Campion, *Phys. Stat. Solid.-Appl. Res.*, 1999, **175**, 233–239.

24. P. Hildebrandt, S. Keller, A. Hoffmann, F. Vanhecke and B. Schrader, *J. Raman Spect.*, 1993, **24**, 791–796.

25. P. Hildebrandt, A. Epding, F. Vanhecke, S. Keller and B. Schrader, *J. Molec. Struct.*, 1995, **349**, 137–140.

26. A. Kudelski and J. Bukowska, *Chem. Phys. Lett.*, 1996, **253**, 246–250.

27. I. T. Shadi, B. Z. Chowdhry, M. J. Snowden and R. Withnall, Spect Acta A-Molec, *Biomolec. Spect.y.*, 2003, **59**, 2213–2220.

28. Z.-Q. Tian, B. Ren and D.-Y. Wu, *J. Phys. Chem. B.*, 2002, **106**, 9463–9483.

29. M. Moskovits, *Rev. Mod.n Phys.*, 1985, **57**, 783–826.

30. E. J. Zeman and G. C. Schatz, *J. Phys. Chem.*, 1987, **91**, 634–643.

31. P. C. Lee and D. Meisel, *J. Phys. Chem.*, 1982, **86**, 3391–3395.

32. P. Hildebrandt and M. Stockburger, *J. Phys. Chem.*, 1984, **88**, 5935–5944.

33. J. A. Creighton, C. G. Blatchford and M. G. Albrecht, *Journal of the Chemical Society, Faraday Transactions 2: Molecular and Chemical Physics*, 1979, **75**, 790–798.

34. K. Kneipp, R. R. Dasari and Y. Wang, *Appl. Spectrosc.*, 1994, **48**, 951–955.

35. B. Ren, X.-F. Lin, Z.-L. Yang, G.-K. Liu, R. F. Aroca, B.-W. Mao and Z.-Q. Tian, *J. Am. Chem.Soc.*, 2003, **125**, 9598–9599.

36. R. J. Stokes, A. Macaskill, P. J. Lundahl, W. E. Smith, K. Faulds and D. Graham, *Small*, 2007, **3**, 1593–1601.

37. G. C. Weaver and K. Norrod, *J. Chem. Educ.*, 1998, **75**, 621.

38. S. M. Heard, F. Grieser, C. G. Barraclough and J. V. Sanders, *J. Colloid Interface Sci.*, 1983, **93**, 545–555.

39. S. P. Mulvaney, M. D. Musick, C. D. Keating and M. J. Natan, *Langmuir.*, 2003, **19**, 4784–4790.

40. W. E. Doering and S. M. Nie, *Anal. Chem.*, 2003, **75**, 6171–6176.

41. L. Sun, C. Yu and J. Irudayaraj, *Anal. Chem.*, 2007, **79**, 3981–3988.

42. M. Gellner, K. Kömpe and S. Schlücker, *Anal. Bioanal. Chem.*, 2009, **394**, 1839–1844.

43. J. Wrzesien and D. Graham, *Tetrahedron.*, 2012, **68**, 1230–1240.
44. M. B. Wabuyele and T. Vo-Dinh, *Anal. Chem.*, 2005, **77**, 7810–7815.
45. H. N. Wang and T. Vo-Dinh, *Nanotechnology*, 2009, 20.
46. K. Faulds, L. Fruk, D. C. Robson, D. G. Thompson, A. Enright, W. Ewen Smith and D. Graham, *Faraday Disc.*, 2006, **132**, 261–268.
47. S. Tyagi and F. R. Kramer, *Nat. Biotechnol.*, 1996, **14**, 303–308.
48. D. G. Thompson, A. Enright, K. Faulds, W. E. Smith and D. Graham, *Anal. Chem.*, 2008, **80**, 2805–2810.
49. C. A. Mirkin, R. L. Letsinger, R. C. Mucic and J. J. Storhoff, *Nature.*, 1996, **382**, 607–609.
50. D. Graham, D. G. Thompson, W. E. Smith and K. Faulds, *Nat. Nanotech.*, 2008, **3**, 548–551.
51. D. G. Thompson, K. Faulds, W. E. Smith and D. Graham, *J. Phys. Chem. C.*, 2010, **114**, 7384–7389.
52. F. McKenzie, K. Faulds and D. Graham, *Nanoscale*, 2010, **2**, 78–80.
53. F. McKenzie and D. Graham, *Chem. Comm.*, 2009, 5757–5759.
54. H. Zhang, M. H. Harpster, H. J. Park, P. A. Johnson and W. C. Wilson, *Anal. Chem.*, 2011, **83**, 254–260.
55. F. Michael, *Curr. Opin. Struct. Biol.*, 1999, **9**, 324–329.
56. E. N. Brody and L. Gold, *Rev. Molec. Biotechnol.*, 2000, **74**, 5–13.
57. S. D. Jayasena, *Clin. Chem.*, 1999, **45**, 1628–1650.
58. L. Griffin, G. Tidmarsh, L. Bock, J. Toole and L. Leung, *Blood*, 1993, **81**, 3271–3276.
59. V. Pavlov, Y. Xiao, B. Shlyahovsky and I. Willner, *J. Am. Chem. Soc.*, 2004, **126**, 11768–11769.
60. D. Graham, R. Stevenson, D. G. Thompson, L. Barrett, C. Dalton and K. Faulds, *Faraday Disc.*, 2011, **149**, 291–299.
61. S. P. Ravindranath, Y. Wang and J. Irudayaraj, Sensors Actuators B, *Chem.*, 2011, **152**, 183–190.
62. T. A. Taton, C. A. Mirkin and R. L. Letsinger, *Science*, 2000, **289**, 1757–1760.
63. T. A. Taton, G. Lu and C. A. Mirkin, *J. Am. Chem. Soc.*, 2001, **123**, 5164–5165.
64. Y. W. C. Cao, R. C. Jin and C. A. Mirkin, *Science*, 2002, **297**, 1536–1540.
65. G. Braun, S. J. Lee, M. Dante, T.-Q. Nguyen, M. Moskovits and N. Reich, *J. Am. Chem. Soc.*, 2007, **129**, 6378–6379.
66. L. R. Allain and T. Vo-Dinh, *Anal. Chim. Act.*, 2002, **469**, 149–154.

CHAPTER 12

Covalent and Non-covalent Conjugates of Oligonucleotides as Artificial Restriction DNA Cutters

MAKOTO KOMIYAMA,*[a] YAN XU[b] AND
JUN SUMAOKA[a]

[a] Life Science Center of Tsukuba Advanced Research Alliance, University of Tsukuba, Ten-noudai 1-1-1, Tsukuba, Ibaraki 305-8577, Japan; [b] Division of Chemistry, Department of Medical Sciences, Faculty of Medicine, University of Miyazaki, 5200 Kihara, Kiyotake, Miyazaki 889-1692, Japan
*Email: komiyama@tsukuba.tara.ac.jp

12.1   Introduction

Significant roles of artificial DNA cutters, which have freely tunable scission sites and higher sequence-specificity than naturally occurring enzymes, are widely recognised. Many attempts to conjugate appropriate catalysts with sequence-recognising oligonucleotides have been made, providing a variety of fruitful results.[1–7] However, practical and useful tools for the manipulation of huge DNA are scarce. Our group has been working on this project for a long time, and has recently developed chemistry-based cutters which selectively cut DNA at targeted sites via hydrolysis of the phosphodiester linkage.[8–10] Our strategy is based on: (1) differentiation of targeted scission sites from other

RSC Biomolecular Sciences No. 26
DNA Conjugates and Sensors
Edited by Keith R Fox and Tom Brown
© The Royal Society of Chemistry 2012
Published by the Royal Society of Chemistry, www.rsc.org

parts in terms of their reactivity and (2) selective scission of the targeted site by a catalyst which sufficiently recognises the difference in reactivity (see Figure 12.1). Factor (1) is accomplished by oligonucleotides for site-selective hydrolysis of single-stranded DNA substrate, or by peptide nucleic acid (PNA)[11] for site-selective scission of double-stranded DNA. On the other hand, factor (2) is fulfilled by using Ce(IV)/EDTA[12] as the catalyst for hydrolytic DNA scission. One of the most important advantages of this strategy is the versatility of the DNA substrate employed. Because all that is needed for site-selective scission is a single-stranded portion to serve as a hot spot and Ce(IV)/EDTA as a catalyst, a variety of DNA (*e.g.* telomeric overhang DNA) can be the substrate. In principle, covalent attachment of the catalyst is not required for site-selective scission. However, the scission can of course be made still more clear-cut by covalently fixing the catalyst to the targeted scission site. These tools cut DNA via hydrolysis of phosphodiester bonds, exactly as naturally occurring nucleases do. The scission products are directly susceptible to transformation by various enzymes, so that these cutters match current tools used in molecular biology and biotechnology.

This article describes the efforts of our laboratory in the preparation of site-selective DNA cutters using various oligonucleotide conjugates. Single-stranded DNA is primarily the substrate for the scission.[13–17] The cutters are valuable tools for molecular biology and biotechnology, because there are no naturally occurring enzymes which cut single-stranded DNA selectively at specific sequences. Furthermore, virtually the same principle is applicable to site-selective scission of double-stranded DNA.[18–20] In this application, PNA is used in place of an oligonucleotide, to bind more strongly with double-stranded

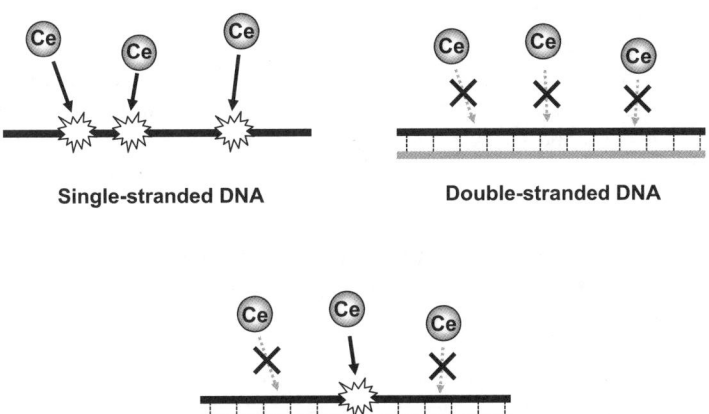

Figure 12.1 The strategy of site-selective hydrolysis of single-stranded DNA by Ce(IV)/EDTA. Only the targeted site is kept single stranded and hydrolysed by Ce(IV)/EDTA. For site-selective scission of double-stranded DNA, a single-stranded portion is formed by use of PNA strands at the predetermined site of each strand and hydrolysed by Ce(IV)/EDTA (see text for details).

DNA and to recognise a predetermined site therein. Recent attempts to prepare a second-generation artificial cutter from Ce(III) are also presented.[21]

12.2   Ce(IV)/EDTA as Molecular Scissors Showing Remarkable Substrate-Specificity

In early 1990s, we found that Ce(IV) effectively hydrolyses DNA. The rate constant for TpT hydrolysis at pH 7 and 50 °C is $1.9 \times 10^{-1}\,h^{-1}$ (the half-life 3.6 h).[22] The scission rate is hardly dependent on the identity of the adjacent nucleobases, and oligonucleotides are hydrolysed almost randomly throughout the main chain. The scission is completely hydrolytic so that the scission products are directly susceptible to transformation by various enzymes used in current molecular biology (*e.g.* alkaline phosphatase, polynucleotide kinase, and terminal deoxynucleotidyl transferase).[23] All these features render the Ce(IV) ion promising for use in the catalytic centre of artificial enzymes designed to hydrolyse DNA. However, this metal ion hydrolyses both single-stranded DNA and double-stranded DNA at almost the same rate, and has rather poor substrate specificity.

In contrast, the Ce(IV)/EDTA complex has a very unique substrate-specificity (Figure 12.1).[13,14] It efficiently hydrolyses the phosphodiester linkages in single-stranded DNA, but its catalytic activity for the hydrolysis of double-stranded DNA is minimal. Apparently, this homogeneous complex is highly potent as molecular scissors for site-selective scission of DNA [thus requirement (2) is satisfactorily fulfilled]. This substrate-specificity of the Ce(IV)/EDTA complex primarily originates from the difference in its binding activity for single-stranded DNA and double-stranded DNA. The K_m value for the Ce(IV)/EDTA-induced hydrolysis of single-stranded DNA is 10 μM, whereas the K_m value for the hydrolysis of double-stranded DNA under the same conditions is greater than 5 mM. Thus, single-stranded DNA is bound to Ce(IV)/EDTA more strongly (> 500-fold) than double-stranded DNA, and is efficiently hydrolysed.

12.2.1   Combination of Oligonucleotide Additives and Ce(IV)/EDTA for Site-Selective Scission of Single-Stranded DNA

The substrate specificity of Ce(IV)/EDTA described above (scission rate: single-stranded DNA ≫ double-stranded DNA) indicates that any target site in a single-stranded DNA substrate should be preferentially hydrolysed by this complex if this site only is kept single-stranded and the other parts are made double-stranded. In order to verify this hypothesis, a gap structure was formed at a predetermined site in single-stranded DNA by using two complementary oligonucleotides (Figure 12.2).[13,14] The products of Ce(IV)/EDTA scission were analysed by polyacrylamide gel electrophoresis. Exactly as designed, only the

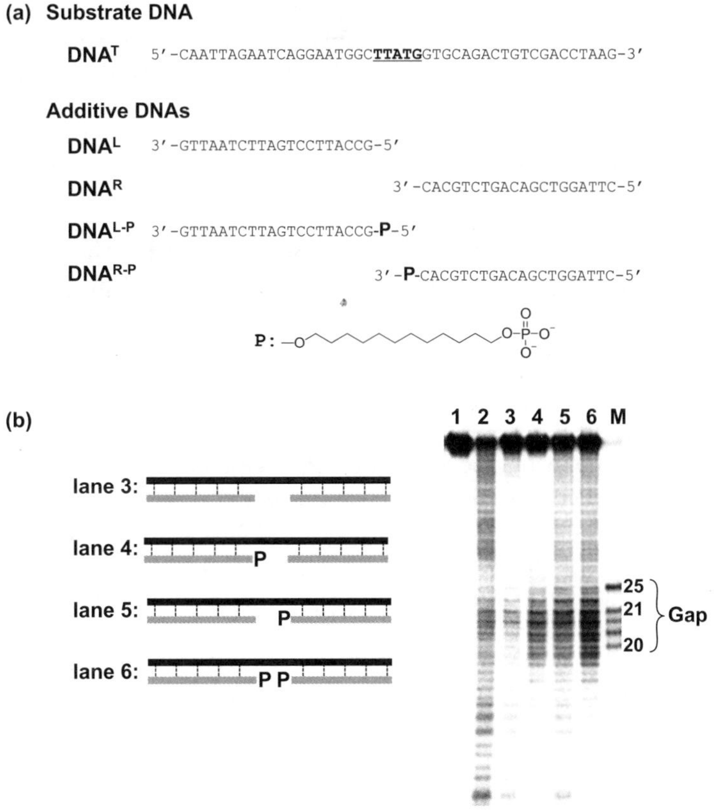

(a) **Substrate DNA**

   DNA^T   5'-CAATTAGAATCAGGAATGGC**TTATG**GTGCAGACTGTCGACCTAAG-3'

Additive DNAs

   DNA^L   3'-GTTAATCTTAGTCCTTACCG-5'

   DNA^R   3'-CACGTCTGACAGCTGGATTC-5'

   DNA^L-P 3'-GTTAATCTTAGTCCTTACCG-**P**-5'

   DNA^R-P   3'-**P**-CACGTCTGACAGCTGGATTC-5'

(b)

Figure 12.2 Site-selective hydrolysis of single-stranded DNA by combining Ce(IV)/ EDTA with various DNA additives. (a) Sequences of target DNA and additive oligonucleotides. The gap site formed in the substrate DNA^T is shown by the underlined bold characters. (b) Typical gel electrophoresis patterns: lane 1, DNA only; lane 2, only with Ce(IV)/EDTA; lane 3, DNA^L/ DNA^R with Ce(IV)/EDTA; lane 4, DNA^L-P/DNA^R with Ce(IV)/EDTA; lane 5, DNA^L/DNA^R-P with Ce(IV)/EDTA; lane 6, DNA^L-P/DNA^R-P with Ce(IV)/EDTA; M, markers. Reaction conditions: $[DNA^T] = 1.0\,\mu M$, [each of the DNA additives] $= 2.0\,\mu M$, [Ce(IV)/EDTA] $= 1.0\,mM$, [HEPES (pH 7.0)] $= 7.5\,mM$ and [NaCl] $= 100\,mM$ at 37 °C for 95 h.
(Reproduced by permission of the American Chemical Society from ref. 16.)

products for the site-selective scission of the gap site were formed (lane 3). Each of the bands corresponded to scission of the five phosphodiester linkages in the gap structure (the fragments for the scission of one phosphodiester linkage were further separated into two bands on the basis of whether the scission terminus was –OH or –phosphate). Other parts of the substrate DNA formed double strands with the additives, and were hardly hydrolysed at all.

Importantly, this selective scission was further promoted by attaching a monophosphate group to the 3'- or 5'-terminus of the oligonucleotide additives (lanes 4 and 5).[15,16] When these modified oligonucleotides formed duplexes with the substrate DNA, the monophosphate group at the termini was placed at the edge of the gap site. Simultaneous introduction of two monophosphates, at both edges of the gap, was still more effective (lane 6). These monophosphate groups recruit the catalytic Ce(IV)/EDTA to the targeted scission site, resulting in enhancements in both the rate of scission and the site selectivity.

12.2.2 Promotion of Site-Selective Scission by Conjugating Oligonucleotide with a Multiphosphonate Ligand

Far more efficient site-selective scission of single-stranded DNA was accomplished when multiphosphonate groups were attached, in place of the monophosphate group, to the oligonucleotide additives.[17] Especially eminent are nitrilotris(methylenephosphonic acid) (NTP) and *N,N,N′,N′*-ethylenediaminetetrakis(methylenephosphonic acid) (EDTP). These ligands were connected to the termini of oligonucleotides by oxime formation (Figure 12.3). In the final

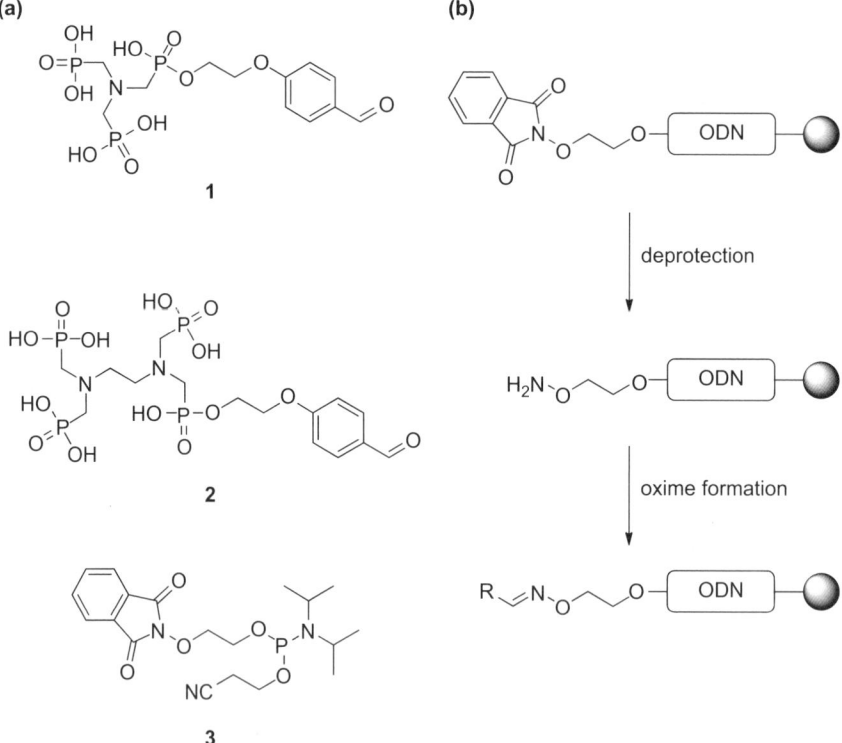

Figure 12.3   (a) 2-(4-Formylphenoxy)ethyl esters of NTP (**1**) and EDTP (**2**), and the protected amino-oxy linker (**3**) used to incorporate **1** and **2** into the oligonucleotide. (b) Preparation method for multiphosphonate–oligonucleotide conjugates on the support (R is NTP or EDTP).

stage of solid-phase synthesis of the oligonucleotides, a protected amino-oxy linker was attached to the terminus, and after the deprotection, reacted on the support with 2-(4-formylphenoxy)ethyl esters of NTP and EDTP. Because of the cooperation of three or four phosphonate groups, these ligands have very high affinity for Ce(IV) and strongly recruit Ce(IV)/EDTA to the target site.

Accordingly, site-selective scission was evident even when the Ce(IV)/EDTA concentration was low and the combination of two monophosphate–oligonucleotide conjugates provided no site-selective DNA scission (Figure 12.4). This argument was further supported by the dependence of the DNA scission efficiency on the Ce(IV)/EDTA concentration (Figure 12.5). With increasing Ce(IV)/EDTA concentration, the rate of scission by two NTP–oligonucleotide system (closed triangles) saturated at a much lower Ce(IV)/EDTA concentration than the rate for the two monophosphate–oligonucleotide system (closed circles). The saturation of the two EDTP–oligonucleotide system (closed squares) occurred at still lower Ce(IV)/EDTP concentration. It is interesting that notable acceleration by these multiphosphonate ligands (EDTP or NTP) was evident especially when two of them were simultaneously placed at the gap

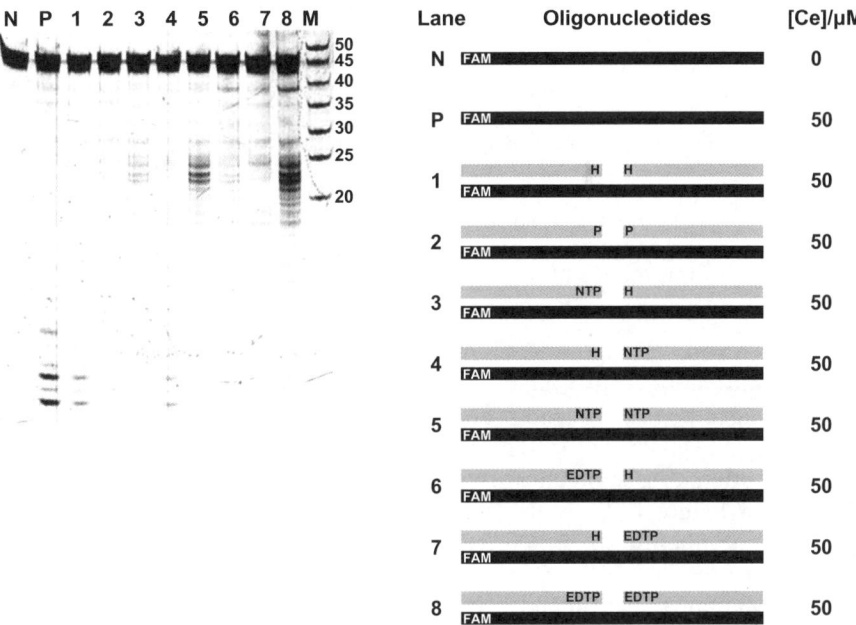

Figure 12.4 Polyacrylamide gel electrophoresis patterns for the hydrolysis of target DNA at a five-base gap by Ce(IV)/EDTA. The ligands used in each lane are presented on the right-hand side. Reaction conditions: [Ce(IV)/EDTA] = 50 μM, [substrate DNA] = 1 μM, [each of oligonucleotide additives] = 2 μM, [HEPES (pH 7.0)] = 7.5 mM and [NaCl] = 100 mM at 50 °C for 72 h.
(Reproduced by permission of The Royal Society of Chemistry from ref. 17.)

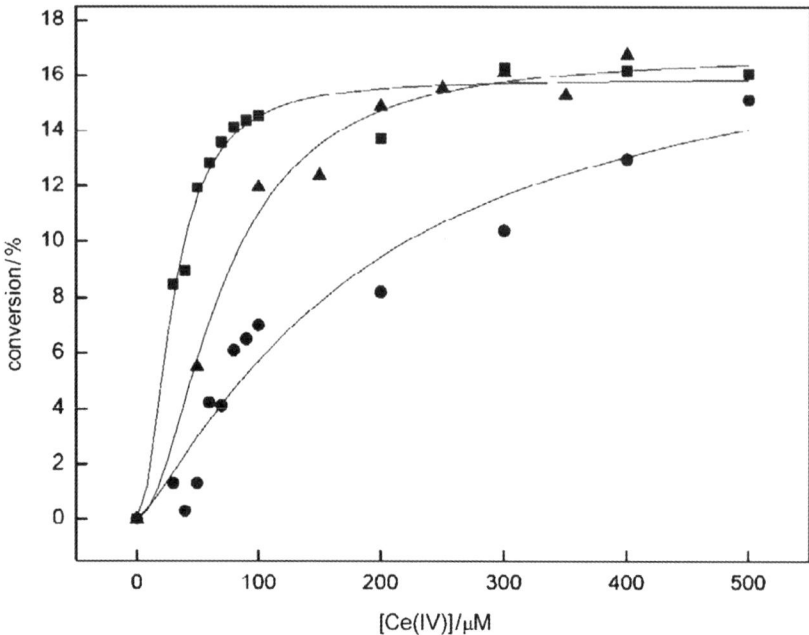

Figure 12.5   Conversion of gap-selective scission by Ce(IV)/EDTA as a function of
Ce(IV)/EDTA concentration: (■) the two EDTP system; (▲) the two
NTP system; (●) the two monophosphate system. A pair of these
ligands was placed at the edge of five-base gap, as presented in
lanes 8, 5, and 2 in Figure 12.4, respectively. Reaction conditions:
[target   DNA] = 1 μM,   [each   additive] = 2 μM,   [HEPES
(pH = 7.0)] = 7.5 mM, [NaCl] = 100 mM at 50 °C for 20 h.
(Reproduced by permission of The Royal Society of Chemistry
from ref. 17.)

site (lanes 5 and 8 in Figure 12.4). A single NTP or EDTP at either side of the
gap was not very effective (lanes 3, 4, 6 and 7). Apparently, these multiphosphonate ligands strongly (and probably cooperatively) draw the Ce(IV) species
to the gap site, giving rise to the ease of DNA scission.

These cutters can be practical tools for the manipulation of single-stranded
DNA.[24] In Figure 12.6, by using two EDTP–oligonucleotide conjugates the
gene of blue fluorescent protein (BFP) was converted to the gene of green
fluorescent protein (GFP). To the termini of two short oligonucleotides (20-
and 21-mer), an EDTP ligand was attached through an oxime linkage. The
oligonucleotide portions are complementary with the C225–G244 and the
T250–G270 sequences in the sense strand of BFP (838-mer), respectively. By
mixing these additives and the target DNA, a 5-base pair gap structure was
formed between T245 and A249, which is just before the chromophore-coding
region. The two EDTP ligands were located at the edges of this gap. When this
system was treated with Ce(IV)/EDTA, two scission bands (around 250- and
590-mer) were formed, as expected from the selective scission at the gap site.

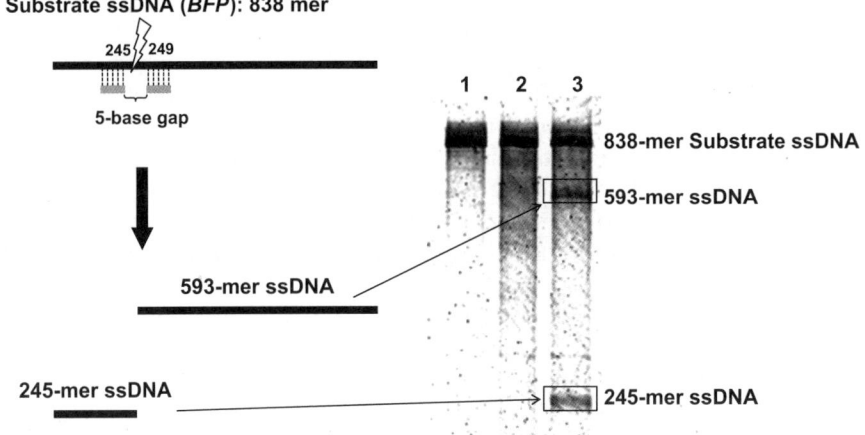

Figure 12.6 Site-selective scission of sense-strand of BFP gene (838-mer) by combining Ce(IV)/EDTA and two EDTP–oligonucleotides. Lane 1, DNA only; Lane 2, DNA + Ce(IV)/EDTA (without EDTP–oligonucleotides); Lane 3, DNA + Ce(IV)/EDTA + EDTP–oligonucleotides. Reaction conditions: [DNA] = 0.1 μM, [each EDTP–oligonucleotide] = 0.5 μM, [HEPES (pH 7.0)] = 5 mM, [NaCl] = 100 mM, [Ce(IV)/EDTA] = 5 μM at 50 °C for 17 h.

It should be noted that most of the substrate DNA on both sides of the gap site (G1–C224 and A271–C838) was single stranded, but these flanking portions were not hydrolysed to a measurable extent by the Ce(IV)/EDTA and only the gap site was hydrolysed. The two EDTP ligands placed at the gap site strongly attract Ce(IV)/EDTA to the target site, and thus undesired off-target scission was effectively suppressed. Consistently, with the use of two oligonucleotides bearing no EDTP, these single-stranded parts were also hydrolysed and the products of site-selective scission were never observed. The upstream fragment of the BFP gene (G1–T245), obtained by the site-selective scission, was ligated with the downstream fragment of GFP (G246–C838) which was separately prepared. The recombinant single-stranded DNA thus prepared was converted to double-stranded form, and integrated into plasmid DNA. The corresponding protein was successfully expressed in *E. coli* and emitted green fluorescence. According to sequencing experiments, no mutation was detected except for the chromophore-coding site. This cutter (the combination of two short EDTP–oligonucleotides and Ce(IV)/EDTA) should be useful in the manipulation of long single-stranded DNA (for example, the genome of adeno-associated virus, which is often used as a vector).

12.2.3 Scission of Human Telomeric Overhang DNA

Human telomeric DNA consists of a duplex region composed of TTAGGG repeats, ending in a G-rich single-stranded overhang of 100–200 nucleotides.[25–30]

The overhang is required for telomere end protection and is regarded as a crucial component of telomere structure.[25,31–33] This overhang is also a substrate of telomerase, which elongates the telomeric sequence by adding G-rich repeats.[25,33,34] In 80–90% of human tumours telomerase is activated, but it is low or undetectable in most normal somatic cells.[35] Thus, the development of approaches to target human telomeric DNA is an area of great interest because of its potential as a target for the discovery of anticancer agents.[35–51] A promising method is to disrupt the telomere–telomerase interaction via targeting the telomere DNA substrate, thereby preventing telomere elongation by telomerase.[52,53]

As is well known, G-rich human telomeric DNA folds into four-stranded G-quadruplex structures.[54,55] Furthermore, a new topology containing the $(3+1)$ G-tetrad core was recently reported for human telomeric G-quadruplex.[56–59] It was also shown that a dimeric G-quadruplex with the same topology is formed from a three-repeat human telomeric sequence and another single-repeat human telomeric sequence (Figure 12.7a).[60] This dimeric G-quadruplex structure suggests a way in which a segment of three G-tracts can bind to a remote G-tract. This pioneering work led us to develop a novel approach toward the

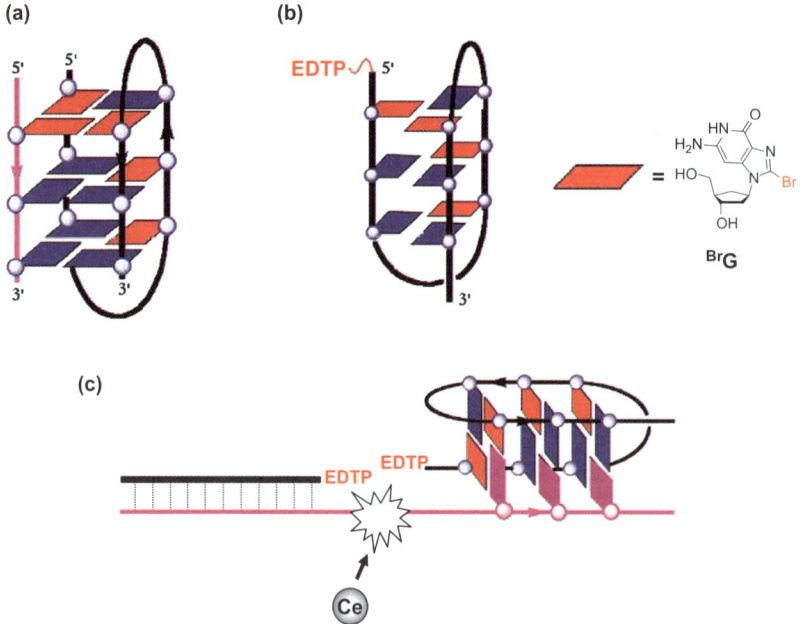

Figure 12.7   (a) Schematic structure of the $(3+1)$ core dimeric G-quadruplex formed by a three-repeat human telomeric sequence and a single-repeat telomeric sequence. (b) Three-repeat oligonucleotide equipped with EDTP at the 5′ end. The red boxes show the substitutions of dG in the *syn* conformation with 8-bromoguanosine (BrG). (c) Sequence-specific cleavage of human telomere by G-quadruplex-forming DNA–EDTP and Ce(IV)/EDTA.

scission of human telomere DNA substrate. Our strategy involves cleavage of the human telomeric DNA sequence by forming a G-quadruplex between an oligonucleotide containing three G-tracts and targeted telomeric DNA (Figure 12.7c).[61] As mentioned above, a three-repeat human telomeric sequence (16-mer) can form a dimeric G-quadruplex with a single-repeat telomeric sequence (6-mer) in a human telomere DNA substrate. To this an oligonucleotide containing three G-tracts and an EDTP ligand, which efficiently recruits Ce(IV)/ EDTA to the targeted scission site, was covalently bound (Figure 12.7b). This conjugate should bind to the targeted human telomere DNA by G-quadruplex formation and cause a sequence-specific strand break. Formation of a stable G-quadruplex between the EDTP-modified oligonucleotide and the target telomere sequence is a key part of the present strategy. Accordingly, in order to stabilise (or 'freeze') the G-quadruplex structure, four dG units, which must take the *syn* conformation in the G-quadruplex, were substituted with *syn*-preferring 8-bromoguanosine (^{Br}G) (see Figure 12.7).[56,57,62] These ^{Br}G substitutions at the *syn* positions in the EDTP-modified oligonucleotide significantly increased the thermal stability of the G-quadruplex when compared with the parent unsubstituted EDTP–oligonucleotide conjugate.[61]

The sequence-specific hydrolysis of human telomeric DNA was examined on a 5′-end FAM-labelled 29-mer DNA substrate containing a 7-mer human telomeric sequence. The EDTP–oligonucleotide conjugates (with ^{Br}G substitutions) and 5′-end labelled DNA were incubated with 10 µM Ce(IV)/EDTA and 200 mM KCl in 10 mM HEPES (pH 7.0) for 50 h (Figure 12.8). As shown in lane 2, the telomere DNA was notably cleaved. In the absence of the EDTP–oligonucleotide conjugate, however, no cleavage occurred even when Ce(IV)/EDTA was added. A mutated random substrate DNA that did not contain the telomere sequence was never cleaved. All these results confirmed that the present DNA cleavage results from the formation of a G-quadruplex structure.

In practical applications, the secondary or tertiary structure of the telomeric DNA substrate could hinder the binding of this cutter to target DNA (for example, the telomere could take an intramolecular G-quadruplex structure via a four-repeat telomere sequence). In order to shed light on this subject, a longer human telomeric fragment [a 45-mer DNA substrate containing (TTAGGG)$_4$ was treated with this cutter (Figure 12.9)]. Even though this substrate could form some complicated structures, one site at the 5′ end of the four-repeat telomeric sequence was successfully cleaved (lane 3). This demonstrates that EDTP-modified oligonucleotides containing ^{Br}Gs can form a stable G-quadruplex even with a long telomere DNA target, by interrupting the self-formed intramolecular G-quadruplex structure in the substrate DNA, and induce telomere sequence-specific cleavage by Ce(IV)/EDTA. Targeting of the human telomere DNA substrate through G-quadruplex formation has been demonstrated for the first time. This approach to prevention of telomere elongation is conceptually different from previous work, in that the formation of a stable G-quadruplex is used directly to cut telomere DNA.[63] In most traditional studies, telomere G-quadruplex structures were stabilised by small molecules.

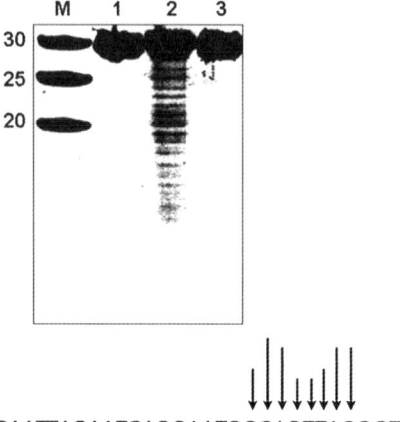

lane 2: 5'FAM-CAATTAGAATCAGGAATGGCAGT<u>TAGGGT</u>-3'

lane 3: 5'FAM-CAATTAGAATCAGGAATGGCAGCGTACAG-3'

Figure 12.8   Sequence-specific cleavage of human telomere by G-quadruplex formation: (lane 1) Ce(IV)/EDTA without EDTP–oligonucleotide; (lane 2) with
EDTP–oligonucleotide and Ce(IV)/EDTA; (lane M) marker oligonucleotides. The substrate is 5'-end FAM-labelled 29-mer DNA containing
a 7-mer human telomere sequence (underlined). In lane 3, a 29-mer DNA
containing no telomere sequence was treated with EDTP–oligonucleotide and Ce(IV)/EDTA.
(Reproduced by permission of the American Chemical Society from
ref. 61.)

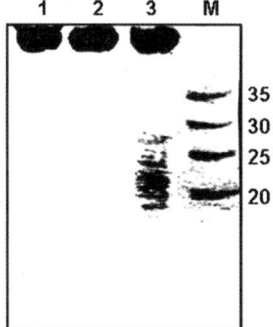

5'FAM-CAATTAGAATCAGGAATGGCA<u>TTAGGGTTAGGGTTAGGGTTAGGG</u>-3'

Figure 12.9   Sequence-specific cleavage of long human telomere strand by G-quadruplex formation: (lane 1) without both EDTP–oligonucleotide and
Ce(IV)/EDTA; (lane 2) without EDTP–oligonucleotide but with Ce(IV)/
EDTA; (lane 3) with both EDTP–oligonucleotide and Ce(IV)/EDTA;
(lane M) marker oligonucleotides. The substrate is 5'-end FAM-labelled
45-mer DNA containing the d(TTAGGG)₄ human telomere sequence at
the 3'-end (underlined).
(Reproduced by permission of the American Chemical Society from
ref. 61.)

12.2.4 Use of Peptide Nucleic Acid (PNA) to Prepare Artificial Enzymes for Site-Selective Scission of Double-Stranded DNA

An artificial restriction DNA cutter (ARCUT) for the site-selective scission of double-stranded DNA has recently been developed by our group (Figure 12.10).[18–20,64–68] This cutter is also based on the substrate-specificity of Ce(IV)/EDTA, which efficiently hydrolyses only single-stranded DNA. In order to form the hot spots for the scission (single-stranded portions) in double-stranded DNA, two strands of pseudo-complementary peptide nucleic acid (pcPNA) are used, because oligonucleotides cannot bind sufficiently well to double-stranded DNA.[69–71] The neutral backbone of pcPNA, which is in contrast with the negatively charged backbone of oligonucleotides, is the primary driving force of this effective binding. Two pcPNA strands are designed to be laterally shifted to one another by five nucleobases. When these two pcPNAs invade double-stranded DNA to form a so-called double-duplex invasion complex, a gap-like structure (unpaired single-stranded portion) is formed at the predetermined site in each of the two strands of the double-stranded DNA. These single-stranded portions are preferentially hydrolysed by Ce(IV)/EDTA, leading to the site-selective scission of the double-stranded DNA. With this new tool, even the whole human genome could be selectively cut at one site.[67] Furthermore, various

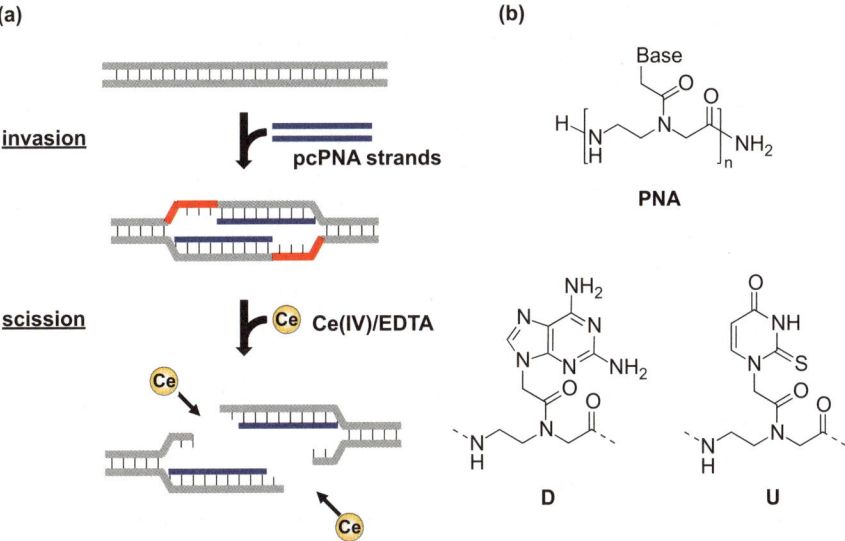

Figure 12.10 (a) Scheme of site-selective scission of double-stranded DNA by ARCUT. Two single-stranded portions (red lines) are formed by the invasion of pcPNAs (blue lines) and preferentially hydrolysed by Ce(IV)/EDTA. (b) The chemical structures of PNA, and pseudo-complementary bases [2,6-diaminopurine (D) and 2-thiouracil (U)] used for pcPNA. Note that D and U are used simply to promote the invasion process (details are shown in refs 69–71).

kinds of biotechnological application, including homologous recombina-tion,[72,73] have been accomplished (a comprehensive review of these applications is beyond the scope of this article and readers are referred to recent reviews[9,10]).

Interestingly and importantly, the scission efficiency of this cutter is sig-nificantly enhanced by attaching a monophosphonate group to the pcPNA strands (a phosphoserine is bound to the pcPNA).[20] These monophosphate groups recruit Ce(IV)/EDTA to the target site, exactly as described for the site-selective scission of single-stranded DNA in Figure 12.2. The mechanistic similarity between the single-stranded DNA cutters and the double-stranded DNA cutters is evident. Thus, the information required to improve the effi-ciency of a double-stranded DNA cutter can be obtained from detailed analysis of the single-stranded DNA cutter (and *vice versa*).

12.3 Preparation of an Artificial DNA Cutter from Ce(III) as a Precursor of Catalytically Active Ce(IV)

In the work described in previous sections, undesired scission at off-target sites was suppressed by localising Ce(IV)/EDTA near the target site with the use of various ligands. High scission yield and specificity, which are sufficient for various applications, have been satisfactorily accomplished. For still more complicated DNA manipulation, especially *in vivo*, however, further improvements of the site selectivity and more complete suppression of off-target scissions are desirable. One of the most promising ways should be to bind catalytically active Ce(IV) covalently to oligonucleotides. In this section, pre-pared a second-generation site-selective DNA cutter is described; the method uses a Ce(III) salt as the starting material and a Ce(IV) complex is attached firmly to oligonucleotide additives.[21] Compared with Ce(IV), Ce(III) is markedly less prone to gel formation, and more tractable. Although Ce(III) itself hardly hydrolyses DNA, it can be promptly oxidised by molecular oxygen under neutral conditions and converted to catalytically active Ce(IV).[74,75] These fea-tures indicate that a well-characterised Ce(IV) complex should be obtainable by oxidation of an appropriate Ce(III) complex. However, in order for this pro-cedure to be successful, the Ce(III) ions [and the resultant Ce(IV) ions] must be sufficiently separated from each other during the oxidation and their mutual aggregation must be avoided.

It was found that EDTP is the most appropriate ligand to separate the Ce(III) ions from each other [and from Ce(IV) ions also]. Site-selective scission of single-stranded DNA by this cutter was performed as follows (Figure 12.11). First, two conjugates of EDTP and oligonucleotide (20-mer) were synthesised. Then, by using them as additives, a five-base gap was formed in the middle of a single-stranded DNA substrate. The DNA cleavage reaction was started by adding $Ce(NO_3)_3$ to the mixture, and incubated at 50 °C for 20 h under air. In the course of the reaction, the Ce(III)/EDTP complex, bound to the oligonu-cleotides, was rapidly oxidised by ambient molecular oxygen to a Ce(IV)/EDTP complex. As depicted in lane 2 of Figure 12.11, the site-selective scission

(a) 5′-FAM......CAATTAGAATCAGGAATGGC**TTATG**GTGCAGACTGTCGACCTAAG......-3′

3′-GTTAATCTTAGTCCTTACCG-**E** **E**-CACGTCTGACAGCTGGATTC-5′

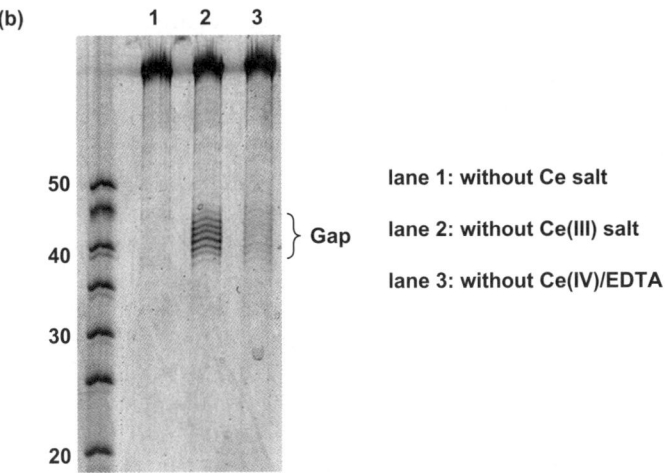

(b)

lane 1: without Ce salt

lane 2: without Ce(III) salt

lane 3: without Ce(IV)/EDTA

Figure 12.11 Site-selective DNA scission by the Ce(III)-derived DNA cutter. (a) The sequences of 5′-FAM labelled DNA (85-mer) and EDTP–oligonucleotide conjugates (20-mer). **E** stands for the EDTP ligand. (b) Denaturing polyacrylamide gel electrophoresis patterns. Lane M, marker; lane 1, target DNA and additive EDTP–oligonucleotides; lane 2, with Ce(NO$_3$)$_3$; lane 3, with Ce(IV)/EDTA. Reaction conditions: [DNA] = 1.0 μM, [each EDTP–oligonucleotide conjugate] = 1.0 μM, [Ce(NO$_3$)$_3$ or Ce(IV)/EDTA] = 4.0 μM, [HEPES (pH 7.0)] = 5.0 mM and [NaCl] = 100 mM at 50 °C for 20 h under air.

notably occurred at the five-base gap site, providing the corresponding five bands. Under the comparable conditions, Ce(IV)/EDTA showed virtually no DNA hydrolysis [the concentration of Ce(IV) was much lower than employed previously]. The efficiency of DNA scission increased with increasing concentration of Ce(NO$_3$)$_3$ and attained a plateau when the ratio of Ce(NO$_3$)$_3$ to EDTP was around 2–4. These results indicate that a well-defined Ce(IV) complex can be successfully prepared on EDTP ligand from the Ce(III) precursor and is highly active for DNA hydrolysis. This new preparation method for generating a site-selective DNA cutter is advantageous for various practical applications, especially *in vivo*, because the amount of unbound Ce(IV) is minimised, and accordingly off-target scission should be more completely suppressed.

12.4 Conclusion

By combining EDTP–oligonucleotide conjugates and Ce(IV)/EDTA, an artificial tool for site-selective hydrolysis of single-stranded DNA has been developed. The site specificity and the target site can be freely modulated, and therefore this cutter should be useful for developing new single-stranded DNA-based biotechnology. Another cutter composed of an EDTP–oligonucleotide conjugate and Ce(IV)/EDTA satisfactorily shows sequence-selective scission of the human telomeric DNA overhang. Here, the telomere sequence is recognised by the formation of a G-quadruplex from the conjugate and a telomeric DNA substrate. Thus the scission occurs only at the telomeric overhang, opening a way for the design of new anticancer agents. Furthermore, artificial restriction DNA cutters for cleaving double-stranded DNA at predetermined sites can also be obtained by combining Ce(IV)/EDTA with PNA. Even the human genome can be cut at a targeted site. DNA scission by all these cutters proceeds through hydrolysis of phosphodiester linkages and is compatible with the enzymes that are used in current molecular biology and biotechnology. Various DNA fragments, even with chemical modifications (*e.g.* methylation, hydroxymethylation, and non-canonical bases), can be clipped from naturally occurring sources, and can be used to construct sophisticated DNA nanomaterials. Improvements in the site selectivity and scission efficiency should further promote their applicability. One of these attempts is to prepare a second-generation cutter by oxidising a Ce(III) complex to the corresponding Ce(IV) complex. Applications of these DNA cutters *in vivo* and *in vitro* are currently being studied in our laboratory.

Acknowledgements

This work was partially supported by Grants-in-Aid for Specially Promoted Research from the Ministry of Education, Science, Sports, Culture and Technology, Japan (18001001 and 22000007) and by the Global COE Program for Chemistry Innovation.

References

1. S. J. Franklin, *Curr. Opin. Chem. Biol.*, 2001, **5**, 201–208.
2. A. Sreedhara and J. A. Cowan, *J. Biol. Inorg. Chem.*, 2001, **6**, 337–347.
3. J. Suh, *Acc. Chem. Res.*, 2003, **36**, 562–570.
4. F. Mancin and P. Tecilla, *New J. Chem.*, 2007, **31**, 800–817.
5. T. A. Shell and D. L. Mohler, *Curr. Org. Chem.*, 2007, **11**, 1525–1542.
6. J. Sumaoka, Y. Yamamoto, Y. Kitamura and M. Komiyama, *Curr. Org. Chem.*, 2007, **11**, 463–475.
7. C. L. Liu and L. Wang, *Dalton Trans.*, 2009, 227–239.
8. M. Komiyama, Y. Aiba, Y. Yamamoto and J. Sumaoka, *Nat. Protoc.*, 2008, **3**, 655–662.
9. H. Katada and M. Komiyama, *Chembiochem.*, 2009, **10**, 1279–1288.
10. Y. Aiba, J. Sumaoka and M. Komiyama, *Chem. Soc. Rev.*, 2011, **40**, 5657–5668.

11. M. Egholm, O. Buchardt, L. Christensen, C. Behrens, S. M. Freier, D. A. Driver, R. H. Berg, S. K. Kim, B. Norden and P. E. Nielsen, *Nature*, 1993, **365**, 566–568.

12. T. Igawa, J. Sumaoka and M. Komiyama, *Chem. Lett.*, 2000, 356–357.

13. Y. Kitamura and M. Komiyama, *Nucleic Acids Res.*, 2002, **30**, e102.

14. Y. Kitamura, J. Sumaoka and M. Komiyama, *Tetrahedron*, 2003, **59**, 10403–10408.

15. W. Chen, T. Igawa, J. Sumaoka and M. Komiyama, *Chem. Lett.*, 2004, **33**, 300–301.

16. W. Chen, Y. Kitamura, J.M. Zhou, J. Sumaoka and M. Komiyama, *J. Am. Chem. Soc.*, 2004, **126**, 10285–10291.

17. T. Lönnberg, Y. Suzuki and M. Komiyama, *Org. Biomol. Chem.*, 2008, **6**, 3580–3587.

18. Y. Yamamoto and M. Komiyama, *Chem. Lett.*, 2004, **33**, 920–921.

19. Y. Yamamoto, A. Uehara, T. Tomita and M. Komiyama, *Nucleic Acids Res.*, 2004, **32**, e153.

20. Y. Yamamoto, M. Mori, Y. Aiba, T. Tomita, W. Chen, J. M. Zhou, A Uehara, Y. Ren, Y. Kitamura and M. Komiyama, *Nucleic Acids Res.*, 2007, **35**, e53.

21. T. Lönnberg, Y. Aiba, Y. Hamano, Y. Miyajima, J. Sumaoka and M. Komiyama, *Chem. Eur. J.*, 2010, **16**, 855–859.

22. M. Komiyama, N. Takeda, Y. Takahashi, H. Uchida, T. Shiiba, T. Kodama and M. Yashiro, *J. Chem. Soc., Perkin Trans. 2*, 1995, 269–274.

23. J. Sumaoka, Y. Azuma and M. Komiyama, *Chem. Eur. J.*, 1998, **4**, 205–209.

24. Y. Aiba, T. Lönnberg and M. Komiyama, *Chem. Asian J.*, 2011, **6**, 2407–2411.

25. E. H. Blackburn, *Cell*, 2001, **106**, 661–673.

26. S. A. Stewart, I. Ben-Porath, V. J. Carey, B. F. O'Connor, W. C. Hahn and R. A. Weinberg, *Nat. Genet.*, 2003, **33**, 492–496.

27. T. R. Cech, *Cell*, 2004, **116**, 273–279.

28. M. A. Blasco, *Nat. Rev. Genet.*, 2005, **6**, 611–622.

29. R. E. Verdun and J. Karlseder, *Nature*, 2007, **447**, 924–931.

30. T. de Lange, *Nat. Rev. Mol. Cell Biol.*, 2004, **5**, 323–329.

31. M. K. Bhattacharyya and A. J. Lustig, *Trends Biochem. Sci.*, 2006, **31**, 114–122.

32. P. M. Lansdorp, *Trends Biochem. Sci.*, 2005, **30**, 388–395.

33. K. Collins and J. R. Mitchell, *Oncogene*, 2002, **21**, 564–579.

34. D. Alves, H. Li, R. Codrington, A. Orte, X. Ren, D. Klenerman and S. Balasubramanian, *Nat. Chem. Biol.*, 2008, **4**, 287–289.

35. N. W. Kim, M. A. Piatyszek, K. R. Prowse, C. B. Harley, M. D. West, P. L. Ho, G. M. Coviello, W. E. Wright, S. L. Weinrich and J. W. Shay, *Science*, 1994, **266**, 2011–2015.

36. W. C. Hahn, S. A. Stewart, M. W. Brooks, S. G. York, E. Eaton, A. Kurachi, R. L. Beijersbergen, J. H. Knoll, M. Meyerson and R. A. Weinberg, *Nat. Med.*, 1999, **5**, 1164–1170.

37. S. Neidle and G. Parkinson, *Nat. Drug Discovery*, 2002, **1**, 383–393.
38. L. H. Hurley, *Nat. Rev. Cancer*, 2002, **2**, 188–200.
39. J. W. Shay and W. E. Wright, *Nat. Rev. Drug Discovery*, 2006, **5**, 577–584.
40. J.-L. Mergny, J.-F. Riou, P. Mailliet, M. P. Teulade-Fichou and E. Gilson, *Nucleic Acids Res.*, 2002, **30**, 839–865.
41. A. De Cian, G. Cristofari, P. Reichenbach, E. De Lemos, D. Monchaud, M. P. Teulade-Fichou, K. Shin-ya, L. Lacoix, J. Lingner and J.-L. Mergny, *Proc. Natl. Acad. Sci. USA*, 2007, **104**, 17347–17352.
42. P. J. Perry and T. C. Jenkins, *Mini. Rev. Med. Chem.*, 2001, **1**, 31–41.
43. J. F. Riou, *Curr. Med. Chem. Anti-Cancer Agents*, 2004, **4**, 439–443.
44. E. M. Rezler, D. J. Bearss and L. H. Hurley, *Curr. Opin. Pharmacol.*, 2002, **2**, 415–423.
45. J. W. Shay, Y. Zou, E. Hiyama and W. E. Wright, *Hum. Mol. Genet.*, 2001, **10**, 677–685.
46. E. Raymond, D. Sun, S. F. Chen, B. Windle and D. D. Von Hoff, *Curr. Opin. Biotechnol.*, 1996, **7**, 583–591.
47. F. Pendino, I. Tarkanyi, C. Dudognon, J. Hillion, M. Lanotte, J. Aradi and E. Segal-Bendirdjian, *Curr. Cancer Drug Targets*, 2006, **6**, 147–180.
48. D. Cairns, R. J. Anderson, P. J. Perry and T. C. Jenkins, *Curr. Pharm. Des.*, 2002, **8**, 2491–2504.
49. S. Huard and C. Autexier, *Curr. Med. Chem. Anticancer Agents*, 2002, **2**, 577–587.
50. E. M. Rezler and D. J. Bearss, *Ann. Rev. Pharmacol. Toxicol.*, 2003, **43**, 359–379.
51. R. Rodriguez, S. Müller, J. A. Yeoman, C. Trentesaux, J.-F. Riou and S. Balasubramanian, *J. Am. Chem. Soc.*, 2008, **130**, 15758–15759.
52. K. Jantos, R. Rodriguez, S. Ladame, P. S. Shirude and S. Balasubramanian, *J. Am. Chem. Soc.*, 2006, **128**, 13662–13663.
53. A. De Cian, E. DeLemos, J.-L. Mergny, M.-P. Teulade-Fichou and D. Monchaud, *J. Am. Chem. Soc.*, 2007, **129**, 1856–1857.
54. Y. Wang and D. J. Patel, *Structure*, 1993, **1**, 263–282.
55. G. N. Parkinson, M. P. Lee and S. Neidle, *Nature*, 2002, **417**, 876–880.
56. Y. Xu, Y. Noguchi and H. Sugiyama, *Bioorg. Med. Chem.*, 2006, **14**, 5584–5591.
57. A. Matsugami, Y. Xu, Y. Noguchi, H. Sugiyama and M. Katahira, *FEBS J.*, 2007, **274**, 3545–3556.
58. K. N. Luu, A. T. Phan, V. Kuryavyi, L. Lacroix and D. J. Patel, *J. Am. Chem. Soc.*, 2006, **128**, 9963–9970.
59. A. Ambrus, D. Chen, J. X. Dai, T. Bialis, R. A. Jones and D. Z. Yang, *Nucleic Acids Res.*, 2006, **34**, 2723–2735.
60. N. Zhang, A. T. Phan and D. J. Patel, *J. Am. Chem. Soc.*, 2005, **127**, 17277–14285.
61. Y. Xu, Y. Suzuki, T. Lönnberg and M. Komiyama, *J. Am. Chem. Soc.*, 2009, **131**, 2871–2874.
62. Y. Xu and H. Sugiyama, *Nucleic Acids Res.*, 2006, **34**, 949–954.
63. Y. Xu, *Chem Soc Rev.*, 2011, **40**, 2719–2740.

64. Y. Yamamoto, A. Uehara, K. Miura, A. Watanabe, H. Aburatani and M. Komiyama, *Nucleosides Nucleotides & Nucleic Acids*, 2007, **26**, 1265–1268.
65. Y. Yamamoto, K. Miura and M. Komiyama, *Chem. Lett.*, 2006, **35**, 594–595.
66. Y. Yamamoto, A. Uehara, A. Watanabe, H. Aburatani and M. Komiyama, *Chembiochem.*, 2006, **7**, 673–677.
67. K. Ito, H. Katada, N. Shigi and M. Komiyama, *Chem. Commun.*, 2009, 6542–6544.
68. Y. Miyajima, T. Ishizuka, Y. Yamamoto, J. Sumaoka and M. Komiyama, *J. Am. Chem. Soc.*, 2009, **131**, 2657–2662.
69. G. Haaima, H. F. Hansen, L. Christensen, O. Dahl and P. E. Nielsen, *Nucleic Acids Res.*, 1997, **25**, 4639–4643.
70. J. Lohse, O. Dahl and P. E. Nielsen, *Proc. Natl. Acad. Sci. USA*, 1999, **96**, 11804–11808.
71. M. Komiyama, Y. Aiba, T. Ishizuka and J. Sumaoka, *Nat. Protoc.*, 2008, **3**, 646–654.
72. H. Katada, H. J. Chen, N. Shigi and M. Komiyama, *Chem. Commun.*, 2009, 6545–6547.
73. H. Katada and M. Komiyama, *Curr. Gene Therap.*, 2011, **11**, 38–45.
74. S. A. Hayes, P. Yu, T. J. O'Keefe, M. J. O' Keefe and J. O. Stoffer, *J. Electrochem. Soc.*, 2002, **149**, C623–C630.
75. P. Yu, S. A. Hayes, T. J. O'Keefe, M. J. O'Keefe and J. O. Stoffer, *J. Electrochem. Soc.*, 2006, **153**, C74–C79.

Subject Index